CELL BIOLOGY RESEARCH PROGRESS SERIES

DAUGHTER CELLS: PROPERTIES, CHARACTERISTICS AND STEM CELLS

CELL BIOLOGY RESEARCH PROGRESS SERIES

Tumor Necrosis Factor
Toma P. Rossard (Editor)
2009. ISBN: 978-1-60741-708-8

Tumor Necrosis Factor
Toma P. Rossard (Editor)
2009. ISBN: 978-1-61668-276-7 (Online book)

Cell Determination During Hematopoiesis
Geoffrey Brown and Rhodri Ceredig (Editors)
2009. ISBN: 978-1-60741-733-0

Handbook of Cell Proliferation
Andre P. Briggs and Jacob A. Coburn (Editors)
2009. ISBN: 978-1-60741-105-5

Handbook of Cell Proliferation
Andre P. Briggs and Jacob A. Coburn (Editors)
2009. ISBN: 978-1-60876-854-7 (Online book)

Daughter Cells of Microalgae
Dilwyn J. Griffiths
2010. ISBN: 978-1-60876-787-8

Handbook of Free Radicals: Formation, Types and Effects
Dimitri Kozyrev and Vasily Slutsky (Editors)
2010. ISBN: 978-1-60876-101-2

Endocytosis: Structural Components, Functions and Pathways
Brynn C. Dowler (Editor)
2010. ISBN: 978-1-61668-189-0

Endocytosis: Structural Components, Functions and Pathways
Brynn C. Dowler (Editor)
2010. ISBN: 978-1-61668-717-5 (Online book)

Cell Respiration and Cell Survival: Processes, Types and Effects
Gijsbert Osterhoudt and Jos Barhydt (Editors)
2010. ISBN: 978-1-60876-462-4

Handbook of Molecular Chaperones: Roles, Structures and Mechanisms
Piero Durante and Leandro Colucci (Editors)
2010. ISBN: 978-1-60876-366-5

Cell Division: Theory, Variants and Degradation
Yuri N. Golitsin and Mikhail C. Krylov (Editors)
2010. ISBN: 978-1-60876-986-5

Basophil Granulocytes
Paul K. Vellis (Editor)
2010. ISBN: 978-1-60741-797-2

Daughter Cells: Properties, Characteristics and Stem Cells
Ayane Hitomi and Masuyo Katoaka (Editors)
2010. ISBN: 978-1-60876-790-8

Cytoskeleton: Cell Movement, Cytokinesis and Organelles Organization
Sébastien Lansing and Tristan Rousseau (Editors)
2010. ISBN: 978-1-60876-559-1

Prostaglandins: Biochemistry, Functions, Types and Roles
Gillian M. Goodwin (Editor)
2010. ISBN: 978-1-61668-272-9

Prostaglandins: Biochemistry, Functions, Types and Roles
Gillian M. Goodwin (Editor)
2010. ISBN: 978-1-61668-645-1

Lipids: Categories, Biological Functions and Metabolism, Nutrition and Health
Paige L. Gilmore (Editor)
2010. ISBN: 978-1-61668-464-8

Lipids: Categories, Biological Functions and Metabolism, Nutrition and Health
Paige L. Gilmore (Editor)
2010. ISBN: 978-1-61668-522-5 (Online book)

CELL BIOLOGY RESEARCH PROGRESS SERIES

DAUGHTER CELLS: PROPERTIES, CHARACTERISTICS AND STEM CELLS

AYANE HITOMI
AND
MASUYO KATOAKA
EDITORS

Nova Biomedical Books
New York

For permission to use material from this book please contact us:
Telephone 631-231-7269; Fax 631-231-8175
Web Site: http://www.novapublishers.com

LIBRARY OF CONGRESS CATALOGING-IN-PUBLICATION DATA

Available upon request.

ISBN: 978-1-60876-790-8

Published by Nova Science Publishers, Inc. ✝ New York

CONTENTS

PREFACE

This book summarizes some of the major processes involved in the production of daughter cells as they have been reported from work mainly with synchronous cultures of microalgae over the 50 years or so since that technique was first developed. It highlights the key findings that have led to our present understanding of cell cycle processes in microalgae with particular reference to those that control daughter-cell production. The authors of this book also systematize the data on molecular genetic system controlling drosophila bristle morphogenesis and proposes an integral scheme of its functioning. In addition, the current understanding of kidney regeneration after injury are examined from the perspective of renotropic factors, renal stem/progenitor cells, and stem cell therapies. In unicellular and multicellular organisms, asymmetric division enables segregation of damaged molecules into one daughter cell. The authors suggest that partition of damaged proteins and organelles and segregation of template DNA may function together to produce long-lived stem cells. Daughter cells' behavior of the chlorarachniophytes are also described and the evolution and biological implications are discussed. Lastly, recent advances in stem cell biology are summarized, and ways in which clinical medicine could take advantage of this fascinating field of biology are examined.

Chapter 1 - This review summarises some of the major processes involved in the production of daughter cells as they have been reported from work mainly with synchronous cultures of microalgae over the 50 or so years since that technique was first developed. It highlights the key findings that have led to our present understanding of cell cycle processes in microalgae with particular reference to those that control daughter-cell production. Among the microalgal species that have been studied are those that divide by simple binary fission, like the diatoms (Bacillariophyceae), dinoflagellates (Pyrrophyta) and the euglenoids (Euglenophyta) as well as those, mainly represented by the green microalgae (Chlorophyta), that divide by various more complex processes of multiple fission. In the latter, the cell cycle includes multiple rounds of duplication yielding 2^n daughter cells per mother cell ($n = 2$-5). Some examples of cell cycle processes in colonial microalgal forms are included for comparison and, for the purpose of this review, the unicellular prokaryotic cyanobacteria (blue-green algae) are also included because they allow comparisons between prokaryotic and eukaryotic cell cycles.

It will be shown that although certain details of the cell cycle will vary between the microalgal species, as affected by morphological and physiological features of the adult cell, the basic control and regulatory mechanisms have much in common not only among the microalgae themselves but between them and the cell cycles of other micro-organisms, animals and plants. The case studies described will highlight the flexibility of the microalgal cell cycle under different conditions and the range of processes involved in the progression of the cell through its natural cycle culminating in daughter cell production and release. They will also address the concept of commitment to divide, the role of cell cycle regulatory proteins and their genetic control and possible links of cell cycle processes to certain endogenous circadian rhythms known to exist in a number of microalgal species. It will be clear that despite the large amount of information now available on various microalgal cell cycles there is much yet to be revealed, especially in relation to the control mechanisms involved.

Chapter 2 - Today's actual question is *creation of the science on stem cells* and development of its theoretical bases, taking into account comparative evolutionary cell biology and the *cardinal problem of the developmental biology – the problem of wholeness* for representatives of various kingdoms. This allows revealing correlations and studying correlative dependences of various structures at different levels of biohierarchy. The creation of every science is impossible without the application of methodology.

In current work the system of nontraditional ideas about the nature and role of stem cells in ontogenesis, reproduction and evolution of plants is proposed.

The main properties of plant stem cells have been developed by us, which has shown the integrity of morphogenous and reproductive processes at all stages of plant's life cycle. A special attention should be paid to the dynamics of the stem cells' pool and its correlative dependence with niche.

System approach and analysis of data on male and female generative structures – egg cell at different stages of development, double fertilization in model objects: syngamy, triple fusion, plasmogamy and karyogamy taking into consideration the cell cycle, zygote, embryogenesis of model objects in representatives of a lot of flowering plants taxa allowed us to express nontraditional idea, that from the embryological point of view zygote is proximate stem cell and gives a rise for all other stem cells in plant ontogenesis, particularly those of shoot and root apices.

Zygote contains information about architectonics and development of organization of an individual, which is connected with gradients and correlations between its parts, and consequently with heterogeneity of different axes of a future organism.

The data of developmental biology and especially comparative embryology of representatives of various taxa of *flowering* plants enable to consider that *formation of stem cells derived from zygote, are probably typical for all organs – flower, stem, leaf, root – and at all stages of the life cycle – sporophyte, gametophyte. Their localization is depending, first of all, on their functioning.*

Not every cell possesses the totipotency and not every totipotent cell is able to become the stem cell, and not every stem cell can produce an individual.

One has to pay espesial attention to the role of somatic cells in organization of embryo's body, in reproduction, and particularly, in changing of morphogenetic programs in ontogenesis and evolution. This view of the role of somatic cells also corresponds to the data of discoveries made in the middle of the XXth century: stem cells, growing plants out of somatic cells, haploidy, parasexual hybridization, and embryoidogeny. These categories are for the first rime considered by us from the position of stem cells.

A new phenomenon, embryoidogeny, was for the first time revealed by us in 1977 – a new type of vegetative propagation, where elementary structural unit is an embryoid (somatic embryo) forming out of somatic cells.

Thus, not only zygote, but somatic cell as well is able to be the initial cell – stem cell of a plant, playing a big role in ontogenesis and evolution.

The diversity of morphogenetic processes comprehends two components: potentially infinite somatic cells **activity that is conditioned by the functioning of dormant centres' cells,** being in intermediate differentiated state, and loss of stemness – senescence and apoptosis.

The plasticity and polyvariety of these processes are conditioned by the diversity of the morphogenesis and origin of vegetative and generative structures in which the peculiar reserve of cells, possessing the stemness property, is situated. Hence in plant morphogenetic processes even third component is present – restoring of stemness in many respects depending on stemness degree.

The arsenal of the data received in the XX-XXI centuries on biology of development including stem cells biology has allowed us to propose and discuss the three main perspective trends in study of stem cells – traditional and nontraditional.

Nontraditional trends:

- gametes – zygote – embryo – individual – population – coenosis;
- different types of somatic cells considering their localization and function in the organism; possibility of homologization of some somatic cells with the zygote;

Traditional trend:

- apical meristem of shoot apex and root apex in which organization somatic cells take part.

Chapter III - Formation of specialized spatial structures comprising various cell types is most important in the ontogenesis of multicellular organisms. An example is the *D. melanogaster* bristle organs. Bristles (micro- and macrochaetes) are external sensory organs, elements of the peripheral nervous system, playing the role of mechanoreceptors. Their comparatively simple organization comprising only four specialized cells and a common origin of these cells make macrochaetes a convenient model for studying cell differentiation.

The four cells forming bristle organ result from two successive divisions of a single cell, sensory organ precursor (SOP) cell. The number of macrochaetes on drosophila body corresponds to the number of SOP cells.

The morphogenesis of macrochaetes comprises three stages, the first two determining a neural fate of the cells. The third stage is cell specialization into components of the bristle organ—neuron, thecogen, tormogen, and trichogen.

Development of each bristle commences from segregation of proneural clusters, of 20–30 cells, from the massif of undifferentiated cells of the wing imaginal disc. At this stage, each cluster cell can potentially become a SOP cell. At the second stage, the only SOP cell and its position are determined within each cluster. Finally, two asymmetric divisions of the SOP cell with subsequent differentiation of the daughter cells gives the bristle organ.

Several dozens genes are involved in the control of macrochaete morphogenesis. The main component of this system is the proneural genes of *achaete-scute* complex (*AS-C*). An increased content of proneural proteins fundamentally distinguished the cells that will follow the neural developmental pathway from the disc epidermal cells.

A local *AS-C* expression, initiated at specified disc sites by specific transcription factors, determines the number and topology of proneural clusters. The expression of *AS-C* genes, continuing in the cells of the cluster, increases the difference in proneural protein content, first, between the cluster cells and then, between the cluster cells and the single SOP cell, where it reaches the maximum level. This process is provided by both the intracellular regulation of *AS-C* gene activity and intercellular events mediated via the EGFR and Notch signaling pathways.

The third stage in macrochaete morphogenesis comprises two successive asymmetric SOP cell divisions, determining the final specialization. The selector genes, in particular, *numb*, *neuralized*, *tramtrack*, and *musashi*, play the key role in cell type specification.

This review systematizes the data on molecular genetic system controlling drosophila bristle morphogenesis and proposes an integral scheme of its functioning.

Chapter IV - Stem cells (SCs) are thought to reside in many tissues within a niche formed by a group of surrounding cells and their extracellular matrices, which provide an optimal microenvironment for the SCs function. In general, the niche consists of a highly organized microenvironment in which various factors, such as signals coming from secreted cytokines, extracellular matrix (ECM), and intercellular adhesions are thought to work cooperatively to maintain the undifferentiated stem cell phenotype. Among SCs, adult SCs are often localized into specific niches where they utilize many, but not necessarily all, the external and intrinsic factors used by the embryonal counterparts to select a specific fate. Within the niche, SCs can maintain their ability for self-renewal as well as their potential so that, consequently, their detachment from the niche compartment induces their differentiation and loss of self-renewal capacity. Thus, when a stem cell begins to divide, it is thought that one daughter cell remains in a niche to replace the original stem cell, whereas another daughter cell is expelled out of the niche and starts its process of differentiation. In this process, a cell retains self-renewal properties and differentiation inhibitory factors, so that it stays as SC, whereas another daughter cell is destined to proliferate during a certain number of divisions for final differentiation along a particular lineage. This latter daughter cell will receive too few *stemness* factors to maintain its identity as a stem cell, and/or will inherit differentiation factors that can overcome its stem cell phenotype. To maintain tissue homeostasis and correct functioning of an organism, the number of daughter cells that retain stem cell identity must be strictly controlled such that differentiated cells can be generated in response to any injury. Likewise, the rate of division of SCs into a niche must be tightly controlled, since an overproduction of daughter cells destined to be differentiated may be harmful because this may result in cancer generation. In the present chapter, the authors will analyze distinct types of niches for distinct tissues as well as the mechanisms that influence the generation of daughter cells from each niche.

Chapter V - Kidney has a capacity to regenerate itself after a variety of insults. For the past few decades, the search for the factor(s) that can promote regeneration of the injured kidney has been a principal and fundamental theme in nephrology. Using various kidney injury models, the role of growth factors in kidney regeneration has been extensively examined and several factors such as hepatocyte growth factor (HGF) have been demonstrated to act as renotropic factor(s). Lately, along with other organs, the identification of adult tissue stem cell is one of the most exciting subjects in this area. A number of cell populations expressing putative stem cell markers or possessing stem cell property has been found within the kidney. Using BrdU labeling methods, it was shown that a slow-cycling tubular cell population was actively engaged in the regeneration processes of the kidney after ischemia and function as an origin of regenerating cells. Like in other tissues, side population cells have been identified in the kidney. Several groups have reported that cell therapies with bone marrow-derived cells (hematopoietic stem cell or mesenchymal stem cell) are highly effective for the treatment of acute or chronic renal failure in animals, which seems to be via the production of regenerative factor(s) rather than direct differentiation into a component of renal cells. In the near future, therapeutic strategy with stem cells might be possible for the treatment of various kidney diseases.

In this chapter, the authors summarize and discuss the current understanding of kidney regeneration after injury from the perspective of renotropic factors, renal stem/progenitor cells, and stem cell therapies.

Chapter VI - Exfoliation is a loss of cellular material retaining the basic cytological features of typical cells (plasma membrane, cytoplasm and nucleus). However, the set of physiological parameters leading to exfoliation is complex and difficult to measure. Exfoliated epithelial cells can be obtained from a wide range of mucosa like mammary glands, oral mucosa by scraping, bronchial mucosa by expectoration and gastric or colonic epithelia by extensive rinsings during routine procedures of endoscopic examination. From an histological point of view, epithelia are organized in functional units containing different cellular compartments (stem, proliferative, mature or functional and senescent cells). Exfoliation is either (1) a natural loss of body cells implying a molecular signal related to the turn-over of terminally differentiated cells and to the progressive mobilization of proliferative as well as stem cells or (2) the result of manual exfoliation by applying mechanical constraints like gentle swabbing or scraping. Exfoliated epithelial cells are believed to be either in apoptosis or in anoïkis. In the latter group, cells may survive by autophagy retaining their growth potential if proper matrix is rapidly provided after exfoliation. Most studies are based on cytological proof using microscopic examination along with a limited number of biomarkers. Only few studies using proteomics or transcriptomics are available and they open discussion about tissue references and normalization. Exfoliated cells may represent valuable source of information on the physiopathological status of the mucosa. In translational research, ethical reasons as well as legal ones are limiting the use of biopsies in favor of measures realized on epithelial cells isolated from biological fluids (milk, urines, gastric aspirates, stools) collected by way of non-invasive technology. Unambiguous knowledge on the advantages and the limitations of exfoliation methodologies could have far reaching impact on chrononutrition or chronopharmacology.

Chapter VII - Plants can show flexible capacity of coping with the impairing factors of their abiotic and biotic environment and their high developmental plasticity, and easily forced cellular totipotency is a general feature. Plants have a notable potential to maintain separate cell groups in their apical meristems to assure growing. In addition to these proembryonic cell division centers, plants posses various manners (e.g., dormant meristemoids and residual stem cells) to acquire cell division capability in their bodies and, as a result, the daughter cells achieve an altered differentiation status there. The initiated growth and regeneration might also be based on newly formed dividing cells entering again into the cell cycle, and the different cell fate of unequal daughter cells is usually fundamental in the realization of altered development. In the latter case, a precondition of the subsequent events is dedifferentiation of mature, specialized cells and remodeling of their genetic program.

The authors results demonstrate that cells may attain disparate dedifferentiation statuses depending on their ancestry, and the initiated redifferentiation processes might be arrested even before entering cell divisions or continued up to complete root formation. Transfer cells of leaf veins share function in short-distance transport under normal conditions and stem cell formation under forced organogenesis. The *Rubia tinctorum* in vitro model system allowed us to detect the main symptoms of the rapid morphological changes during dedifferentiation. Modification of cell wall structure, enhanced autophagic activity, asymmetric plastid fission and unequal cell division seem to be basic steps of meristemoid (stem cell) formation. The theoretical background and the importance of the cytological characteristics mentioned above are also discussed.

Chapter VIII - To escape senescence and dying, cells endure continual proliferation, as only dividing cells can self-renew. In the course of evolution, asymmetric division was used by multicellular organisms to solve the problem of replicative aging of stem cells. There are data showing that aged tissue-specific stem cells retain their intrinsic proliferative potential and that age-related changes in the systemic environment and niche in which stem cells reside can modulate the molecular signaling pathways critical to the function of stem cells. In unicellular and multicellular organisms, asymmetric division enables segregation of damaged molecules (biological garbage) into one daughter cell. This appears to be an evolutionarily conserved mechanism preventing transmission of damaged proteins to the longest-lived daughter cell. Thus, dividing stem cells have a simple physiological mechanism to free themselves of toxic proteins destined for degradation during asymmetric cell division. The authors suggest that partition of damaged proteins and organelles and segregation of template DNA may function together to produce long-lived stem cells.

Chapter IX - Stem cells (SC) research is an important part of biotechnology studies that could lead to the development of new therapeutic strategies in different fields of medicine.

It is known that stem cells exist in early embryos as well as in some individual adult tissues. A lot of efforts have been made to understand biology of stem cells, particularly the features concerning their proliferation, self-renewal and differentiation abilities (the three major characteristics of SC) and subsequently to find out the role of the genes and proteins involved in these processes.

Research on stem cells is advancing knowledge on how healthy cells replace damaged cells in adult organisms. This promising area of science is also leading scientists to investigate the possibility of cell-based therapies to treat disease, which is often referred to as regenerative or reparative medicine.

Stem-cell biology is in a phase of dynamic expansion and is forming connections with a broad range of basic and applied disciplines. Stem cell research has offered the promise of effective cell-based therapies in treating many debilitating diseases such as diabetes, cardiovascular and neurodegenerative diseases, and cancer. In the last few years, the therapeutic potential of stem cells has inspired the imagination, intense interest, and targeted investment of scientists, clinicians, and the general public toward this area of biology.

Thus, in this chapter the authors would like to summarize recent advances in stem cell biology, as well as to discuss how clinical medicine could take advantage from this fascinating field of biology.

Chapter X - Chlorarachniophytes constitute a group of photosynthetic cercozoans (one of the ecologically significant protists) living in marine environments. They possess plastids of secondary endosymbiotic origin, which have a remnant nucleus of the green algal endosymbiont (nucleomorph) located in the periplastidial compartment, between the inner and the outer pairs of the plastid membranes. With respect to the plastid evolution, the chlorarachniophytes constitute therefore one of the most interesting unicellular algal groups. Since this algal group was found in 1930, little is known about their cell cycle as well as their taxonomy and geographical distribution. During the authors taxonomic studies of the chlorarachniophytes, an interesting difference in behavior between the daughter cells has been observed: once the cell has divided, one of the amoeboid daughter cells constricts till it is as narrow as one of its pseudopodia; then, the cytoplasm of this new cell flows in a thin trickle along the pseudopodium, collecting to form another cell body some distance from the stay-at-home nucleus. This recently-discovered phenomenon raises some questions about the mechanism and the evolution of the unique behavior. Here I describe in detail the daughter cells' behavior of the chlorarachniophytes, and discuss about the evolution and biological implications.

In: Daughter Cells
Editors: A. Hitomi et al., pp. 1-50
ISBN: 978-1-60876-790-8

Chapter I

DAUGHTER CELLS OF MICROALGAE

Dilwyn J. Griffiths*
James Cook University, Townsville, Australia.

ABSTRACT

This review summarises some of the major processes involved in the production of daughter cells as they have been reported from work mainly with synchronous cultures of microalgae over the 50 or so years since that technique was first developed. It highlights the key findings that have led to our present understanding of cell cycle processes in microalgae with particular reference to those that control daughter-cell production. Among the microalgal species that have been studied are those that divide by simple binary fission, like the diatoms (Bacillariophyceae), dinoflagellates (Pyrrophyta) and the euglenoids (Euglenophyta) as well as those, mainly represented by the green microalgae (Chlorophyta), that divide by various more complex processes of multiple fission. In the latter, the cell cycle includes multiple rounds of duplication yielding 2^n daughter cells per mother cell (n = 2-5). Some examples of cell cycle processes in colonial microalgal forms are included for comparison and, for the purpose of this review, the unicellular prokaryotic cyanobacteria (blue-green algae) are also included because they allow comparisons between prokaryotic and eukaryotic cell cycles.

It will be shown that although certain details of the cell cycle will vary between the microalgal species, as affected by morphological and physiological features of the adult cell, the basic control and regulatory mechanisms have much in common not only among the microalgae themselves but between them and the cell cycles of other micro-organisms, animals and plants. The case studies described will highlight the flexibility of the microalgal cell cycle under different conditions and the range of processes involved in the progression of the cell through its natural cycle culminating in daughter cell production and release. They will also address the concept of commitment to divide, the role of cell cycle regulatory proteins and their genetic control and possible links of cell cycle processes to certain endogenous circadian rhythms known to exist in a number of

* Correspondence concerning this article should be addressed to: Dilwyn J. Griffiths, School of Biological Sciences, James Cook University, Townsville, Queensland 4811, Australia. *dilwyng@ozemail.com.au.*

microalgal species. It will be clear that despite the large amount of information now available on various microalgal cell cycles there is much yet to be revealed, especially in relation to the control mechanisms involved.

INTRODUCTION

"Daughter cells" are here defined by reference to the cell cycle i.e. the sequence through which the cell passes as it progresses through its growth phase until it reaches and completes the processes involved in fission of the protoplast to produce and release daughter cells to initiate the next cell cycle. In its simplest form, the typical eukaryotic cell cycle can be described by the sequence:

$$G1 \rightarrow S \rightarrow G2 \rightarrow M \rightarrow G1$$

where G1 is the growth phase during which daughter cells increase in size, S is the phase of DNA synthesis, G2 is a pre-mitosis phase, M is mitosis, which includes cytokinesis and release of new daughter cells which then enter the G1 phase of the next cycle. There are many variations in the detail of such a sequence; in some cases, for example, the G2 and M phase may overlap or be otherwise difficult to distinguish by conventional analytical techniques.

This review focuses on the cell cycle as it has been studied in microalgae mostly using synchronous cultures, where the individual cells of the culture have been brought into phase so that the behaviour of the culture as a whole mimics that of the individual cell. This makes available enough material to allow bulk analyses of different parameters. The extent to which this can be achieved will depend on the degree of synchrony attained. For microalgal cultures, two types of procedures have generally been used to produce synchronous cultures namely selection and induction. In the former, cells at a particular stage of the cycle are selected and separated from a normal asynchronous culture and used to initiate the new synchronous culture. In the latter, an asynchronous culture is subjected to some treatment that temporarily arrests development at a particular stage of the cell cycle. Removal of the blockage may, under suitable conditions, produce cells from which a new synchronous culture can be initiated. Another type of induction synchrony is that achieved by subjecting cultures, sometimes already partially synchronised by selection, to repeated environmental shifts. This is particularly suitable for the microalgae, where repeated light and dark cycles, for example, have been successfully used to bring cultures into synchrony.

Case studies of work with synchronous cultures and other microalgal-based experimental systems will be described below, with special emphasis on those processes involved in the production and release of daughter cells. The control mechanisms involved in daughter cell production in microalgae are likely to be somewhat different from those in, for example, higher plants where the sessile growth habit, requires a special form of cytokinesis involving formation of a cell plate between the newly constituted daughter nuclei. In the unicellular microalgae, by contrast, cell division is of the non-vegetative type and is concerned solely with reproduction. The daughter cells (zoospores, autospores or gametes) become completely invested by new cell walls, and the parent cell wall does not form part of the wall of the daughter cells as in vegetative cell division. In the case of those species capable of sexual reproduction the daughter cells occur as gametes which, after sexual fusion, produce zygotes around which the new cell wall develops.

The microalgal genera most commonly used in studies of the cell cycle represent a wide range of different unicellular morphologies and cell division patterns. These range from simple binary fission where each cell divides once to produce two daughter cells to various forms of multiple fission yielding 2^n daughter cells per mother cell ($n = 2 – 5$, depending on species and/or growth conditions). Multiple fission is characterised by a growth phase during which the cell can double its mass several times compared with the one doubling per cell cycle in cells dividing by binary fission. During simple binary fission as is found, for example, in the euglenoids, diatoms and dinoflagellates, the generation time is identical with the actual doubling time. The shortest possible generation time would therefore determine the upper limit of productivity during balanced growth. Microalgal genera that divide by multiple fission, like *Chlorella* and *Chlamydomonas,* are capable of enhanced production and the doubling time can then be determined by the ratio t/n (where t is the generation time and n is the number of doublings). For species of *Chlorella*, for example, doubling times of 5 to 6h are usual although values as low as ca. 3h have been reported for a high temperature strain (Sorokin, 1959).

Many of the microalgae that have been studied have cell cycles strongly influenced by some special features of the parent cell morphology. The diatoms, for example, have a specific requirement for the element silicon, a vital component of their rigid cell wall that will restrict and in other ways influence various aspects of cell growth and division and hence a range of cell cycle processes. The protozoan-like, flagellate euglenoids, by contrast, have no true cell wall; instead the outer membrane of the cell is modified to form a periplast (pellicle) which may be either rigid or plastic (as in some species of *Euglena* which change shape when they move). Longitudinal division of the protoplast occurs after the mother cell has become enclosed by a gelatinous matrix and, under certain conditions the cells may undergo repeated division to form the palmelloid condition. The other group of flagellates that has been widely used in studies of the cell cycle are the dinoflagellates. They have a unique form of mitosis and nuclear division that may be regarded as being intermediate between prokaryotic and eukaryotic cell cycles. The unicellular prokaryotic cyanobacteria (blue-green algae) are included here with the microalgae because they allow interesting points of comparison between their cell cycle control mechanisms and those of the eukaryotic microalgae. The colonial chlorophyte *Scenedesmus* is also included because its cell cycle, where each of the four constituent cells of the coenobium divides to form daughter coenobia, also offers interesting comparisons with the unicellular microalgal cell cycles.

For all these reasons then, the various case studies are grouped, for the purpose of this review, on a taxonomic basis although, as will be shown, many of the cell cycles have features in common. This will apply particularly to cell cycle control mechanisms and their relationship, where they have been shown to exist, to certain underlying endogenous rhythms.

THE CELL CYCLE IN *CHLAMYDOMONAS* AND *DUNALIELLA*

Chlamydomonas, with its unicellular spherical to ovoid morphology and its simple form of reproduction by longitudinal binary or multiple fission, is generally regarded to have the most primitive or basic morphology of all the green algae. Cell division is preceded by loss of the two flagella and the protoplast then undergoes successive divisions to form two, four, eight or sometimes more motile daughter cells which are released when the cell wall of the parent cell breaks down and is discarded. Under certain conditions, the daughter cells do not develop flagella but undergo repeated divisions within a mucilaginous envelope to form what has been described as the "*Palmella*" stage. During sexual reproduction the daughter cells become haploid gametes which may be isogamous or anisogamous, and may occur as + or – strains with the opposites fusing to form the diploid zygote. Germination of the zygote (zygospore) involves fission of the protoplast to form new daughter cells (zoospores or aplanospores).

Dunaliella, although sharing many of the gross Volvocalean features of *Chlamydomonas,* is placed in a different family (Polyblepharidaceae). The cells of *Dunaliella,* unlike those of *Chlamydomonas,* have no outer wall, only a firm membrane which constitutes the outer boundary of the protoplast. Its method of reproduction by binary fission, therefore, more closely resembles that found in another large group of naked green flagellates, the Euglenophytes (which are described in a later section). Species of *Dunaliella* occur in a wide range of salinities including marine and hypersaline environments.

Cell Cycle Stages in Synchronous Cultures of *Chlamydomonas*

Based on the pioneering work of Bernstein (1960, 1964), Lorenzen and his group in Gottingen (eg Kuhl & Lorenzen, 1964, Lorenzen & Hesse, 1974) and Tamiya's group in Tokyo (e.g. Mihara & Hase, 1971), synchronous cultures of *Chlamydomonas* have been extensively used to study factors influencing the cell cycle. Lien & Knutsen (1979), for example, investigated the optimal conditions for the synchronous growth of *C. reinhardtii* Dangeard under a range of different growth conditions and light-dark regimes. They found that under a 12:4h, L(20,000 lux):D regime at 35° C, with serial dilution at the start of each cycle, the cells produced an average of 8 spores (daughter cells) per sporangium when the starting density was 4 x 10^6 cells.ml^{-1} but this increased to ca. 30 when the starting density was reduced to 5 x 10^4 cells.ml^{-1}. Reduction in starting cell density, and the consequent increased light availability per cell, caused not only an increase in the number of progeny but also in their average volume. However, cell division and sporulation were noted by Lien & Knutsen (1979) to occur at about the same time from the start of the light period independently of the starting cell density. They interpreted this to indicate that the growth conditions allowed the cells to experience their shortest generation time (as had previously been defined for *Chlorella* by Lorenzen – see Lorenzen & Schleif, 1966).

Increases of cell volumes recorded during the cell cycle were exponential over the first 8 hours of light with a doubling time of 2.5h, followed by a two hour period of a slightly lower growth rate and a final two hours at the maximum growth rate. During the subsequent sporulation, the mean cell volume decreased to match that of the zoospores of the preceding cycle. Mean cell volume increased by a factor of 25-30 during the light period; the difference between this and the estimated cell volume of the average number of spores (16 under these conditions) produced from each sporangium could be accounted for, according to Lien & Knutsen (1979), by the intercellular spaces between the spheroid zoospores when packed within the sporangium.

A number of other parameters were monitored through the cell cycle. RNA, for example, accumulated throughout the light phase, at first exponentially (0-8h), followed by a short period of almost no RNA accumulation and a final period when RNA accumulated at the highest rate recorded during the whole synchronous cycle. The drop in the rate of RNA accumulation between about 7 and 10 h coincided with the start of DNA synthesis and cell division. The increase in DNA that occurred over the period 8 to 12h of the light phase, when expressed on a per cell basis, matched that which would be expected from the average number of 16 daughter cells produced. Lien & Knutsen (1979) were further able to show that for strains producing 16 spores per sporangium, maximum frequencies of cells with 2, 4, 8, and 16 nuclei occurred at ca. 10.0, 10.6, 11.1 and 11.5h of the light phase, respectively. Maximum frequencies of cells in process of dividing into 2, 4, 8, and 16 protoplasts occurred at ca. 10.3, 10.8, 11.3, and 11.7h, respectively indicating that the maximum frequencies of the different division stages appeared 30-40 minutes apart with nuclear division preceding the corresponding cytoplasmic division by 12-18 minutes. It was further estimated that each cell needs approximately 1h, on average, to complete the sequence from DNA replication, karyokinesis to cytokinesis for each round, leading finally to the 16 daughter cells. Evaluation of synchrony based on observation of sporulation, DNA synthesis, karyokinesis and cytokinesis showed that the degree of synchrony is highest for cells growing under optimal conditions.

Immunofluorescence studies of *Chlamydomonas* in the period leading into mitosis showed that while the cell's complement of actin protein is arrayed around the nucleus at interphase, it undergoes dramatic reorganisation during mitosis and cytokinesis (Harper et al. 1992). It becomes involved, it was reported, in metaphase plate formation and in the formation of the cleavage furrow during telophase and cytokinesis.

The final stage of the cell cycle of *Chlamydomonas*, release of the daughter cells, has been shown to involve a lytic enzyme "sporangin" that mediates breakdown of the sporangial wall (Schlösser, 1981). Sporangin, which is released into the medium by synchronous cultures of *C. reinhardtii* when the daughter cells are released, has been purified and identified as a glycoprotein (Matsuda et al. 1995). It was shown to be capable of breaking the peptide bonds of several synthetic peptides.

In its purified form, sporangin has been found to react under gel electrophoresis as a single band (125kDa) polypeptide which, under certain conditions yields two degradation products of 76 and 62kDa (Kubo et al. 2009). When sporangin mRNA was monitored over a 20h cell cycle it was found that it first became detectable by 14h when the cells shifted into S/M phase at the start of mitosis and cytokinesis, but rapidly disappeared as the cells entered the next G1 phase. It was concluded that expression of the sporangin protein during the cell cycle was regulated at the transcription level. Further studies showed that sporangin is stored in sporangia as an inactive 127kDa proenzyme and released as an active 125kDa enzyme into the culture medium concurrently with digestion of the sporangial wall.

Hatching of the sporangia was shown to be correlated with flagellation. When mature sporangia subjected to pH shock - to detach the flagella of the daughter cells - were incubated in fresh medium to allow regeneration of the flagella it was found that they did not hatch until after 50 min. incubation and then only slowly (Kubo et al. 2009). Untreated sporangia (with full length flagella), by contrast, hatched after only 10 min. of incubation. It was further shown that it was the absence of flagella *per se* rather than the pH shock that affected hatching of the sporangia. Immunofluorescence imaging showed that prosporangin becomes localised in the flagella of the daughter cells prior to hatching in agreement with previous findings (e.g. Goodenough & St. Clair, 1975) that flagella appear to be essential for efficient hatching.

Commitment to Division in *Chlamydomonas*

John (1984) identified in the *Chlamydomonas reinhardtii* cell cycle, a key control point for cell division, similar to that which, at the time, was already well known for yeast cells (Lee & Nurse, 1988). It occurred in the late G1 phase of the cycle and, as in the case of yeast, acted as the start point for progress to DNA synthesis. It was characterised as that point beyond which cell division could be completed with little or no requirement for further growth (Donnan & John, 1984). From that point the cells were brought quickly to the S phase, usually within 0.1 of a cell cycle. Another essential feature of this control mechanism, according to John (1984) was the requirement for a critical minimum cell size. Thus, in cells growing very slowly, the critical mass is reached at a time when there is already a commitment to form two daughter cells. At higher growth rates doubling of the cell mass occurs before the cell has experienced the requirement for a minimum timed interval in G1. Division, under these conditions, is delayed beyond one doubling in cell mass and then, according to McAteer et al. (1985), daughter cell size will be stabilised by additional commitments that can recur as long as the mass per committed daughter cell remains above a critical minimum. In cells undergoing multiple divisions (to give four or eight daughter cells), the additional commitments (each leading to a further doubling of cell number) were observed by Donnan et al.(1985) to follow straight after the first.

The mechanism controlling the number and timing of the consecutive commitments as described by John et al. (1989) was further investigated using another species of *Chlamydomonas* (*C. eugametos*) (Zachleder & van den Ende, 1992). By growing synchronous cultures of *C. eugametos* under various conditions of light intensity or temperature they were able to show that the length of the pre-commitment period in this species was strictly dependent on the rate of assimilation. This contrasts with the findings of Donnan & John (1983) referred to above, who observed that the length of the pre-commitment period was constant under different growth rates. Zachleder & van den Ende (1992) were further able to show that while mitosis commences only after the cells have attained a certain critical mass, the rate at which the cell accumulates that mass determines the overall timing of the cell cycle. The occurrence of more commitments, they concluded, is dependent on the input of energy after the first commitment. To explain the apparent discrepancy between their findings and those of John's group, Zachleder & van den Ende

(1992) suggested that while in *C. reinhardtii* the commitments that determine multiple cell divisions lie close together, those in *C. eugametos* (and in *Scenedesmus* – see Šetlík & Zachleder, 1984 - to be described later) are more widely spaced. This, according to Zachleder & van den Ende (1992), may be the reason why in *C. reinhardtii* the first commitment point seems to be the major point at which the cell cycle is controlled in response to cell size and external conditions whereas in *Scenedesmus* and in *C. eugametos* this control is distributed over later stages of the cell cycle (as in fission yeast). In the latter case the critical cell volume to trigger the reproductive sequence of events was attained repeatedly during a single cell cycle. The mechanism, it was suggested, appears to be related to an activation of the regulatory proteins of the cell cycle, which themselves are controlled by cell growth.

Oldenhof et al. (2007) took the question of the timing of the commitment to divide one step further by confirming that in *C. reinhardtii,* the point of commitment was dependent on the growth rate and coincided with the moment at which the cells had approximately doubled in size. The timing of cell division, they found, was temperature dependent and occurred at a fixed time from the start of the light period, irrespective of the light intensity or of the timing of the commitment point. From this, they concluded that the commitment point is that point when all the prerequisites for progression through the cell cycle are checked. It is not, they maintained, the point at which cell division is initiated but rather a check-point to ensure that the cells have passed the minimum size required for division.

Noting that in animal cells, passage of cells from G1 to S and through the S phase is gated by a process involving the retinoblastoma (RB) family of tumor suppressors, Umen & Goodenough (2001), studied a mutant strain of *C. reinhardtii* in which an RB homolog had been identified. They observed that the mutant produces daughter cells considerably smaller than normal, initiates the cell cycle at below normal size with the inclusion of extra rounds of mitosis. Unlike mammalian RB mutants, however, the *C. reinhardtii* mutant did not have a shortened G1, did not enter S phase prematurely but could complete the cell cycle normally.

Light Quality and the Division Process in *Chlamydomonas*

A number of studies (e.g. Lien & Knutsen, 1979) have shown that when cultures synchronized by light:dark cycles are kept in the light beyond the commitment point, they continue to grow and can double in size several times, thereby passing additional rounds of division before later entering the division phase. Münzner & Voigt (1992) showed that when synchronized cultures of *C. reinhardtii* growing under heterotrophic conditions in the dark were exposed to light during the second half of the growth period, cell division was delayed. This was interpreted as indicating that entry into the division phase is regulated by the transition from light to darkness. They then found that while blue light was as effective as white light in delaying cell division, red and far-red light exposure was ineffective, pointing to the involvement of a blue light receptor.

Oldenhof et al. (2004, 2006) observed that in the presence of blue light, the commitment point for cell division in *C. reinhardtii* is shifted later, coinciding with a larger cell size and hence allowing for two rounds of division. In red light, on the other hand, the commitment point coincides with the minimum cell size for cell division, when cells have approximately doubled in size. The delay in cell division is less when the cells are exposed to shorter periods of blue light before being transferred to red light. Interestingly, Oldenhof et al. (2006) found that red light could reverse the inhibitory effect of blue light on the initiation of cell division at the minimum cell size for cell division, but darkness could not. The ability of red light to reverse the inhibitory effect of blue light was evident even in the absence of growth, that is, when cells that had doubled in size were treated with the photosynthetic inhibitor DCMU at the time of the red light treatment.

Cell Cycle Control Proteins in *Chlamydomonas*

The presence of *Chlamydomonas* proteins specifically related to cell division or to the preparatory steps for cell division, were first tested for by John et al. (1989). Their role was inferred from their functional similarities to proteins already known to play such a role in yeast and animal cells. One such protein p34cdc2 had been shown to be essential for the start of the division process in yeast (Simanis & Nurse, 1986) and a protein similar in molecular size and immunological characteristics, was then demonstrated for *C. reinhardtii* by John et al, (1989). Its specific link with cell division processes was indicated by noting its level of phosphorylation in proliferating synchronous cultures of *C. reinhardtii* compared with that in cultures held in quiescence by nitrogen starvation. Moreover, cells supplied with ^{32}P label before the commitment to cell division (i.e. during the first 4h of G1) showed no labelling of their p34 protein while those receiving the label between 5h and 8h of G1 (i.e. when the majority of the cells are committed to division) contained a high degree of phospho-p34 labelling. It was estimated that at least a sevenfold increase in the rate of incorporation of ^{32}P into p34 occurred at the time of commitment to division.

Quantitative estimation of the amount of the p34 protein at different stages of the cell cycle indicated an abrupt increase in accumulation at the time of commitment to divide (John et al., 1989). Phosphorylation of p34 protein was first detected at the time of the first commitment. 2.5h later, when the cells were in mitosis, there was additional incorporation into a second protein distinguished from the first by its slower migration under electrophoresis. By the time mitosis was complete and daughter cell numbers were beginning to increase, phosphate incorporation into p34 protein had stopped. Both forms of phospho-p34 were specifically recognised by antibody to the protein kinase (p34cdc2) known to be essential for the start of the division process in yeast.

This work was the first to provide evidence that *Chlamydomonas,* like yeast (and mammalian cells), has a p34 protein that is somehow involved in the late G1 control point of the cell cycle. Its mechanism of action, it was suggested, may be controlled by interaction with other proteins (eg. cyclins) rather than by changes in the actual amount of p34. It was concluded that the two forms of p34 detected in *C. reinhardtii,* acted successively at late G1 and in the G2 phase (when mitosis is initiated).

In their study of synchronous cultures of a mutant strain of *C. reinhardtii* lacking a normal cell wall, Zachleder et al. (1997) monitored changes in the activity of histone H1 kinase, a p34 protein like that mentioned above and a homologue of the p34cdc2 protein kinase known to be active in yeast and mammalian cells. The activity of histone H1 kinase was shown to vary markedly over the cell cycle. One of the peaks of activity was always found at the end of the cell cycle when multiple rounds of DNA replication, nuclear division and cleavages occurred (to give eight daughter cells). There were also peaks of histone H1 kinase activity during the light period, three in cultures exposed to high irradiance and two in those exposed to low irradiance, each peak coinciding with periods of active increases of cell volume. The increases in histone H1 kinase activity, it was further shown, always preceded the commitment points.

Cultures exposed to 4, 7, or 12h light experienced, respectively, one, two and three peaks of increased histone kinase activity during the light period. In the first case, there was only one subsequent duplication of DNA, followed by one nuclear division and the formation of two daughter cells; the maximum kinase activity occurring just prior to the first commitment point and later declining during the dark period. Under a 7h light period there were two rounds of DNA replication and two nuclear divisions which occurred during the dark period to give four daughter cells. Under 12h light the three peaks of kinase activity were each followed by a commitment to trigger the associated reproductive processes yielding, at the end of the cycle, eight daughter cells.

Changes in H1 kinase activity and associated growth characteristics of cultures under the light:dark regime were then compared with those of cultures maintained in continuous light. With both treatments, kinase activity showed the same relationship to changes in cell volume and attainment of the commitment point despite the difference in the timing of these processes. This led Zachleder et al.(1997) to conclude that the rhythm of the oscillation of kinase activity was not simply a function of the alternating light:dark regime but rather related to the special role of the cyclin-dependent kinases in triggering and maintaining the reproductive process of DNA replication, nuclear division and cytokinesis.

Because of the more frequent sampling regime used by Zachleder et al. (1997) they were able to distinguish two consecutive commitment points characterised by two types of cell cycle specific maxima of histone H1 kinase activity. One maximum was always coupled with multiple rounds of DNA replication followed a short time later by nuclear divisions, at the end of the cell cycle. The other maximum showed a rate-limiting dependence and was observed at the end of each of the multiple growth steps.

In their discussion of the observed oscillation of histone H1 kinase activity during the cell cycle of *Chlamydomonas,* without any apparent direct relationship to DNA replication, Zachleder et al. (1997) questioned why the cells, once committed, do not immediately initiate DNA replication. They speculated that this might be due to the action of specific inhibitors which block progression from the G1 to S phase as has been described for budding yeasts (King et al., 1996). The repeated stopping and starting of the increase in cell volume, without an external signal was not explained nor was the decrease in the histone H1 kinase activity after the commitment to cell division had been reached.

Genetic Control of Cell Cycle Regulators in *Chlamydomonas*

Bisova et al. (2005) identified and profiled the expression of the core regulators of the cell cycle of *Chlamydomonas.* They found that cells of *C. reinhardtii* encode orthologs of the major CDK(cyclin-dependent protein kinases) and cyclin families found in other organisms as well as two CDK subtypes and two cyclin subtypes not found in higher plants, fungi or animals. Orthologs of a number of other cell cycle regulators were also identified, most of them apparently coded by single copies of their respective genes.

C. reinhardtii was found to encode a single ortholog for five of the known plant-type CDKs as well as four new members of the CDK family. mRNA for CDKA1 was found to be present constitutively during the cell cycle with expression increasing as the cells entered the growth phase at the beginning of the light period and increasing further around the time of S/M. Abundance of presumed Chlamydomonas CDK protein was relatively constant during the cell cycle with phosphorylation-induced isoforms appearing during S/M. mRNA for another CDK (CDKB1) showed two peaks of expression, one corresponding to passage through commitment and a second, very strong peak during S/M. The Chlamydomonas-specific CDKs G1 and H1 were also expressed constitutively at a low level, with CDKG1 message levels rising somewhat after commitment.

Single orthologs of each of the A- and B-type cyclins, known from other systems to function in S-phase and mitosis, respectively, have been identified in *C. reinhardtii* (Bisova et al. 2005). Orthologs of another three D-type cyclins (presumed to function in G1) were also identified. *C. reinhardtii* also has some other divergent cyclins and it is presumed that these, like the other cyclins, combine with the appropriate CDKs in their various cell cycle regulatory activities. A CKS (cyclin-dependent kinase subunit) protein that is presumed to mediate interactions involving the CDKs was also detected as well as a transcription factor known from other systems to play a part in regulating DNA replication.

In describing the suite of regulators they had identified in *C. reinhardtii*, Bisova et al. (2005) remarked on their overall similarities to those known from higher plants. It was noted, however, that whilst most of the *C. reinhardtii* cell cycle regulatory genes are single copy, most of their higher plant equivalent have undergone multiple duplication. Bisova et al. (2005) raise the question whether the relatively small number of *C. reinhardtii*-specific cell cycle regulators identified by them may be related to the capacity of that alga for multiple fission. They further speculate that the capacity for multiple fission may have evolved through altering the regulation of core conserved cell cycle genes and not by utilization of novel genes. They also propose a possible explanation for the observed transient increase in expression of S/M-phase genes in *C. reinhardtii.* The G1 delay period in that alga (to accommodate multiple fission), they suggest, renders the expression of S/M-phase genes at commitment either unnecessary or even detrimental so their expression is suppressed.

Circadian Rhythms and the Cell Cycle in *Chlamydomonas*

Many of the basic biochemical mechanisms underlying the functioning of the *Chlamydomonas* cell have been shown to be governed by an endogenous rhythm of approximately 24h period (i.e. a circadian rhythm) driven by a time measuring device termed the circadian clock. Such a mechanism is common to a wide range of organisms and is assumed to allow anticipation of daily changes in environmental conditions thus optimising cellular functions. In nature, the cue for setting the phase of the endogenous circadian clock is light.

One manifestation of a strong circadian rhythm in *Chlamydomonas* is that controlling phototactic activity for which the receptor is thought to be a retinal-based pigment with an absorption maximum at 503nm (Foster et al. 1984). Kondo et al. (1991) grew *Chlamydomonas* in bright light followed by a 24h period of darkness to synchronize the cellular clock. They then examined the photobiology of the clock by applying brief light pulses during the 24h dark period and monitored the resulting phase shifts in the rhythm. The action spectrum for the phase shift exhibited two prominent peaks (at 520nm and 660nm), both eliciting a 4h phase shift. Significantly, the red light (660nm)-induced phase shift was not diminished by subsequent far-red (730nm) exposure.

For microalgal cultures growing under alternating light/dark cycles it is often difficult to distinguish processes controlled by the endogenous circadian rhythm from those inherent in the cell cycle itself. In an attempt to resolve this problem, Jacobshagen et al. (2001) tested for the abundance in *C. reinhardtii* of mRNAs for a range of proteins (including some thought to play a role in regulation of cell division) to see if their expression met the criteria required for a circadian rhythm namely: (1) persistence under constant conditions with a period of ca. 24h; (2) entrainment to daily environmental cycles of light or temperature; (3) temperature compensation of the period to keep it close to 24h. They concluded that although many of the genes they tested exhibited circadian gene expression, many did not. In this respect *C. reinhardtii*, they suggest, differs from the prokaryote *Synechococcus* (to be discussed in a later section) where circadian gene expression is reported to be a common, if not universal phenomenon.

Among the genes of *Chlamydomonas* whose expression has been described as following a circadian rhythm are the *FtsZ* and *MIN* genes coding for proteins involved in chloroplast division (see below). The mRNA expression of these genes maintained an oscillating circadian rhythm even under continuous darkness or continuous light albeit with a considerable dampening in the latter case.

Morphological Changes over the Cell Cycle in *Chlamydomonas*

Studies of structural changes within the cells of *Chlamydomonas* as they progress through the cell cycle have largely centred on the two major organelles, the chloroplast and the mitochondria and their respective genomes (the organellar nucleoids). In synchronous cultures of *C. reinhardtii*, for example, Kuroiwa et al., (1981) reported that the chloroplast nucleoids increase in size and quadruple in number just before nuclear division. Ehara et al., (1990) observed changes in the structure and frequency of chloroplastic nucleoids during the division cycle. At different times, they could occur as ten granular nucleoids (at the beginning of the light period) or as threads or fibrils dispersed throughout the chloroplast (towards the end of the light period). Soon after chloroplast division, the thread-like nucleoids were transformed into about 20 granular forms which gradually combined to form about ten larger granular bodies in the zoospores immediately before they were released from the mother cells. During gametogenesis, the number of nucleoids decreases and, after gametic union to form zygotes, their number continues to decline.

A more detailed study of living cells stained with vital dyes (SYBR Green I and $DiOC_6$) confirmed that, at the start of the light phase, the new daughter cells contained about 30 mitochondrial (mt) and 10 chloroplastic (cp) nucleoids (Hiramatsu et al., 2006). They could be distinguished from the cell nucleus (ca.2μm diam.), by their much smaller size. The smaller stained bodies, situated near the cell surface were judged to be mt nucleoids and the slightly larger ones, cp nucleoids. After 3h of the light phase the number of mt nucleoids increased to about 40 and the cp nucleoids increased to about 20. Later (9h light) the cells, which were now larger, contained more than 100 mt nucleoid particles and more than 50 cp nucleoids. At the beginning of the dark period (D0h) the cell nucleus became slightly larger indicating nuclear DNA synthesis while the nucleoids became dispersed uniformly throughout the cell. At D3h, the time of the first karyokinesis, and during the subsequent cytokinesis, the nucleoids were distributed equally among the daughter cells, about 50 in each daughter cell.

More detailed studies of a mutant strain that has only one cp nucleoid throughout the cell cycle showed that of the four daughter cells contained in the mother cell at D6-9h, only one contained the large cp nucleoid. After 3h of the subsequent light phase, however, all contained a large cp nucleoid. Quantitative analysis of organelle DNA showed that for both mt and cp nucleoids, DNA synthesis occurs throughout the light in contrast to the synthesis of nuclear DNA which occurs only towards the end of the light period.

Based on an evaluation of the various morphological forms assumed by the nucleoids over the cell cycle Hiramatsu et al.(2006) traced the probable sequence of events involved in the transfer of the organelle genomes from one generation to the next.

Chloroplast Division in *Chlamydomonas*

Division of the chloroplast in *C. reinhardtii* has been shown to involve specific division proteins similar to those known to be active in bacterial cell division (Wang et al. 2003). One such protein is FtsZ which has been shown to form the Z-ring that constricts the bacterial cell membrane under the guidance of other proteins concerned with spatial arrangement of the ring (Bi & Lutkenhaus, 1991). Genes coding for FtsZ protein in *C. reinhardtii* were identified by Wang et al. (2003) and the genes coding for another two proteins MIND and MINE1 were identified by Hu et al. (2008). Although both of the latter proteins were thought to be located in the chloroplast, their respective genes were reported as being encoded in the nucleus.

Cultures of *C. reinhardtii* synchronized under 12:12h L:D cycles showed step-wise increases in mRNA of *FtsZ* and *MIN* genes corresponding with the step-wise synchronous cell divisions. The mRNA levels showed an oscillating rhythm over successive 24h cycles, increasing gradually in the light phases, peaking at the light-dark transitions, decreasing in the dark phases and reaching minimal levels at the dark-light transition points. Under light:dark conditions, peak levels of gene expression correlated closely with cell division.

In continuous darkness, the cells reproduced more slowly but maintained an oscillating circadian rhythm of mRNA gene expression. Under continuous light, synchrony of cell division was lost after 28h and the circadian rhythm of mRNA gene expression was also very much dampened from that point onward. These results pointed to a circadian expression of *FtsZ* and *MIN* genes of *C. reinhardtii* correlated with cell division (and hence, plastid division).

Changes in Photosynthetic Performance during the Cell Cycle in *Chlamydomonas*

A range of metabolic processes affecting photosynthetic performance of synchronous cultures of a number microalgal species have been shown to vary significantly over the cell cycle. Polypeptides of the PSII reaction centre of *C. reinhardtii*, were shown by Herrin & Michaels (1984) to be, for the most part, synthesised during the first few hours of the light period. Polypeptides of the light-harvesting chlorophyll *a,b* complex of PSII (LHCII) were, on the other hand, synthesised between the 7th and 9th hours of the light period. Cell cycle dependent changes of photosynthetic activity in the same species have been attributed both to changes in PSII activity (Marcus et al., 1986) and to changes in the level of RuBisCO (e.g. Herrin & Michaels, 1984).

Phosphorylation of thylakoid proteins, especially the LHCII polypeptides, was also shown by Marcus et al. (1986) to change markedly during the cell cycle of *C. reinhardtii,* reaching a maximum after 8 to 9 hours of light, decreasing to its minimum value towards the end of the light period and remaining at that level throughout the dark period. Changes in the activity of carbonic anhydrase, a component of the CO_2 concentrating mechanism facilitating supply of CO_2 (from bicarbonate), also varied during the cell cycle in a way similar to the other photosynthetic parameters (Marcus et al. 1986). Whether any or all of these cell cycle-related changes are the direct result of the light:dark regime or of an independent internal rhythm was not fully resolved by Marcus et al. (1986) although some of their observations would seem to favour the former explanation.

The Sexual Life Cycle of *Chlamydomonas*

The sequence of events involved in the sexual life cycle of *Chlamydomonas* has been comprehensively reviewed by Harris (2001) and it has been shown that three of the steps involved have a specific requirement for light. These are: gamete formation, mating of the gametes and zygote germination (Treier et al. 1989; Pan et al. 1997; Gloeckner & Beck, 1995; Saito et al. 1998). Gametogenesis takes place under conditions of nitrogen starvation but for maturation of the gametes there is an additional requirement for blue light (Weissig & Beck, 1991). Blue light is also required for maintaining the mating competence of the gametes. The final stage in the sexual life cycle, zygote germination, requires an initial ca. 3h light exposure for the quantitative induction of meiosis. All steps subsequent to meiosis may occur in the dark over a 20h period.

Using strains of *C. reinhardtii* with reduced levels of the photoreceptor phototropin – one of the photoreceptors known in a range of organisms to respond to the blue region of the irradiance spectrum – Huang & Beck (2003) were able to show that diminished levels of this receptor partially impaired all three of the light-dependent steps of the sexual life cycle. It has been suggested (Pan et al. 1997) that one of the possible physiological responses downstream of blue-light perception may be some chemical modification of the flagella involved in sexual agglutination. Some of the stages in the development of the zygote resulting from gametic fusion in *Chlamydomonas* have been summarised by Suzuki et al. (2000). Particular attention was paid to the zygotic cell wall which develops rapidly after fusion and which, within about 2.5 to 3h forms an insoluble structure through cross-linkages between certain zygote specific cell wall proteins (Minami & Goodenough 1978, Waffenschmidt et al. 1993). The wall itself appears dense and multilayered (Grief et al., 1987) and the cells are held together within a fibrillar gelatinous material, the pellicle matrix (Minami & Goodenough 1978). The long fibrils of the pellicle matrix are thought to be the result of binding of glutathione- sensitive lectin-like residues to sugar molecules to form the pellicle and it is further suggested that pellicle formation is particularly important when the zygotes form in a liquid medium rather than in their more usual soil habitat (Suzuki et al. 2000). The adhesiveness of the pellicle may, according to Suzuki et al. (2000), be effective in concentrating the dispersed zygotes so that when germination occurs, there is an increased likelihood of outcrossing among the meiotic products.

Zygotes resulting from union of gametes of *Chlamydomonas* induced by incubation in a nitrogen-free medium have been observed undergoing meiosis using the DNA-specific fluorochrome SYBR Green I (Aoyama et al. 2006). The time course of meiosis was estimated by measuring the frequencies of zygotes having two or four cells following exposure to light and it was judged that Meiosis I probably occurred after 16-18h (Aoyama et al. 2008). Changes in chromosome morphology and behaviour were observed throughout the meiotic cycle and the total number estimated to be 18 with the DNA content of the largest chromosome being 3-6 times greater than that of the smallest.

Germination of zygotes of *C. reinhardtii* usually occurs only after a period of dormancy. The standard protocol for inducing germination is to expose the zygotes, soon after mating, to constant light for one day, followed by constant darkness for about 6 days before final exposure to constant light again (Harris, 1989). After about 12-16h, the zygospores undergo meiosis followed soon afterwards by their release when the thick zygotic cell wall breaks down. Germination efficiency was shown by Suzuki & Johnson (2002) to be influenced by the total amount of light exposure but the cue for germination was a long-day photoperiod, sensed just prior to mating and during the first days of zygospore development. Because of the greater resistance of zygospores to freezing injury, suppression of germination under short-days, it was proposed, might be regarded as an adaptive response to over-wintering.

Cell Cycle Proteins in *Dunaliella*

In addition to the ever-growing family of cyclin-dependent kinases (CDKs) now known to play a central role in the cell cycle control mechanisms of microalgae (as they do in other systems), there is also a group of proteins that regulate various other components of the cell cycle and are detectable in individual cells by immunohistochemistry and immunofluorescence. An analog of the so-called proliferating cell nuclear antigen (PCNA) was detected in *Dunaliella tertiolecta* and found to be exclusively located in the nucleus (Lin et al. 1995). PCNA was found to be most abundant in S-phase cells, less abundant in some other phases but undetectable in early G1 or late M-phase. This pattern would be consistent with the presumed role of PCNA as an accessory protein to DNA polymerase-δ and hence essential for DNA replication. In exponentially growing and partially synchronised cultures, the percentage of PCNA-stained cells increased during the light period to reach a peak (75%) before the onset of the dark period when most of the cells (71%) were in the S-phase. The DNA synthesis inhibitor, hydroxyurea, depressed PCNA abundance whereas the mitosis inhibitor colchicine was without effect.

PCNA-like proteins have also been shown to occur in a number of other microalgal species namely *Isochrysis galbana, Thalassiosira weissflogii* and *Skeletonema costatum* (Lin et al. 1994) where their electrophoretic mobility and their reaction to PCNA-specific antibody has confirmed their affinity with PCNA in a wide range of organisms ranging from viruses to animals.

The precise role of PCNA in microalgal cell cycles is yet to be elucidated although by analogy with what has been described from other systems, an involvement with certain cyclins and with some CDK inhibitors in their respective roles in switching on and off the transition from the G1 to the S phase is strongly suggested.

THE CELL CYCLE IN *CHLORELLA* AND *NANNOCHLORIS*

These two non-motile, unicellular members of the Chlororcoccales have cell cycles that involve division of the protoplast to form non-motile autospores that are released following break down of the mother cell wall. *Chlorella* was by far the most commonly used microalga in the earlier studies using synchronous cultures and some of the major findings arising from that work are described below. The genus *Nannochloris*, although sharing with *Chlorella* a capacity for autosporulation, also contains species that divide by binary fission or by budding (Krienitz et al. 1996; Yamamoto et al. 2003). The taxonomic status of the non-autosporic species has been the subject of considerable debate (Yamamoto et al. 2003) and there have also been suggestions that some of the autosporic species are more closely related to some species of *Chlorella* (Krienitz et al. 1996; Yamamoto et al. 2001).

Early Work with Synchronous Cultures of *Chlorella*

Synchronous cultures of *Chlorlella* began to be extensively used in studies of the cell cycle in the late 1950s and the early sixties. This early work has been well summarised in two comprehensive reviews (Tamiya, 1966; Pirson and Lorenzen, 1966). Using synchronous cultures of *C. ellipsoidea*, Tamiya and his co-workers were able to identify seven distinct phases of the cell cycle (recognisable from microscopic observation of cell size and nuclear appearance) starting with recently released autospores and ending when they, in turn, release the next generation of autospores (four from each mother cell). This work led them to propose the probable existence of certain division-inducing factors which, it was further suggested, might be largely independent of the synthesis of nuclear and cellular material during the period of cellular growth.

The work summarised by Pirson & Lorenzen (1966) introduced the concept of the "shortest possible life cycle" (Lorenzen, 1957, Senger, 1961). By examining the behaviour of synchronous cultures of *Chlorella* under a range of conditions, they were able to identify growth conditions under which further enhancement of the factors affecting production, (cell mass and the number of autospores), did not affect the timing of autospore release.

They proposed a clock with a starting mechanism, the *Zeitgeber* (Pirson & Lorenzen, 1958; Senger, 1961), as a photoinducible system, superimposed on photoproduction. Thus, when an asynchronous culture was transferred to a dark –light regime with light sufficient to permit only one full cycle for those cells already primed to divide, the population became separated into two synchronous groups receiving their Zeitgeber impulses in turn and hence dividing alternately, both with stretched cycles (Senger, 1961).

These early studies also noted that light intensities supporting high photosynthetic rates during the growth phase of synchronous cultures of *Chlorella* could be inhibitory for certain developmental steps towards the end of the life cycle (Sorokin & Krauss, 1959, 1965). A consequence of this would be a stretching of the life cycle unless the inhibitory effect is removed by darkening. Tamiya (in Zuethen, 1964) suggested that this photoinhibition effect may be the result of competition between cellular syntheses depending on light energy and light independent formation of specific substances required for cell division. Hase et al. (1959) demonstrated the existence of specific substances, one of which they designated as an S-containing deoxyribopolynucleotide whose accumulation was causally related to cellular division, anticipating the later discovery of cell division proteins and other cell division factors now known to play a regulatory role in the cell cycle of a range of organisms from bacteria to higher plants and animals.

Nitrogen Metabolism of Synchronous Cultures of *Chlorella*

The enzyme nitrate reductase (NR) is known in a variety of higher plant tissues to loose its activity quickly in the dark and to regain it in the light (e.g. Hageman & Flesher, 1960). Similar fluctuations have been shown to occur in synchronous cultures of *Chlorella* maintained in light/dark cycles (Hodler et al. 1972; Tischner, 1976; Griffiths, 1979).

Two species of *Chlorella* (*C. vulgaris,* Brannon No.1, 211/11d and *C. emersonii,* 211/11n) grown autotrophically in LD cycles showed a rapid increase in NR activity over the first few hours of the light period followed by a slower rate of increase in activity over the remainder of the light period (Griffiths, 1979). During the subsequent dark period there was an immediate and rapid decline in NR activity, back to the level in the original daughter cells at the start of G1. Cultures not transferred to the dark but maintained in the light, showed an initial partial reduction in NR activity followed by a recovery such that the released autospores, unlike autospores produced during the dark period, had NR activities almost identical with that of the mother cells from which they had been derived. Under continuous illumination, the increase in NR activity keeps pace with the overall increase in biomass. It is only when the cultures are grown in an alternating LD cycle that they have bursts of very rapid increases in NR (in early G1) followed by rapid declines in activity during autospore production and release (during the dark phase).

There was evidence that while part of the increase in NR activity during G1 in LD cultures may be due to synthesis of new enzyme proteins, part may be due to other causes such as a reactivation of existing protein. Tischner & Hüttermann (1978) have shown that the rapid increase in NR activity at the start of the light period in synchronous cultures of another strain of *Chlorella* (*C. vulgaris*, 211/8k) is due to a light-mediated activation rather that a *de novo* synthesis of the enzyme.

"Giant" Cells of the Emerson Strain of *Chlorella*

When synchronous cultures of the Emerson strain of *Chlorella* (*C. emersonii*, 211/11n) were grown autotrophically under a 20:12 LD cycle with dilution at the end of each dark period, peak cell volumes increased from an initial 63μm^3 to a final 440μm^3 during the light period and the population density increased from 0.7×10^6 ml^{-1} to 2.5×10^6 ml^{-1} during the succeeding dark period (av. autospore no./cell = 3.6) (Griffiths 1970). When the cultures were transferred to a 1% glucose medium in the dark for a period of 7 days, their subsequent behaviour depended on the stage of the original LD cycle at which the transfer was made. Cells transferred towards the end of the light period (t = 18h) increased in number from an initial 1.7×10^6 ml^{-1} to a final 5.28×10^6 ml^{-1} over the 7 day period. The cell volumes at the end of the period of heterotrophic growth covered a wide range with a small proportion having volumes >4,000 μm^3. When the 7 day heterotrophic cultures were initiated with cells at the start of the original LD cycle, there was no significant increase in cell numbers during the period of heterotrophic growth and the cell volumes were distributed as a narrow peak around 2,500 μm^3 with some cells reaching volumes >6,000 μm^3.

It was concluded that for cells of the Emerson strain, it is only those that have already reached the latter stages of the growth period of the LD cycle (i.e. late G1) that are able to divide under heterotrophic conditions on glucose. Heterotrophic cultures initiated with cells from early G1 yield a homogeneous population of "giant" cells. The "giant" cells, when washed and resuspended in inorganic medium and returned to a normal LD cycle, recovered their capacity to divide, producing autospores with peak cell size distribution close to that of autospores in a normal autotrophic culture (ca. 150 μm^3) during the first recovery cycle.

In addition to their larger cell volumes (up to 20-30 times larger than late G1 cells of an autotrophic culture) the "giant" cells showed a number of ultrastructural differences from autotrophically grown cells (Griffiths & Griffiths 1969). There was a massive accumulation of starch in the chloroplast and the synthesis of thylakoids and of photosynthetic pigments had clearly not kept pace with the increase in cell volume during the period of heterotrophic growth. Recovery of the "giant" cells after return to autotrophic conditions was accompanied by pigment synthesis and restoration of photosynthetic capacity (Thinh & Griffiths 1970). The two major components of the light-induced recovery of giant cells, cell division and chloroplast development, could be separated by treatment with the protein-synthesis inhibitors chloramphenicol (which has its effect at the chloroplastic ribosomes) and cycloheximide (which has its effect at the cytoplasmic ribosomes). Best separation was obtained with the former which severely inhibited chlorophyll synthesis and the development of a photosynthetic capacity without affecting cell division (Thinh & Griffiths 1971, 1972a, 1972b). Evidence was also obtained that photosynthesis is not an essential requirement for the light-recovery of "giant" cells but recovery is strongly dependent upon respiratory mobilisation of reserve substrates (Thinh & Griffiths 1973a). The light-induced recovery occurs without significant DNA synthesis but with substantial RNA synthesis which occurs in two phases, an early 8- or 9-fold increase and a later phase coinciding with autospore production (Thinh & Griffiths 1973b).

Supplying heterotrophic cultures of the Emerson strain with L-arginine (in addition to the NO_3 of the standard medium) allowed the cells to reach their maximum "giant" dimension much sooner (ca. 6 days compared with the ca. 12 days in a 1% glucose medium not supplied with arginine) (Thinh & Griffiths 1974). The maximum "giant" size reached was the same with or without arginine and the number of autospores produced per "giant" cell was also the same but the "giants" produced in heterotrophic cultures supplied with arginine were "green giants" with well developed chloroplast thylakoids and photosynthetic pigments. Their production was preceded by an arginine-stimulated DNA and RNA synthesis. The arginine effect on autospore production in the Emerson strain may have a parallel in the findings of Kanazawa and his co-workers (Kanazawa, 1964; Kanazawa et al. 1965) that certain arginine-containing peptides may play an important part in the preparation of *Chlorella* cells for division.

Further studies have indicated that the arginine effect may be mediated through synthesis within the chloroplast of specific polypeptides necessary for nucleic acid synthesis and autospore production (Thinh & Griffiths 1977). Studies of the competition between arginine and its guanidoxy structural analogue, canavanine, in the special role of the former in its effect on autospore production in heterotrophic cultures of *C. emersonii,* have indicated that the analogue interferes with the synthesis of proteins essential for cell division and chloroplast development (Thinh & Griffiths 1980). Evidence was also obtained that protein synthesis that takes place after DNA replication is particularly sensitive to canavanine and that arginine-proteins may have a general role in controlling cell division in *Chlorella.*

Circadian Rhythms of Daughter Cell Production in *Chlorella*

Cells of *Chlorella fusca* var *vacuolata* growing under a 12:12h LD regime, were shown by Wu et al. (1986) to produce 18 autospores per mother cell at the end of every dark period. The cultures, after dilution, were then maintained in darkness for either 12 or 24h, (called the waiting time, WT) after the dark period and then returned to the 12:12h LD schedule to check for autospore production. It was found that after 12h WT, autospore production was 50% whilst after 24h WT, it was much higher, close to that obtained with 0h WT. Minimum production occurred again after 36h WT and this oscillating pattern continued for 4 days.

Cells exposed to weak light ($0.07 w.m^{-2}$) during WT exhibited the lowest autospore production independently of the length of WT but interruption of the weak light by a dark period (3h) increased autospore production by an amount that varied with the phase of the circadian rhythm. Interruption of WT (total darkness) by light pulses (0.5h light used for normal growth) also affected autospore production in a way showing phase shifts varying with the time of the pulse. These results were interpreted by Wu et al (1986) to indicate that autospore production in *C. fusca* is regulated by coupling to circadian oscillations.

The technique of subjecting cultures of *Chlorella* to a waiting time (WT) in the dark to show up any underlying endogenous rhythms was also employed by Chen & Lorenzen (1986). They noted that the rates, but not the duration of syntheses of chlorophyll, DNA and soluble protein changed rhythmically with additional WT after a growth cycle (LD 14:10). In particular, the rates of carbohydrate synthesis during the test cycle was completely different in cells subjected to a WT of 0 and 24h compared with those subjected to a WT of 12h. It was also noted that cells subjected to a WT of 12h remained longer in each multinuclear phase and had a much lower productivity.

Cell Division in *Nannochloris*

Cell division in seven strains of *Nannochloris* (from six species) was examined by Yamamoto et al. (2003) and compared with that in some other species belonging to the Chlorococcales. Using fluorescence microscopy (including the use of the cell wall specific fluorescent dye Fluostain I) they were able to confirm previous findings (Naumann, 1921) that *N. bacillaris* divides by binary fission while *N. coccoides* divides by budding. Two other species, *N. atomus* and *N. eucaryotum* were shown to be typically autosporic, like the three species of *Chlorella* examined and the one species of *Trebouxia.*

Analysis of possible phylogenetic relationships among the different dividing forms of *Nannochloris* and between them and the *Chlorella* species and some *Trebouxia* species was carried out based on DNA sequencing of actin genes and of 18S rRNA genes. This analysis pointed to *Nannochloris* being polyphyletic within the Trebouxiophyceae. The two nonautosporic species of *Nannochloris* (*N. bacillaris* and *N. coccoides*) were deemed to be monophyletic and positioned distally. Based on these and other analyses, Yamamoto et al. (2003) concluded that autosporulation is the ancestral mode of cell division in the genus *Nannochloris* and that binary fission and budding have evolved secondarily.

Division of the chloroplast during binary fission in *N. bacillaris* was shown by Sumiya et al. (2008) to involve special division proteins (FtsZ) similar to those involved in the ring structures that form during division of eubacteria (Bi & Lutkenhaus, 1991). The FtsZ proteins were identified by indirect fluorescent antibody staining and shown to play a part in regulating the position and action of the plastid division ring. The number of FtsZ rings was shown by Sumiya et al. (2008) to increase under phosphate enrichment of the medium but the doubling time of the cells was similar in phosphate-limited and phosphate enriched media. Although up to six FtsZ rings per chloroplast could be formed in cells grown in a phosphate enriched medium, and the chloroplast DNA content was 2.3 times that in a phosphate limited medium, the number of plastid division rings remained the same. In the presence of the DNA replication inhibitor (5-fluorodeoxyuridine) only one FtsZ ring was formed, even under phosphate enrichment suggesting that excess chloroplast DNA replication induces multiple FtsZ ring formation under phosphate enrichment.

THE CELL CYCLE IN *SCENEDESMUS*

Scenedesmus, like many other members of the Chlorococcales, but unlike *Chlorella* and *Nannochloris* (described above) is colonial with each colony being composed of four, eight or rarely 16 ellipsoid, oblong or fusiform cells, usually grouped in one plane with the long axes of the cells parallel to one another. In *S. quadricauda* reproduction occurs by two successive crosswise divisions of the protoplast of each cell and the daughter cells, after the necessary adjustments, form a new colony which is released by rupture or gelatinisation of the parent cell wall (Trainor et al. 1976). Synchronous cultures of various species of *Scenedesmus* were first described by Komarek (reported in Tamiya, 1966) and subsequent work by Šetlík et al. (1972) and Zachleder et al. (1975) went on to define the cell cycle of this microalga as comprising overlapping reproductive processes which interact in a variety of ways depending on the growth conditions.

Cell Cycle Patterns in *Scenedesmus*

Synchronous populations of *S. quadricauda* subjected to various irradiance levels, photoperiods, treatment with growth inhibitors and growth under heterotrophic conditions, were shown by Zachleder & Šetlík (1990) to follow a range of different cell cycle patterns. At low level irradiance levels, the cells attained consecutively two commitments to nuclear division, and two cell reproductive sequences in the one cell cycle. The mother cells yielded exclusively four daughter cells. Because of the low growth rates, the growth steps were long, allowing the reproductive steps of the first cell reproductive sequence to be completed before commitment to the next reproductive step. Under these conditions, the only variable components of the cell cycle were the growth steps; the reproductive component occupied fairly constant time periods regardless of growth rates. At high irradiance levels, the cells divided exclusively into eight daughter cells with four overlapping cell reproductive sequences in each cell cycle. The higher irradiance levels had the effect of shortening the growth phases causing an increase in the extent of overlapping reproductive steps and a consequent prolongation of the duration of these steps. Under treatment by chloramphenicol, an inhibitor of chloroplastic protein synthesis, cell growth was reduced thus prolonging the growth steps and preventing overlap of the reproductive steps of successive cell reproduction sequences.

It was concluded from these and other related results (e.g. those obtained using cultures grown in a glucose medium in the dark) that light has no direct effect on the duration of the reproductive steps. The role of light, it was suggested, was limited to its trophic function in photosynthesis.

Further studies of another species of *Scenedesmus* (*S. armatus*) (Tukaj et al. 1996) showed that RNA and total protein increased stepwise during the cell cycle with short time intervals between individual steps during which the rate of accumulation of RNA and protein decreased. At each of the steps, there was an approximate doubling of both RNA and protein and the number of steps increased with irradiance. At each of the growth steps there was a commitment to trigger the sequence of reproductive events (DNA replication, nuclear division and protoplast fission). The response to different irradiance levels was similar to that reported for *S. quadricauda* (see above) confirming that the length of the pre-commitment periods was reduced with increasing irradiance (as a result of an increasing rate of growth). The lengths of the post-commitment periods were independent of light as long as the number of growth steps (and reproductive sequences) did not change. At higher irradiances, the number of sequences of reproductive events increased resulting in a lengthening of post-commitment periods. Tukaj et al. (1996) concluded that the cell cycle of *Scenedesmus* is controlled by the activation of regulatory proteins which, in turn, are controlled by cell growth.

Changes over the cell cycle in the photosynthetic activity of synchronous cultures of *Scenedesmus* follow the same basic pattern shown by microalgae generally. In *S. quadricauda*, for example, photosynthetic activity, as measured by the rate of oxygen evolution, increased two-fold during the first 3-4h of the light period, remained high for the next 3 to 4 h and then declined during the last half of the light period (Kaftan et al. 1999). Photosynthesis was at a minimum during cell division, which occurred at the beginning of the dark period. Heil & Senger (1986) showed that in *S. obliquus,* the increase in photosynthetic activity during the early light phase was accompanied by an increase in phosphorylation of thylakoid membrane proteins suggesting some structural and physiological changes at that time in certain of the light-harvesting proteins. Then, Setlik et al. (1981) showed that the minimum photosynthetic activity in *S. quadricauda* coincides with the appearance of short stacks of thylakoid membranes (pseudograna). Using a photoacoustic spectroscopy technique, Szurkowski et al (2001) found that photosynthetic energy storage by synchronous cultures of *S. armatus* increased from 27% in young autospores to 35% in mother cells starting to release the next generation of daughter cells.

Kaftan et al. (1999) measured the flash-induced oxygen yield by cells of *S. quadricauda* as an indication of the maximum capacity of photosystem II (PSII) centres independently of subsequent electron transport reactions. They noted that over a period when the capacity of PSII was unchanged on a chlorophyll basis, the maximum rate of steady-state O_2 evolution increased more than 100%. This was interpreted as indicating that the increase in steady-state O_2 evolution was not due to an increased PSII capacity but to an increased rate of electron flow subsequent to PSII. The steep decline in photosynthetic activity during the latter part of the light period was deduced to be due to limited PSII activity. Kaftan et al. (1999) concluded that during the time of high photosynthetic capacity, PSII is used at its maximum capacity but that, after the cell has accumulated enough reserves for the next cell division (which occurs in the middle of the light period), the photosynthetic activity declines. During this period of decline, when energy inputs are not a priority, there is a dramatic reorganisation of the thylakoid membranes and the number of active PSII centres steadily declines, perhaps through inhibition of the PSII repair cycle. Restoration of PSII activity, a light-dependent process, occurs only after the daughter cells have entered the next light period.

Special Features of the Cell Cycle Related to the Colonial Habit

The combination of its colonial habit and its tendency for morphological variability (including alternate unicellular and colonial stages under certain conditions) confers upon *Scenedesmus* considerable flexibility in the form of its cell cycle. This has been well illustrated by Zachleder et al. (2002). They synchronized cultures of *S. quadricauda,* under a L(100Wm$^{-2)}$:D, 14:10h regime and then transferred cells at different stages of the light phase into the dark. It was observed that over the first 3 or 4h of the light period the cells doubled their RNA and total protein content. This period also allowed the cells to reach what was described as the first commitment point as indicated by the observation that during the subsequent dark period, although there was no further growth, one round of DNA replication and one nuclear division (to give two nuclei) was accomplished. It was further shown that in *S. quadricauda,* this first sequence of the reproductive process is completed by division not into two daughter cells but by the formation of a binucleate cell.

When the light period was extended to 6h, two growth steps were permitted and the amount of RNA and protein increased by a factor of four (in two steps), and two commitment points were reached, one at the end of each of the growth steps. These triggered two replications of DNA, each followed several hours later (and during the dark phase) by nuclear division. Shortly after completion of the second nuclear division (to produce four nuclei), the protoplast divided twice to produce four daughter cells which then formed the new daughter colony.

If the light period was extended to 15h (the time of protoplast cleavage in the normal L:D cycle) the cells completed three successive growth steps and passed through three commitment points. The one cell cycle, under these conditions, went through three rounds of DNA replication and the subsequent mitoses and protoplast cleavages to give a final eight-celled colony. Under continuous light (without the interruption to the cycle resulting from transfer to the dark), the cells became committed to further sequences with those initiated later in the cell cycle not being completed in the cell cycle in which they started but in the next one. Newly born daughter cells from cultures grown under different conditions may therefore differ markedly depending on how far along the cell cycle they have progressed.

Zachleder et al. (2002) then went on to investigate the effect upon the cell cycle of *S. quadricauda* of two inhibitors known to affect different aspects of cellular growth and division namely cycloheximide (a specific inhibitor of protein translation) and fluorodeoxyuridine (a specific inhibitor of the replication of nuclear DNA). When synchronous cultures were treated with cycloheximide (2 mg l^{-1}) for 3h between the second and third commitment points, protein synthesis stopped immediately. Accumulation of RNA, replication of DNA and division of the nuclei were also inhibited. By noting the recovery after washing off of the inhibitor, it became apparent that the reproductive processes had been more drastically affected than the growth processes. Cycloheximide primarily caused prolongation of the cell cycle (by about 10h) and a consequent increase in the number of commitment points. The timing of DNA replication, nuclear division and protoplast cleavage were also affected and there was extensive overlap of adjacent cell cycles with the final production of binucleate daughter cells.

Addition of fluorodeoxyuridine (25mg l^{-1}) at the beginning of the cell cycle prevented DNA replication and hence any follow-up nuclear division, protoplast fission or daughter cell formation. Growth processes were not, however, substantially affected and RNA and protein continued to accumulate. Washout of the inhibitor allowed the cells to trigger and complete the reproductive processes comprising DNA replication, nuclear division and protoplast cleavage without any intervening gap phases.

Co-ordination of Cellular and Organellar Events in the *Scenedesmus* Cell Cycle

Tight regulation of the different components of the growing and dividing cell would be a particular requirement for an organism like *Scenedesmus* having such extreme flexibility in the various pathways that the cell cycle can follow. The nature of this regulation has been studied using specific inhibitors that interfere with either chloroplastic or cytosolic events or with events centred on the nucleus or nucleoids (Cepák & Zachleder, 1988; Zachleder & Šetlik, 1990; Zachleder et al. 2004) and also by subjecting synchronous cultures to different light intensities (Zachleder & Cepák, 1987). The latter studies showed that at the lower light intensities there was a decrease in the size of the mother cells as well as a decrease in the number of daughter cells per mother cell, with the size of the daughter cells remaining unchanged.

Cepák et al. (2007) grew synchronous cultures of *S. obliquus* under a range of different temperatures (15° C – 33° C) and noted the effects on attainment of commitment points, on division of cell nuclei and chloroplast nucleoids and on the length of different cell cycle phases. They were able to show that lower temperatures caused a lengthening of the cell cycle (from 12h at 33° C to 70h at 15° C) without changing the number of daughter cells per mother cell. The size of the mother cells and consequently the size of the daughter cells increased with decreasing temperature. Temperature was without effect on the shape or size of the chloroplast nucleoids which were distributed randomly within the chloroplast during the entire cell cycle and divided asynchronously at all temperatures. Their numbers increased proportionally with cell and chloroplast size. Division of the chloroplast nucleoids preceded division of the cell nucleus.

Clearly, temperature and light (see above) had contrasting effects on various events of the cell cycle. In particular, while the mother cells at lower light levels were smaller (and had fewer daughter cells) those subjected to lower temperatures were larger but with no change in the number of daughter cells (which were, of course, larger). These differences were interpreted by Cepák et al. (2007) as being linked to effects on the length of the cell cycle and the time required to bring the mother cells to the point of commitment to divide. They were further interpreted as arguing against a role for a temperature-independent endogenous timer in regulation of cell division in *Scenedesmus*.

By using nalidixic acid (NAL), a specific inhibitor of chloroplastic DNA synthesis in algae generally, Zachleder et al. (2004) were able to confirm that, if the inhibitor was applied to synchronous cultures of *S. quadricauda* at the beginning of the cell cycle, chloroplast DNA did not replicate and nucleoids did not divide. Co-ordination between chloroplast division and cytokinesis was, however, maintained even though chloroplastic DNA replication was blocked. The inhibitor was without effect on the timing and number of rounds of reproductive processes in the nucleocytosolic compartment and their co-ordination with growth rate of the cell was unaffected. Transfer of cells to the dark during the cell cycle confirmed that the inhibitor affected none of the reproductive events in the nucleocytosolic compartment.

In their studies of another colonial member of the Chlorococcales, *Desmodesmus armatus,* Matusiak-Mikulin et al. (2006) drew particular attention to the significance of the ratio of dry matter to cell volume (DM:CV) as an indicator of key events in the cell cycle. They noted that synchronous cultures of *D. armatus* under a 14h:10h, light:dark regime attained three commitment points to division, at 3,6and 11h of the light period resulting in the formation of eight autospores which were released from each parent cell during the dark period. The successive commitment points, they further showed, coincided with times when the DM:CV ratio reached extreme values; minimum values at 3 and 6h, maximum value at 10h. The end of the pre-commitment period (3h) coincided with the maximum efficiency of energy capture by PSII reaction centres whereas the effective quantum efficiency of PSII was maximal at the time of the second commitment point (6h). Ultrastructural observations distinguished two functionally different phases of the cell cycle: the growth phase occupying the greater part of the cycle and the later reproductive phase. The initial decrease of DM:CV to its minimum value at 3h may be due to limited photosynthetic activity during acclimation of the photosynthetic apparatus or to a more rapid increase in cell volume resulting from recovery of the cytoplasm from its denser form in newly released autospores. The ultrastructural evidence also showed that cell density peaked at the 5-6h point due to the high photosynthetic activity. At the time of the switch from the growth to the reproductive phase, starch and lipids started to accumulate so that the young autospores, just before their release from the parent cell, were almost completely packed with starch and lipids.

Cell Cycle Regulatory Proteins in *Scenedesmus*

The involvement in the *Scenedesmus* cell cycle of cyclin-dependent kinases (CDKs) acting in heterodimeric complex with cyclins has been confirmed by Bišová et al. (2000). They reported that there were at least two different complexes expressing CDK-like activity in *S. quadricauda,* one relating to the attainment of the commitment point, the other to mitotic activity. The CDK-like proteins, like those previously reported for *Chlamydomonas* (Zachleder et al. 1997), also displayed histone H1 kinase activity, the basic reaction commonly used for assessing their activity. By using different illumination regimes (and other treatments) Bišová et al. (2000) were able to manipulate the timing of commitment points and cell division with consequent matching of the commitment points to a histone H1 kinase activity and of mitosis to a specific CDK activity.

Further information on these regulatory proteins has been provided by Vitová et al. (2008). Cyclin B was detected in *S. quadricauda* using anti-cyclin B1 antibody raised against recombinant hamster cyclin B1 and its accumulation analysed by immuno-blotting. Mitotic CDK-like kinase activity was monitored as the activity of *suc*1-bound H1 kinases.

Vitová et al. (2008) showed that during the course of a 12:6h L:D cycle, kinase activity peaked at 13h (i.e. at the start of the dark period) while expression of mitotic cyclin was detected as a specific band of about 66kDa from 10-17h, being most intense in the 13h sample. Both the kinase activity and cyclin decreased to their minimum levels by the end of the cell cycle, consistent with the reports (e.g. Murray et al. 1989) that mitotic CDK-cyclin complex activity is inactivated during mitosis due to degradation of cyclin through ubiqitin-mediated proteolysis.

In cultures treated with the specific inhibitor of nuclear DNA replication (5-fluorodeoxyuridine, FdUrd) both the kinase activity and the amount of cyclin B continued to increase during the dark period and were still increasing several hours after the decreases noted in the untreated cultures. Growth and attainment of the commitment points were unaffected by FdUrd while DNA replication, nuclear division and protoplast fission were blocked. Division of the chloroplasts was not blocked but delayed by about 2h, and over a longer time span compared with the untreated cells.

Division of the chloroplasts in a wide range of microalgae is known to involve a special division protein FtsZ which in the *S. quadricauda* experiments reported by Vitová et al. (2008) increases during the growth phase, reaches a maximum after 12h and then decreases. This finding was consistent with observations that a Z-ring is formed before chloroplast division and disassembles during chloroplast constriction (Miyagishima et al. 2001). In FdUrd- treated cultures, FtsZ increased more slowly and, after 23h did not decrease which was interpreted as being consistent with the comparative resistance of chloroplast division to the inhibitor.

The Cell Cycle in Diatoms

Diatoms, unlike most other microalgae, require the element silicon (Si) for their growth and it has long been known that cell division will not occur in a medium depleted of silicic acid (Lewin, 1955; Lewin and Cheng, 1968; Werner, 1966). Silicon is specifically required for the deposition of new valves which occurs between mitosis and daughter cell separation and this creates a strong link between the cell cycle and silicon metabolism (Brzezinski 1992). Deposition of new valves during daughter cell formation involves the sequential formation of the girdle bands of the cingulum which, it has been shown, contain a large fraction of the Si in many species (Round 1972b). In *Navicula pelliculosa,* new valves are deposited after cytokinesis followed by deposition of the first and sometimes the second girdle band while the daughter cells are still attached (Chiappino & Volcani 1977). Formation of the third and final girdle band follows so that deposition of the entire frustule is completed during that segment of the cell cycle beginning just before cytokinesis and ending just before daughter cell separation (Sullivan 1977). In addition to the Si dependency at the G2/M phase associated with the construction of new valves, there is also evidence of an Si-linked arrest point at the G1/S boundary (Brzezinski et al. 1990). The latter, it has been suggested, may be linked with a special mechanism regulating DNA synthesis (Vaulot, 1985).

Cell Synchrony and the Si-requirement in Diatoms

The specific requirement for silicon has been used, usually in conjunction with light-dark cycles, to induce synchrony in a range of diatoms. In synchronised cultures of *Cylindrotheca fusiformis*, for example, cells undergo DNA synthesis between 0.5h and 3.5h into the recovery period from silicon starvation and between 8h-13h after transition from the last dark period into continuous light in light-dark synchronised cultures (Paul & Volcani, 1976; Borowitzka & Volcani, 1977). Daughter cell production is completed between 4h-7h or 13h-17h with the respective synchronies. The silicon requirement for production of daughter cells in *C. fusiformis* was shown by Okita & Volcani (1978, 1980) to be due, at least in part, to a silicon requirement for the synthesis of proteins, including DNA polymerase. By analysing the timing of mRNA and polypeptide accumulation in synchronous cultures of this diatom, Reeve & Volcani (1985) were able to identify changes in individual mRNAs and polypeptides that were common to both silicon-depletion- and light-dark-induced synchronies, pointing to a probable link to cell cycle events. Although the majority of the changes in mRNA and protein accumulation in synchronous cultures of *C. fusiformis* were shown to be due to the synchronisation procedure applied, eleven mRNAs and three polypeptides accumulated between mid-S phase and daughter cell production and were interpreted as being cell cycle stage specific.

Limitation of the growth rate of diatoms by the availability of light or of essential nutrients may lead to an increased incorporation of silica into the cell wall, the result, at least in some cases, of prolongation of the cell cycle phase in which most cell wall synthesis occurs (Martin-Jézéquel et al. 2000). Silica starvation has been shown to arrest the cell cycle at two specific locations, namely, G1 (or the G1-S phase boundary) or at G2+M (Brzezinski et al., 1990). The arrest in early G1, and further progress through the cell cycle after silicate replenishment, was studied in the marine centric diatom *Thalassiosira pseudonana* by Hildebrand et al. (2007). They were able to confirm that Si-starved cultures arrested in early G1, when replenished with silicate proceeded to girdle band synthesis which was confined to a particular period of G1. By measuring silicic acid uptake, silica incorporation into the cell wall and fluorescence visualisation of newly synthesised cell wall structures, they were able to measure incorporation of silica into individual girdle bands and valves.

Cell Cycle-related Metabolic Changes in Diatoms

By growing the marine diatom *Thalassiosira pseudonana* in continuous culture under N- or light-limitation, Claquin et al. (2002) showed that both types of limitation resulted in similar regulation of the cell cycle. At the lower growth rates, there was an increase in the duration of G1, S, and G2+M phases and the percentage of cells in the G1 phase decreased but the percentage of cells in the G2+M phase increased. Cultures with lower growth rates had significantly higher amounts of biogenic silica per cell and per cell surface, reflecting the longer time spent in the G2+M phase and confirming the relative independence of silicon uptake from regulation by other elements. The apparent uncoupling of Si metabolism from C or N metabolism implies, according to Claquin et al. (2002), that it is possible to infer the growth rate of the culture from frustule thickness.

Subjecting diatom cultures to silica starvation or to a prolonged dark period, although effective in inducing synchrony, may not necessarily distinguish true cell cycle effects from those due to the treatments themselves. Claquin et al. (2004) sought to resolve this problem by inducing synchrony in cultures of *Cylindrotheca fusiformis* by using two cell cycle inhibitors (nocodazole and aphidicolin) both of which could be easily removed from the culture medium after the treatment,. Using this technique they were able to obtain clear evidence that variation and regulation of photosynthetic carbon metabolism in *C. fusiformis* was linked to the cell cycle.

They showed that in unsynchronised cultures of *C. fusiformis* under a light:dark cycle (6:6h) regime, the percentage of cells in the successive cell-cycle phases remained almost constant over the 12 hours, being around 56, 23 and 21% for the G1, G2+M and S phases respectively. Both maximal photosynthetic capacity (PB_m) and the maximum electron transport rate (ETR_m) were independent of light and darkness. In cultures synchronised by the two inhibitors and subjected either to continuous light or to a light:dark cycle, PB_m varied by a factor of 2 or 3 as a function of cell cycle stage and ETRm followed a similar pattern. Both PBm and ETRm were constant and low before division, and increased during division to reach the highest values at the end of division.

The high photosynthetic activity after cell division may, according to Claquin et al., (2004), serve to increase the synthesis of carbon products and thus cellular growth during the growth phase. Then, after having accumulated enough reserves for the next division, photosynthetic activity declines perhaps through feed-back regulation. The link between ETRm and cell cycle events was interpreted by Claquin et al. (2004) as being partly due to mitochondrial respiration although Mehler activity (i.e. reduction of O_2 with subsequent production of H_2O_2 rather H_2O) accounted for 50-60% of the oxygen uptake under the growth irradiance chosen by them and the relative Mehler activity was independent of the cell cycle.

Cell cycle-related effects on nitrate uptake and assimilation have been described for a number of diatoms. In *Thalassiosira weissflogii,* diurnal fluctuations of nitrate reductase activity have been shown to involve control at the transcriptional level (Vergara et al. 1998). The same is also true for another key component of nitrogen metabolism, the nitrate transporter (NAT) which in light/dark synchronised cultures of *Cylindrotheca fusiformis* is strongly linked with the cell cycle (Hildebrand & Dahlin 2000). NATmRNA levels were shown in this diatom to be high in early G1 phase and, after decreasing through the remainder of G1, increased again during DNA synthesis in S phase and into G2, only to finally decrease again after M phase. In cultures synchronised by silicon starvation, NATmRNA levels were high at the G1/S phase boundary, high throughout S and G2 and finally low again after M phase.

Diurnal expression of genes encoding for enzymes active in nitrogen assimilation in *Thalassiosira pseudonana* was examined by Brown et al. (2009). Transfer of cells from an NH_4^+ medium to a NO_3^- medium resulted in accumulation of *nia* transcripts (encoding for nitrate reductase, NR). *Nia* mRNA levels varied diurnally and the diurnal oscillations were abolished when the cells were transferred to continuous light. Transcript levels for genes encoding for chloroplastic and cytosolic NR increased within 2h of NO_3 addition with diurnal variation. There was also diurnal variation in transcript levels for genes encoding for chloroplast-localized glutamine synthetase and for cytosolic-localized glutamine synthetase, but the latter was out of phase with the others. The diurnal variation in all these transcript levels was abolished when the cells were transferred to continuous light pointing to control by metabolic events triggered by the L:D transitions rather than by an endogenous circadian rhythm.

Sexual Reproduction in the Diatoms

Diatoms, both the centric and pennate forms, can undergo sexual reproduction involving formation of gametes which, during fertilization, fuse to produce zygotes (auxozygotes) (Drebes 1977). Most of the centric diatoms undergo an oogamous form of reproduction in which a large non-motile gamete (egg) is fertilized by a small motile flagellated gamete (sperm). In pennate diatoms the gametes are generally morphologically isogamous and physiologically anisogamous. The gametes are non-flagellate (aplanospores) that are brought together by prior pairing of their mother cells (gametangia). Meiosis occurs in the final stages of gametogenesis.

Populations of the marine centric diatom *Ditylum brightwellii* in coastal waters at Puget Sound, Washington, USA have been described by Rynearson & Armbrust (2000) as exhibiting high levels of genetic diversity suggesting that sexual reproduction is a key component of the life history of this diatom as is presumed to be the case for most other species. Culture studies and field observations of *D. brightwellii* (Koester et al. 2007) have shown that this species is homothallic, producing and releasing two naked, spherical eggs from each oogonium and 64 uniflagellate sperm, 4 from meiosis of each of 16 spermatogonia while they are enclosed within a mother cell (spermatogonangium). Cell size restoration, an essential requirement to counter the progressive decrease in size resulting from repeated binary fission under the restriction of the rigid salicaceous shell (Round, 1972a) occurs from a putatively fertilized egg via auxospore formation.

The Cell Cycle in *Euglena*

Leedale (1959) noted that the microalgal flagellate *Euglena*, growing under autotrophic conditions and subjected to diurnal light-dark cycles tended to divide at night. Later, Cook & James; Edmunds; and Padilla & Cook – reported in Edmunds, 1964) were able to select for different strains of *Euglena* and for different temperatures, the correct light-dark regime to allow close to 100% synchronisation of laboratory cultures. Cultures of the flagellate grown heterotrophically on a complex organic medium could also be synchronised by the application of heat shocks, cold shocks or diurnal temperature changes (Pogo & Arce; Padilla & Cook. – reported in Zeuthen, 1964).

Hagiwara et al. (2001) showed that when a mixed culture of *E. gracilis* is transferred from continuous light to continuous darkness, some cells in the G1, S or G2 phases are able to proceed for a further one or two cell cycle transitions. Eventually they, like the remaining group that could not so proceed, become arrested in the S, G2 or G1 phase confirming that once begun, mitosis can be completed in the absence of light. It was further shown that cells in the later stages of the G1, S or G2 phases are more likely to proceed to the next cell cycle transition in continuous darkness than those in the earlier stages of the respective phases. This apparent requirement for maturation within the phase was interpreted as being consistent with the suggestion that ca. 25% of the steps that occur in the S phase are obligatory for the S/G2 transition in continuous darkness. It was further shown that cells exposed to higher irradiances were more likely to undergo cell cycle transition during the subsequent period of continuous darkness. Stronger light, it was suggested, results in earlier commitment at each cell cycle phase, pointing to the importance of photosynthetic production in the photo-induced commitment to cell cycle transition.

In cultures maintained in continuous light, progression through the cell cycle of cells in the post G1 phases was found to be unaffected by light intensity (ranging from just above compensation to saturation). Progression through the cycle of G1 phase cells was slowed by decreasing light intensity.

Circadian Gating of Cell Division in *Euglena*

Cell division in *Euglena gracilis* was shown by Edmunds (1966, 1978), and Edmunds & Adams (1981) to be subject to periodicities influenced by circadian rhythms. Then Edmunds et al. (1982) noted phase shifts in the circadian rhythm of cell division in *Euglena* following light perturbations. Cultures grown at 25° C in a repetitive 12:12h, L:D regime were entrained to a precise 24h period (τ = 24h, where τ is the time between onsets of cell division in the population). Under a 3:3h L:D regime, however, the culture free-runs but with a τ value of approximately 30h when plotted over several days. The τ value under the latter regime varied slightly between different strains but was always virtually independent of the percentage of cells dividing during any given burst. 3h light perturbations were imposed at different circadian times throughout the 30 h division cycle by giving light during one of the intervals when dark would have fallen in the 3:3h L:D regime. This caused a shift of the rhythm; maximum shifts (approx 11 to 12h advance or delay) being obtained at circadian time (CT) 20 to 22h (the "breakpoint"), and little or none if the light signal was given between CT6h and CT12h. These findings allowed Edmunds et al., (1982) to construct phase response curves pointing to the existence of a circadian oscillator that modulates cell division in *Euglena.*

The mechanism by which the circadian oscillator described above regulates cell division has been described as "gating". The term implies that cell division can occur only when the cycle is in a particular circadian phase and for *Euglena,* as is the case for microalgae generally, the circadian clock is set by light. When cultures of *E. gracilis* entrained to a 24h light-dark cycle (14hL) were transferred to continuous darkness at the eighth hour of the final L:D photoperiod, cell cycle transition was arrested for all cells in phases G1, S or G2 (Hagiwara et al. 2002). Subsequent exposure of the dark-arrested cells to a 6h light period allowed them to continue through the cell cycle in continuous darkness but in a way controlled by the circadian cycle. Light treatment applied at the subjective dusk was most effective in inducing commitment to divide, whereas light treatment around the subjective dawn had no effect.

Hagiwara et al. (2002) went on to suggest that the light signal for commitment of G2 cells to divide originates in the non-cyclic electron transport component of photosynthesis (PET), particularly in electron flux through the cytochrome b_6-*f* complex, and does not involve overall photosynthesis, photophosphorylation or $NADP^+$ reduction. The fact that the circadian rhythm for photoinduction of cell division runs out-of-phase from that of PET was interpreted by Hagiwara et al. (2002) as indicating that the site of light-induced commitment to divide was downstream of non-cyclic PET. Circadian gating of the photo-induction of G2 cells to divide is thought to form an important part of the mechanism for 24h LD cycle-induced synchrony of cell division.

The primary site within the cell cycle of *Euglena* at which the circadian rhythm exerts its control was investigated by Bolige et al. (2005). They noted the report of Carré & Edmunds (1993) that cultures of an achlorophyllous mutant strain if *E. gracilis* grown heterotrophically contained a large number of cells in G2+M phase even in the subjective day when there was no increase in population density. This, they suggest, begs the question whether those G2+M phase cells are developmentally ready for cell division and if so, were they perhaps prevented from completing cell division by circadian control.

Bolige et al. (2005) sought to answer this question by transferring asynchronously dividing (LL) cultures of *E. gracilis* into 2h L:D cycles (1h light:1hdark), a regime previously shown by Edmunds (1988) to be one in which the cell cycle becomes regulated by a circadian rhythm. Under the 2h LD cycle regime, gated increases in population densities occurred only during the subjective night periods (i.e. starting at 24,50,76 and 102h respectively) with no increases in population density occurring during the successive subjective day periods. By combining microscopic observations with measurement of cellular DNA content Bolige et al. (2005) were able to confirm that the circadian rhythm prevents G2 cells from progressing to mitosis so regulating the timing of the G2→M transition and allowing cell population growth to occur only during subjective night. The observations further showed that committed G2 cells, although developmentally ready (i.e. mature) at the time of commitment are nevertheless prevented from progressing to mitosis. Although much remains to be explained, the most likely hypothesis is that gating of population growth is the result of the combined effects of circadian timing of commitment and of developmental control.

THE CELL CYCLE IN DINOFLAGELLATES

Because of their relatively large size and easily recognisable morphological states corresponding to different stages of the cell cycle, the dinoflagellates can for some purposes be studied as individual cells without need of the synchronous culture techniques that are usually necessary for the smaller microalgae. They have other interesting and in many ways unique characteristics including a number of what are generally regarded to be primitive features. Their chromosomes occur in a permanently condensed state and lack the histones that are typically involved in the packaging of eukaryotic DNA (Rizzo 1987). The nuclear envelope remains intact during mitosis and nuclear division involves longitudinal division of the chromosomes, with the chromatids remaining attached to the nuclear envelope. Division of the nucleus is achieved by a number of irregular invaginations of the nuclear envelope (and their associated arrays of microtubules) distributing the paired chromatids into their respective daughter nuclei. Because of these features, many having strong parallels with bacterial cell division, the dinoflagellates are generally described as being mesokaryotic, that is, intermediate between the prokaryotes and eukaryotes(Rizzo 1987).

Circadian Rhythms and the Cell Cycle in Dinoflagellates

Some dinoflagellate species were among the earliest microlagae for which it was suggested that the cell cycle may be linked to certain endogenous circadian rhythms. In *Gonyaulax polyedra*, for example, the rhythm of cell division was shown to persist after the cells had been transferred from an alternating light:dark regime to constant dim light with maxima in the number of dividing cells occurring approximately every 24 hours although the generation time under these conditions was 3 to 4 days (Sweeney & Hastings, 1958). It was suggested that some phase of the cell division cycle may be **"gated"** by the circadian oscillator so that only cells that have reached a certain stage of the cycle at a given circadian time can progress to division, all others being held back until the next opening of the gate. Such an interpretation would be consistent with the observation that for *G. polyedra* the generation time of individual cells was an even multiple of 24h.

In an attempt to explore the suggested link between the circadian cycle and the cell cycle Sweeney (1982) studied a range of cell cycle parameters in cultured individual cells of the marine planktonic dinoflagellate *Pyrocystis fusiformis.* This organism was chosen because it displays a clear circadian rhythmicity of some of its physiological functions (eg. bioluminescence) but has a long generation time (5 to 6 days) easily distinguishable from the circadian cycle. Different stages of the cell cycle were distinguished from the observed morphologies and the S (DNA synthesis) phase identified by measuring nuclear DNA content by fluorescence microspectrophotometry. By examining the relationship between the cell cycle and the circadian cycle it was clear that the transition from one morphological stage to the next (i.e. the transition from $G_1 \rightarrow S \rightarrow G_2 \rightarrow M \rightarrow C$) occurred at specific times of the circadian cycle. Cells changed from one morphological stage to the next only during the night phase of the circadian cycle, under both light:dark cycles and in continuous darkness. Cells in all segments of the cell division cycle displayed the usual circadian rhythm of bioluminescence. Sweeney (1982) concluded that transitions from one morphological stage to the next (and presumably from one stage of the cell cycle to the next) are all under circadian control.

Natural populations of *Prorocentrum triestum* have a cell cycle of approximately 24h and a division rate of 1.00 day^{-1} resulting in synchronous division of all cells late in the dark phase (Costas et al. 1992). In *Gonyaulax polyedra*, which has a longer cell cycle (<1 division per day), only a fraction of the population divides each day, but within the narrow window (the first hours of the light period) dictated by the diurnal cycle (Karentz 1983, Cetta & Anderson 1990). The same is true for other slow-growing species like *Gymnodinium nagasakiense* which divides only during the latter half of the dark phase (Yamaguchi, 1992) and

Gambierdiscus toxicus (division rate 0.15 day^{-1}) with only 15% of the population progressing to mitosis in one photoperiod (van Dolah et al. 1995). Other species, like *Gyrodinium uncatenatum* appear to be less tightly coupled to the photoperiod and may divide throughout the latter half of the light phase and the first half of the dark phase (Cetta & Anderson 1990).

When the red tide dinoflagellate *Gymnodinium breve* was cultured in a 16:8h L:D cycle at 25° C (van Dolah & Leighfield, 1999), the earliest S-phase cells were first apparent at ca. 6h after the onset of light and continued for ca. 10h. The maximum number of cells in the G2 + M phases occurred at 18-20h and mitosis was completed ca. 4h later. A similar pattern of cell cycle progression was observed in a number of other cultured isolates of this species and in samples taken at different times over one diurnal cycle of a natural bloom. It was further shown that a 4h delay in the dark/light transition (i.e."dawn") caused a matching delay of entry into the S phase and a corresponding delay in the appearance of G2 + M phase cells. Bringing the L/D transition ("dusk") forward by 4h did not result in a significant shift in the occurrence of G2 + M phase cells, nor did it alter the time at which cells entered the S-phase during the following diel cycle. Van Dolah & Leighfield (1999) concluded from this that the cue responsible for phasing the *G. breve* cell cycle to the diel cycle is the D/L transition point.

The Role of Light in the Circadian Cycle in Dinoflagellates

The role of the light:dark cycle in providing the major input signal for entraining the circadian rhythm has been studied in another dinoflagellate *Karenia brevis.* Van Dolah & Leighfield (1999) showed that in that organism the transition from dark to light serves to entrain the cell cycle by controlling the entry into the S phase. Brunelle et al. (2007) confirmed this by noting that when cultures of *K. brevis* were grown in a 16:8 L:D cycle, the S phase began between 10 and 13h (and mitosis occurred late in the dark phase). This pattern of entry into the various cell cycle stages did not change when the cultures were placed in continuous light even though, by day three, the percentage of cells proceeding through the cell cycle had almost doubled.

Brunelle et al. (2007) went on to show that the rate at which cells of *K. brevis* progress through the cell cycle is strongly influenced by light quality. Thus, when cultures were transferred from the 16:8 L:D cycle into red, blue or white light of similar intensity, they proceeded into the S phase earlier in the blue light compared with the white light and a greater percentage of the cells entered the cell cycle. Red light caused no such temporal change compared with white light.

The receptor responsible for the blue light effect on cell cycle progression was characterized by Brunelle et al. as being a cryptochrome (Kb CRY) belonging to the cryptochrome DASH group of pigments (CRY DASH) known to be present in a number of organisms. Like the CRY DASH known for *Arabidopsis* (Cruciferae), Kb CRY is localised in the chloroplast. Another source of CRY DASH proteins is the cyanobacterium *Synechocystis* where it is thought to bind to DNA and serve as a transcriptional repressor (Brundler et al. 2003). Kb CRY, as the only blue light receptor in *K. brevis* is considered by Brunelle et al. (2007) to be a likely candidate as the photoreceptor involved in the input signalling pathways of the circadian clock in that organism.

Cell Cycle Regulatory Proteins and the Circadian Cycle in the Dinoflagellates

In their studies of the dinoflagellate *Cryptothecodinium cohnii,* Rodriguez et al. (1994) identified a compound resembling the cyclin-dependant kinase cdc2 kinase known to be a general participant in eukaryotic cell cycle progression. Then, van Dolah et al. (1995) showed that this cdc2-like kinase was present in *Gambierdiscus toxicus* and, moreover, was active *in vitro* against a known substrate of cdc2 kinase, histone H1 (from calf thymus). By measuring the mitotic index (i.e. the percentage of cells undergoing mitosis) at different stages of the 18:8h L:D cycle, van Dolah et al. were able to show that in *G. toxicus,* mitosis occurred only during a 3h window beginning 6h after the onset of the dark phase, with maximal numbers of mitotic cells present 8h after start of the dark period. Cell lysates prepared at each sampling point of the cell cycle were then analysed for cdc2 kinase activity and its ability to phosphorylate calf thymus histone H1. It was found that while the cdc2 kinase-like protein (ca. 34 kDa) was expressed constitutively throughout the cell cycle, its activation was cell-cycle dependent, maximal histone H1 phosphorylation coinciding with mitosis.

When cell lysates of *Gymnodinium breve* subjected to electrophoretic protein separation were probed with a conserved amino acid sequence fragment present in all known members of the CDK protein family, a 34 kDa immunoreactive protein was identified (van Dolah & Leighfield, 1999). Its involvement in cell cycle progression was confirmed by treatment with a specific CDK inhibitor at different phases of the cycle. Cyclins, known from studies with higher plants and animals to have catalytic activity in the cell cycle when they are bound to CDKs, have also been found to occur in dinoflagellates (Barbier et al. 1995).

Molecular Genetics of Circadian Regulation in Dinoflagellates

Making the assumption that the first step in circadian physiological regulation will involve gene expression of some key component of the system, Okamoto & Hastings (2003) made use of the large cDNA library available for *Pyrocystis lunula* to test whether the transcription of any of the genes was under circadian control. About 3% of the unique genes screened were identified as being under circadian control and more than 50% of them could be identified with diverse known genes. Most, it was further shown, were preferentially expressed in either early subjective day or late subjective night. Light exposure at times which might be expected to induce phase shifts in the rhythm revealed 30 differentially expressed genes. Genes identified as being under circadian control, it was concluded, are most likely to code for proteins important in those processes peaking at times most relevant to their function. Light-harvesting genes, for example, may contribute to the diurnal rhythm of photosynthesis (Hastings et al. 1961) while differential actin expression may be involved in the rhythm of chloroplast movement (Sweeney 1984).

THE CELL CYCLE IN UNICELLULAR CYANOBACTERIA

Cell cycle behaviour of the unicellular planktonic cyanobacteria *Synechococcus* and *Prochlorococcus* (Prochlorophyta) has received considerable attention partly because these two genera are important primary producers in many marine and freshwater environments and partly because they offer interesting comparisons with the bacteria on the one hand and the eukaryotic microalgae on the other. Special attention has been directed toward the timing of DNA replication and the factors influencing cell division and daughter cell production.

Studies with Synechococcus

Studies of the cell cycle of the WH8101 strain of *Synehococcus* (and a number of other marine strains) have shown that under a range of different conditions, the frequency distribution of cellular DNA is typically bimodal, the peaks corresponding to one and two chromosomes per cell, respectively (Armbrust et al. 1989, Binder & Chisholm 1995, Vaulot et al. 1996). This indicates that there are no overlapping rounds of DNA replication as are characteristic of fast-growing prokaryotes (and some other strains of *Synechococcus*) where the cells contain multiple genome copies even at very low growth rates (Mann & Carr, 1974; Binder & Chisholm 1990, 1995) and where asynchronous initiation of replication occurs at multiple chromosome origins.

Cultures of the coastal-dwelling WH8101 strain grown under light-limited or nitrogen-limited conditions giving growth rates of between 1.83 d^{-1} and 0.25 d^{-1} all showed the typical bimodal DNA frequency distribution (Binder 2000). Binder calculated the relative duration of various components of the generation time (τ) namely, the time between cell birth and initiation of chromosome replication (B), the time between initiation and termination of replication (C) and the time between termination of replication and cell division (D) and found that, even at the highest growth rates, τ was always greater than (C + D). This was interpreted to indicate that strain WH8101 is not capable of the overlapping rounds of chromosome replication that occur in fast growing bacteria.

Over the range of growth rates examined, chromosome replication time (C) in this strain and in another (open ocean) strain (WH8103 – Burbage & Binder 2007) was constrained to a fairly narrow range of values while the lengths of pre- and post-replication periods (B and D, respectively) both increase with decreasing growth rate. In the WH8103 strain, the length of the post-replication period (D) was maximal (10-20h) at its lowest growth rate. As growth rate increased, D declined rapidly reaching minimal values of ca. 2.2h. The length of the pre-replication period (B) was maximal (29-33h) at the slowest growth rates decreasing to a minimum value of 4.3h at the highest growth rate. The combined length of time required for DNA replication and cell division varied 5-fold over the growth rates tested and the cell mass at the point of initiation of chromosome replication ranged from 127 to 285 $fgC.cell^{-1}$.

These and other parameters measured were interpreted to indicate that these strains conform with the pattern of behaviour ($\tau > C + D$) typical of slow-growing prokaryotes and more like what is found in the eukaryotic cell cycle where a period of DNA synthesis is preceded and followed by a gap in DNA synthetic activity. Their cell cycle is described as starting with daughter cells having a single chromosome, inherited at birth, which begins to be replicated at some point thereafter and then, some time after completion of replication, the cell divides to form new daughter cells. The timing of initiation if chromosome replication, Binder found, appears to be tightly coupled to cell volume with the critical cell size remaining approximately constant except at very slow growth rates.

Synechococcus strain WH7803 (an open ocean strain) and strain PCC6301 (a freshwater strain) both have a multimodal DNA distribution pattern indicating the existence of more than two genome copies per cell (Binder & Chisholm 1990, 1995) although no such bimodal DNA distributions were observed by Liu et al. (1999) in N-limited populations of the former strain.

Studies with *Prochlorococcus*

The prochlorophyte *Prochlorococcus* has been shown to have a cell cycle that is tightly phased to the light:dark regime of its warm, mostly tropical and sub-tropical oceanic environments, indicating a maximum daily division rate of ca. 1.0 (Vaulot et al. 1995, Moore et al. 1995). However, cultures of this prochlorophyte isolated from different locations and maintained under a range of light intensities and different light:dark regimes have yielded growth rates well in excess of one division per day (Shalapyonock et al. 1998). Thus, at light intensities of 40 μmol m^{-2} s^{-1} and higher, DNA frequency distributions indicate that some of the daughter cells released towards the end of the light period undergo a second division during the subsequent dark period. At lower light intensities, this second wave of cell division is absent. This feature of *Prochlorococcus*, with strict timing of both DNA replication and cell division and the very short period of growth needed before the second division, has been described as one that combines the efficient use of light for photosynthesis with the potential for maximum cell division during the dark, so that cells that have fixed enough carbon during the light are able to divide more than once (Shalapyonok et al. 1998).

The relationship between growth rate, various cell cycle parameters and cell size of the MIT9312 strain of *Prochlorococcus* were found by Burbage & Binder (2007) to be very similar to those described above for the WH8103 strain of *Synechococcus.* The DNA frequency distributions for both strains were bimodal at all growth rates, chromosome replication times (C) were similar and in neither strain did C appear to vary systematically with growth rate.

Jacquet et al. (2001) have confirmed that in *Prochlorococcus,* the key parameter for synchronized cell cycling is the "light on" signal, which initiates the S (i.e. DNA synthesis) phase. Moving *Prochlorococcus* cultures from a low- to high-light:dark regime produced an increase in the number of cells in both the S phase and in the G_2 (i.e post DNA replication) growth phase, with a consequent increase in growth. The reverse shift (from high to low-light regime) reduced the growth rate of the population, confirming close coupling between irradiance levels and cell cycling.

The close synchronization of the cell cycle with light:dark regimes simulating those in the upper layers of the ocean was studied by Holtzendorff et al. (2001), using turbidostat cultures of the PCC 9511 strain of *Prochlorococcus*. They cloned and sequenced two genes (*dnaA* and *ftsZ*) coding for enzymes known to be involved in cell cycle-related processes and observed that both genes exhibited clear diel expression patterns with mRNA maxima during the replication (S) phase. Peak FtsZ concentration occurred in the dark period (at the time of cell division).

Circadian Rhythms and Cyanobacterial Cell Division

A circadian rhythm of cyanobacterial cell division was first confirmed in the marine *Synechococcus* strain WH7803 (Sweeney & Borgese, 1989). This work sought to avoid the confusion between circadian and cell cycle control and to establish the independence of the cycle period from temperature (one of the key requirements of a circadian clock) by adjusting the growth conditions so that the generation time was more than 24h and by testing the cultures under constant low irradiance at three temperatures. They found that distinct maxima in cell division continued to occur at about 24h intervals in constant light and temperature for at least four cycles even though the percentage of dividing cells was never high (because of the low light intensity chosen deliberately to avoid cell cycles of 24h). When cell division was tested in low light at three different temperatures, maxima in cell division occurred at almost the same time in all temperatures (with periods close to 24h). These observations and the fact that the circadian rhythm in cell division is entrained by a light:dark cycle, pointed to cell division in *Synechococcus* as being under the control of a circadian "clock".

Cyanobacterial circadian rhythms have been comprehensively reviewed by Golden et al. (1997) and by Iwasaki & Kondo (2000). Their apparent independence of cyanobacterial cell division cycles is now well established and has been shown to apply irrespective of whether the doubling time is close to 24h, greater than, or even less than that. Mori et al. (1996) showed that *Synechococcus* cultures growing with average doubling times of less than 24h (e.g. one division every 10h), continued to exhibit a circadian rhythm of bioluminescence and of expression of the *psbA1* gene (which encodes the D1 protein of the phototsystem II reaction centre). A circadian rhythm of transcription abundance of both the *psbA1* and *psbAII* genes was evident in *Synechococcus* cultures exposed to two different intensities of continuous light supporting doubling times of 12h and 5-6h, respectively (Golden et al. 1997). Although the division cycle appears to be independent of the circadian cycle, it was clear that the cell division cycle is "gated" by the circadian cycle. This brought cell division and DNA synthesis in continuously diluted cultures into a circadian rhythm but certain phases of the circadian cycle appeared not to allow cell division. Mitsui et al. (1993) showed that temperature-induced alterations in the doubling time of *Synechococcus* also affected oscillation of other cell cycle events of photosynthesis and nitrogen fixation. This was interpreted as suggesting that the cell division cycle sets the other oscillations (Mitsui et al. 1995).

The possible adaptive significance of circadian rhythms in cyanobacteria, even under conditions allowing generation times of less than a day, was addressed by Johnson et al. (1998). They conclude that the advantage of having the capacity to anticipate, for example, the onset of solar illumination, with its concurrent benefits and deficits, would accrue to all light-sensitive organisms, even if they divide more rapidly than the daily cycle. Studies of the possible role of gene expression in the circadian control mechanisms in *Synechococcus* have shown that, remarkably, virtually all promoters examined appear to be under circadian control (Liu et al. 1995).

CONCLUSION

The microalgal cell cycle has a purely reproductive function during which the body mass of the parent cell, usually minus the cell wall, becomes transformed into that of the daughter cells. In its simplest form, it could be represented by a *Chlorella* cell growing under a light:dark regime that allows it to pass through only one mitotic division and subsequent cytokinesis per cycle to produce two daughter cells (autospores) – a rarity for that genus and indeed for the chlorophytes generally. The fact that some of the microalgae most commonly used for physiological studies, like *Chlorella, Chlamydomonas* and *Scenedesmus,* undergo the more complex process of multiple fission may, in part, account for the fact that studies of cell cycle control mechanisms in the microalgae have generally not kept pace with those using animal cells, yeast or plant cells.

It may be significant that some of the earlier successes at achieving close to 100% synchronization of the microalgal cell cycle, and the most convincing evidence for endogenous circadian control of cell cycle events, were obtained with *Euglena* whose division by binary fission, its comparatively large size and "protozoan-like" features would all have contributed to its suitability for cell cycle studies. Studies with *Euglena* were the first to introduce the concept of "gating", suggesting that cell division can occur only when the cycle is in a particular circadian phase determined by a light dependent oscillator. Follow-up studies have explored the nature of the light signal for the circadian oscillator and the possible nature of circadian control over the timing of events/phase transitions of the cell cycle.

Another microalgal group dividing by binary fission, and for which there is clear evidence of control of various cellular processes by endogenous circadian rhythms, is the dinoflagellates. Some of the species used, because of their long generation times, are particularly suited for studies of possible control of cell cycle events by circadian oscillations. The evidence from such species shows that although only a proportion of the population may divide in a particular light:dark cycle, they are able to do so only within a narrow window dictated by the diurnal cycle for which the cue is the dark to light transition. More recent studies have identified the receptor pigment(s) thought to be responsible for the input signal setting the circadian clock in these organisms. It has moreover been shown that in some dinoflagellates growing under a light:dark regime, mitosis is restricted to a 3h window (a few hours after the start of the dark period), coinciding with maximal phosphorylation of a cyclin-dependent kinase protein belonging to the family of such molecules known to function in cell division in plant and animal cells and in yeast.

Cell division in the unicellular cyanobacteria, as might be expected from their prokaryotic nature, displays many of the features of bacterial cell division but without the overlapping rounds of DNA replication characteristic of fast-growing bacteria. In this respect, therefore, cyanobacterial cell cycles appear to be more typical of slow-growing bacteria and more like what is found in the eukaryotic cell cycle where there is always a gap between one round of DNA synthesis and the next. It is now well established that circadian rhythms, quite independent of cell cycle patterns, are a universal feature of the unicellular cyanobacteria that have been studied and there is evidence that the division cycle may also be "gated" by the circadian cycle.

The special requirement for silica displayed by the diatoms gives their cell cycle processes some unique features. Moreover, the presence of a rigid, silica-containing cell wall and the inevitable gradual reduction of cell size following repeated binary fission over successive generations, gives the processes of sexual fusion and resulting auxospore formation particular significance in relation to the restoration of cell size. The requirement for silica in cell division has been exploited to induce synchrony and, in conjunction with the use of inhibitors, has provided evidence that in the diatoms, diurnal changes in various metabolic processes are cell cycle-related rather than being under the control of endogenous circadian rhythms.

Species dividing by multiple fission are characterized by having successive rounds of mitosis (and consequent cytokinesis) within the one cycle. In view of this, it is interesting to note that much of the information now available to us on various aspects of the microalgal cell cycle has come from studies of *Chlamydomonas, Chlorella* and *Scenedesmus,* species where cycles incorporating multiple fission are by far the most common. Studies of synchronous cultures of these species have explored the factors involved in determining the number (and size) of daughter cells produced in each cycle, the pattern of metabolic and ultrastructural changes throughout the growth and reproductive phases of the cycle and the processes involved in release of the daughter cells whether they be zoospores, autospores or gametes.

The earlier studies with synchronous cultures of *Chlorella,* especially those concerned with the timing of cell division and autospore production, led to the concept of the "shortest possible light cycle" and an associated timing mechanism the *Zeitgeber* which sets a presumed endogenous rhythm for the cell cycle. The coupling of autospre production in *Chlorella* to circadian oscillations was also proposed in those early studies.. Anticipating the role that was later defined for certain regulatory proteins important in cell division, some of the early work with *Chlorella,* including the studies with "giant" cells of certain strains, highlighted the key role of some aspects of nitrogen metabolism in controlling the cell cycle.

Other studies, particularly with *Chalmydomonas,* have explored the concept of commitment to divide and how it is related to the action of specific cell division proteins. Evidence is now available that the basic cell cycle control processes in the microalgae have much in common with processes serving a similar function in the cell cycles of plant and animal cells and yeast. In particular, various forms of cyclin molecules or cyclin-like proteins and a range of cyclin-dependent protein molecules (kinases) (CDKs), the latter acting in complex with the cyclins, have been identified. Their role in initiating DNA synthesis and mitosis and in subsequent cell cycle events has been confirmed. Cell cycle-related variations in the expression of genes encoding these proteins have been shown to have distinct overall similarities with that of their counterparts in higher plant cell cycles but with some important differences. The pattern of activation/phosphorylation of cyclin-CDK complexes has also been shown to be of particular relevance to the regulation of specific cell cycle events. Expression of mRNA for a range of proteins associated with cell cycle processes, like chloroplast division, has been well documented and, significantly there is, in some cases, strong evidence of an oscillating rhythm over successive 24h cycles.

Studies with the colonial chlorophyte *Scenedesmus,* and of the way the overlapping cell reproductive processes interact under different light intensities, have confirmed that in that species, and perhaps more generally in the chlorophytes, the effects on the timing of cell cycle events could be explained without invoking the concept of an endogenous circadian oscillator. Instead, the regulatory role of light, for example, on the duration of the cycle would appear to be restricted to its effect on photosynthetic production. The results obtained would also argue against the concept of a light-dependent pacemaker (Zeitgeber). It would appear that regulation by an endogenous oscillator, well known for a range of cellular processes, especially in species dividing by binary fission, may either not exist or may be difficult to demonstrate in species undergoing multiple overlapping reproductive steps. It is no surprise, in view of the complexity of the processes involved in multiple fission that the focus of research in those particular species has now moved on from the earlier studies of cell cycle dynamics towards a more biochemical investigation of regulatory proteins and their mode of action.

REFERENCES

Aoyama, H., Hagiwara, Y., Misumi, O., Kuroiwa, T. & Nakamura, S. (2006). Complete elimination of maternal mitochondrial DNA during meiosis resulting in the paternal inheritance of the mitochondrial genome in *Chlamydomonas. Protoplasma 228*:231-242.

Aoyama, H., Kuroiwa, T. & Nakamura, S. (2008). Observations of chromosomal behaviour in living meiotic zygotes of *Chlamydomonas reinhardtii* (Chlorophyceae). *Eur. J. Phycol. 43*:389-394.

Armbrust, E. V., Bowen, J. D., Olson, R. J. & Chisholm, S. W. (1989). Effect of light on the cell cycle of a marine *Synechococcus* strain. *Appl. Environ. Microbiol. 55*:425-432.

Barbier, M., Albert, M., Geraud, M-L., Bhaud, Y., Picard, A. & Soyer-Gobillard, M-O. (1995). Cell-cycle regulation in the primitive dinoflagellate *Crypthecodinium cohnii* Biechler: evidence for a homolog of cyclin B. *Biol. Cell 84*:35-42.

Bernstein, E. (1960). Synchronous division in *Chlamydomonas maewusii. Science 131*:1528-1529.

Bernstein, E. (1964). Physiology of an obligate photoautotroph (*Chlamydomonas moewussi*). I. Characteristics of synchronously and randomly reproducing cells and an hypothesis to explain their population curves. *J. Protozool.11*:56-74.

Bi, E. & Lutkenhaus, J. (1991). FtsZ ring structure associated with division in *Escherichia coli. Nature 354*:161-164.

Binder, B. (2000). Cell cycle regulation and the timing of chromosome replication in a marine *Synechococcus* (Cyanobacteria) during light- and nitrogen-limited growth. *J. Phycol. 36*:120-126.

Binder, B. (1995). Cell cycle regulation in marine *Synechococcus* sp. strains. *Appl. Environ. Microbiol. 61*:708-717.

Binder, B. & Chisholm, S. W. (1990). Relationship between DNA cycle and growth rate in *Synechococcus* sp. strain PCC 6301. *J. Bacteriol. 172*:2313-2319.

Bišová, K,, Krylov, D. M. & Umen, J. G. (2005). Genome-wide annotation and expression profiling of cell cycle regulatory genes in *Chlamydomonas reinhardtii. Plant Physiol. 137*:475-491.

Bišová, K., Vítová, M. & Zachleder, V. (2000). The activity of total histone H1 kinases is related to growth and commitment points while the p13^{suc1} –bound kinase activity relates to mitoses in the alga *Scenedesmus quadricauda. Plant Physiol. Biochem. 13*:755-764.

Bolige, A., Hagiwara, S-y., Zhang, Y. & Goto, K. (2005). Circadian G2 arrest as related to circadian gating of cell population growth in *Euglena. Plant Cell Physiol. 46*:931-936.

Borowitzka, L. J. & Volcani, B. E. (1977). Role of silicon in diatom metabolism. VIII. Cyclic AMP and cyclic GMP in synchronized cultures of *Cylindrotheca fusiformis. Archiv. Microbiol. 112*:147-152.

Brown, K. L., Twing, K. I. & Robertson, D. L. (2009). Unraveling the regulation of nitrogen assimilation in the marine diatom *Thalassiosira pseudonana* (Bacillariophyceae): Diurnal variations in transcript levels for five genes involved in nitrogen assimilation. *J. Phycol. 45*:413-426.

Brudler, R., Hitomi, K., Daiyasu, H., Toh, H., Kucho, K., Ishiura, M., Kanehisa, M., Roberts, V. A., Todo, T., Tainer, J. A. & Getzoff, E. D. (2003). Identification of a new cryptochrome class: structure, evolution and function. *Mol. Cell 11*:59-67.

Brunelle, S. A., Hazard, E. S., Sotka, E. E. & Van Dolah, F. M. (2007). Characterization of a dinoflagellate cryptochrome blue-light receptor with a possible role in circadian control of the cell cycle. *J. Phycol. 43*:509-518.

Brzezinski, M. A. (1992). Cell-cycle effects on the kinetics of silicic acid uptake and resource competition among diatoms. *J. Plankton Res. 14*:1511-1539.

Brzezinski, M. A., Olson, R. J. & Chisholm, S. W. (1990). Silicon availability and cell cycle progression in marine diatoms. *Mar. Ecol. Prog. Ser. 67*:83-96.

Burbage, C. D. & Binder, B. J. (2007). Relationship between cell cycle and light-limited growth rate in oceanic *Prochlorococcus* (MIT9312) and *Synechococcus* (WH8103) (Cyanobacteria). *J. Phycol. 43*:266-274.

Carré, I. A. & Edmunds, L. N. Jr. (1993). Oscillator control of cell division in *Euglena*: cyclic AMP oscillations mediate the phasing of the cell division cycle by the circadian clock. *J. Cell Sci. 104:1163-1173.*

Cepák, V., Přibyl, P., Vítová, M. & Zachleder, V. (2007). The nucleocystolic and chloroplast cycle in the green chlorococcal alga *Scenedesmus obliquus* (Chlorophyceae, Chlorococcales) grown under various temperatures. *Phycologia 46*:263-269.

Cepák, V. & Zachleder, V. (1988). Regulation of chloroplast and cytoplasmic rRNA accumulation by blue light energy and its relation to reproductive events during the cell cycle of the alga *Scenedesmus quadricauda. Plant Science 57*:205-213.

Cetta, C. M. & Anderson, D. M. (1990). Cell cycle studies of the dinoflagellates *Gonyaulax polyedra* Stein and *Gyrodinium uncatenum* Hulburt during asexual and sexual reproduction. *J. Exp. Mar.Biol. Ecol. 135*:69-84.

Chen, P-C. & Lorenzen, H. (1986). Changes in the productivity and nuclear divisions in synchronous *Chlorella* and circadian rhythm. *Pl. Cell Physiol. 27*:1423-1427.

Choappino, M. L. & Volcani, B. E. (1977). Studies on biochemistry and fine structure of silica shell formation in diatoms. VII. Sequential cell wall development in the pennate *Navicula pelliculosa. Protoplasma, 93*:205-221.

Claquin, P., Kromkamp, J. C. & Martin-Jezequel, V. (2004). Relationship between photosynthetic metabolism and cell cycle in a synchronized culture of the marine alga *Cylindrotheca fusiformis* (Bacillariophyceae). *Eur. J. Phycol. 39*:33-41.

Claquin, P., Martin-Jézéquel, V., Kromkamp, J. C., Veldhuis, M. J. W. & Kraay, G. W. (2002). Uncoupling of silicon compared with carbon and nitrogen metabolisms and the role of the cell cycle in continuous cultures of *Thalassiosira pseudonana* (Bacillariophyceae) under light, nitrogen and phosphorus control. *J. Phycol. 38*:922-930.

Costas, E., Gonzales-Gil, S., Aquilera, A. & Lopez-Rodas, V. (1992). A morphometrical analysis of ultrastructural changes during the cell cycle of *Prorocentrum triestinum* using stereology. *Bot. Mar. 35*:429-436.

Donnan, L., Carvill, E. P., Gilliland, T. J. & John, P. C. L. (1985). The cell cycles of *Chlamydomonas* and *Chlorella. New Phytol. 99*:1-40.

Donnan, L. & John, P. C. L. (1983). Cell cycle control by timer and sizer in *Chlamydomonas. Nature 304*:630-633.

Donnan, L. & John, P. C. L. (1984). Timer and sizer controls in the cell cycles of *Chlamydomonas* and *Chlorella.* In: Nurse, P. & Streiblova, E (Eds) *The Microbial Cell Cycle,* Boca Raton; CRC Press, pp. 231-251.

Drebes, G. (1977). Sexuality. In: Werner, D. (Ed.), *The Biology of Diatoms.* Blackwell Scientific Publications, Oxford, 1977 pp. 250-283.

Edmunds, L. N. Jr. (1964). Replication of DNA and cell division in synchronously dividing cultures of *Euglena gracilis. Science 145*:266-268.

Edmunds, L. N. Jr. (1966). Studies on synchronously dividing cultures of *Euglena gracilis* Klebs (strain Z). III. Circadian components of cell division. *J. Cell Physiol. 67*:35-44.

Edmunds, L. N. Jr. (1978). Clocked cell cycle clocks: implications toward chromopharmocology and aging. In: Samis, H. V. Jr. & Capobianco (Eds.) *Aging and Biological Rhythms.* Plenum, New York. pp. 125-184.

Edmunds, L. N. Jr. (1988). Cell division cycles and circadian oscillators. In: *Cellular and Molecular Bases of Biological Clocks.* Springer-Verlag, New York. pp.119-165.

Edmunds, L. N. Jr. & Adams, K. J. (1981). Clocked cell cycle clocks. *Science, 211*: 1002-1013.

Edmunds, L. N. Jr., Tay, D. E. & Laval-Martin, D. L. (1982). Cell division cycles and circadian clocks. Phase-response curves for light perturbations in synchronous cultures of *Euglena. Plant Physiol. 70*:297-302.

Ehara, T., Ogasawara, Y. & Osafune, T. (1990). Behaviour of chloroplast nucleoids during the cell cycle of *Chlamydomonas reinhardtii* (Chlorophyta) in synchronized culture. *J. Phycol. 26*:317-323.

Foster, K. W., Saranak, J., Patel, N., Zarilli, G., Okabe, M., Kline, T. & Nakanishi, K. (1984). A rhodopsin in the functional photoreceptor for phototaxis in the unicellular eukaryote *Chlamydomonas. Nature 311*:756-759.

Gloeckner, G. & Beck, C. F. (1995). Genes involved in light control of sexual differentiation in *Chlamydomonas reinhardtii. Genetics 141*:937-943.

Golden, S. S., Ishiura, M., Johnson, C. H. & Kondo, T. (1997). Cyanobacterial circadian rhythms. *Ann. Rev. Plant Physiol. Plant. Mol. Biol. 48*:327-354.

Goodenough, U. W. & St. Clair, H. S. (1975). Bald2: a mutation affecting the formation of doublet and triplet sets of microtubules in *Chlamydomonas reinhardtii. J. Cell Biol. 103*:405-417.

Grief, C., O'Neill, M. A. & Shaw, P. J. (1987). The zygote cell wall of *Chlamydomonas reinhardtii*: a structural, chemical and immunological approach. *Planta, 170*:433-435.

Griffiths, D. A. & Griffiths, D. J. (1969). The fine structure of autotrophic and heterotrophic cells of *Chlorella vulgaris* (Emerson strain). *Plant Cell Physiol. 10*:11-19.

Griffiths, D. J. (1970). The growth of synchronous cultures of the Emerson strain of *Chlorella vulgaris* under heterotrophic conditions. *Arch. Mikrobiol. 71*:60-66.

Griffiths, D. J. (1979). Factors affecting nitrate reductase in synchronous cultures of *Chlorella. New Phytol. 82*:427-437.

Hageman, R. H. & Flesher, D. (1960). Ntrate reductase activity in corn seedlings as affected by light and nitrate content of nutrient media. *Pl. Physiol. 35*:700.

Hagiwara, S-y., Bolige, A., Zhang, Y., Takashi, M., Yamagishi, A. & Goto, K. (2002). Circadian gating of photoinduction of commitment to cell-cycle transitions in relation to photoperiodic control of cell reproduction in *Euglena. Photochemistry & Photobiology 76*:105-115.

Hagiwara, S-y., Takahashi, M., Yamagishi, A., Zhang, Y. & Goto, K. (2001). Novel findings regarding photoinduced commitments of G1-, S- and G2-phase cells to cell-cycle transitions in darkness and dark-induced G1-, S- and G2-phase arrests in *Euglena. Photochem. Photobiol. 74*:726-733.

Harris, E. H. (1989). The *Chlamydomonas* sourcebook. Academic Press, San Diego, Calif.

Harris, E. H. (2001). *Chlamydomonas* as a model organism. *Ann. Rev. Pl. Physiol and Plant Molecular Biology, 52*:363-406.

Hase, E., Otsuka, H., Mihara, S & Tamiya, H. (1959). Role of sulfur in the cell division of *Chlorella,* studied by the technique of synchronous culture. *Biochim. Biophys. Acta 35*:180-189.

Hastings, J. W., Astrachan, L. & Sweeney, B. M. (1961). A persistent daily rhythm in photosynthesis. *J. Gen. Physiol. 45*:69-76.

Heil, W. G. & Senger, H. (1986). Thylakoid-protein phosphorylation during the life cycle of *Scenedesmus obliquus* in synchronous culture. *Planta 167*:233-239.

Herrin, D. & Michaels, A. (1984). Gene expression during the cell cycle of *Chlamydomonas reionhardtii.* In: Stein, G. & Stein, J. (Eds.) *Recombinant DNA and Cell Proliferation.* Academic Press, New York, pp. 87-106.

Hildebrand, M. & Dahlin, K. (2000). Nitrate transporter genes from the diatom *Cylindrotheca fusiformis* (Bacillariophyceae): mRNA levels controlled by nitrogen source and by the cell cycle. *J. Phycol. 36*:702-713.

Hildebrand, M, Frigeri, L. G. & Davis, A. K. (2007). Synchronized growth of *Thalassiosira pseudonana* (Bacillariophyceae) provides novel insights into cell-wall synthesis processes in relation to the cell cycle. *J. Phycol. 43*:730-740.

Hiramatsu, T., Misumi, O., Kuroiwa, T. & Nakamura, S. (2006). Morphological changes in mitochondrial and chloroplast nucleoids and mitochondria during the *Chlamydomonas reinhardtii* cell cycle. *J. Phycol. 42*:1048-1058.

Hodler, M., Morgenthaler, J.-J., Eichenberger, W. & Grob, E. C. (1972). The influence of light on the activity of nitrate reductase in synchronous cultures of *Chlorella pyrenoidosa. FEBS Letters, 28*:19.

Holtzendorff, J., Partensky, F., Jacquet, S., Bruyant, F., Marie, D., Garczarek, L., Mary, I., Vaulot, D. & Hess, W. R. (2001). Diel expression of cell cycle-related genes in synchronized cultures of *Prochlorococcus* sp. strain PCC9511. *J. Bacteriol. 183*:915-920.

Hu, Y., Chen, Z-W., Liu, W-Z., Liu, X-L. & He, Y-K. (2008). Chloroplast division is regulated by the circadian expression of *FTZS* and *MIN* genes in *Chlamydomonas reinhardtii. Eur. J. Phycol. 43*:207-215.

Huang, K. & Beck, C. F. (2003). Phototropin is the blue-light receptor that controls multiple steps in the sexual life cycle of the green alga *Chlamydomonas reinhardtii. Proc. Natl. Acad. Sci. USA. 100*:6269-6274.

Iwasaki, H. & Kondo, T. (2000). The current state and problems of circadian clock studies in cyanobacteria. *Plant Cell Physiol. 41*:1013-1020.

Jacobshagen, S., Whetstine, J. R. & Boling, J. M. (2001). Many but not all genes in *Chlamydomonas reinhardtii* are regulated by the circadian clock. *Plant Biol. 3*:592-597.

Jacquet, S., Partensky, F., Lennon, J. F. & Vaulot, D. (2001). Diel patterns of growth and division in marine picoplankton in culture. *J. Phycol. 37*:357-369.

John, P. C. L. (1984). Control of the cell division cycle in *Chlamydomonas. Microbial Sci. 1*:96-101.

John, P. C. L., Sek, F. J. & Lee, M. G. (1989). A homolog of the cell cycle control protein p34*cdc2* participates in the division cycle of *Chlamydomonas* and a similar protein is detectable in higher plants and remote taxa. *Plant Cell. 1*:1185-1193.

Johnson, C. H., Golden, S. S. & Kondo, T. (1998). Adaptive significance of circadian programs in cyanobacteria. *Trends in Microbiology 6*:407-410.

Kaftan, D., Meszaros, T., Whitmarsh, J. & Nedbal, L. (1999). Characterization of Photosystem II acticity and heterogeneity during the cell cycle of the green alga *Scenedesmus quadricauda. Plant Physiol. 120*:433-441.

Karentz, D. (1983). Patterns of DNA synthesis and cell division in marine dinoflagellates. *J. Protozool. 30*:581-588.

Kanazawa, T. (1964). Changes of amino acid composition of *Chlorella* cells during their life cycle. *Plant Cell Physiol. 5*:333-354.

Kanazawa, T., Kanazawa, K. & Morimura, Y. (1965). New arginine-containing peptides isolated from *Chlorella* cells. *Plant Cell Physiol. 6*:631-643.

Karentz, D. (1983). Patterns of DNA synthesis and cell division in marine dinoflagellates. *J. Protozool. 30*:581-588.

King, R. W., Deshaies, R. J. Peters, J. M. & Kirschner, M. W. (1996). How proteolysis drives the cell cycle. *Science 274*:1652-1658.

Koester, J. A., Brawley, S. H., Karp-Boss, L. & Mann, D. G. (2007). Sexual reproduction in the marine centric diatom *Ditylum brightwellii* (Bacillariophyta). *Eur. J. Phycol. 42*:351-366.

Kondo, T., Johnson, C. H. & Hastings, J. W. (1991). Action spectrum for resetting the circadian phototaxis rhythm in the CW15 strain of *Chlamydomonas.* I. Cells in darkness. *Plant Physiol. 95*:197-205.

Krienitz, L., Huss, V. A. R. & Hümmer, C. (1996). Picoplanktonic *Choricystis* species (Chlorococcales, Chlorophyta) and problems surrounding the morphologically similar "*Nannochloris*-like" algae. *Phycologia 35*:332-341.

Kubo, T., Kaida, S., Abe, J., Saito, T., Fukuzawa, H. & Matsuda, Y. (2009). The *Chlamydomonas* hatching enzyme, sporangin, is expressed in specific phases of the cell cycle and is localized to the flagella of daughter cells within the sporangial cell wall. *Plant Cell Physiol. 50*:572-583.

Kuhl, A. & Lorenzen, H. (1964). Handling and culturing of *Chlorella.* In: Prescott, D. M. (Ed.) *Methods in Cell Physiology.* Academic Press, New York and London, *1*:159-188.

Kuroiwa, T., Suzuki, T., Ogawa, K. & Kawano, S. (1981). The chloroplast nucleus: distribution, number, size and shape, and a model for the multiplication of the chloroplast genome during chloroplast development. *Plant Cell Physiol. 22*:381-396.

Lee, M. G. & Nurse, P. (1988). Cell cycle control genes in fission yeast and mammalian cells. *Trends Genet. 4*:287-290.

Leedale, G. F. (1959). Periodicity of mitosis and cell division in the Euglenineae. *Biol. Bull.* 116:162-174.

Lewin, J. C. (1955). Silicon metabolism in diatoms. *J. Gen. Physiol. 39*:1-10.

Lewin, J. C. & Chen, C. H. (1968). Silicon metabolism in diatoms. VI. Silicic acid uptake by a colourless marine diatom, *Nitzscia alba. J. Phycol. 4*:161-166.

Lien, T. & Knutsen, G. (1979). Synchronous *growth* of *Chlamydomonas reinhardtii* (Chlorophyceae): a review of optimal conditions. *J. Phycol. 15*:191-200.

Lin, S., Chang, J. & Carpenter, E. J. (1994). Detection of proliferating cell nuclear antigen analog in four species of marine phytoplankton. *J. Phycol. 30*:449-456.

Lin, S., Chang, J. & Carpenter, E. J. (1995). Growth characteristics of phytoplankton determined by cell cycle proteins: PCNA immunostaining of *Dunaliella tertiolecta* (Chlorophyceae). *J. Phycol. 31*:388-395.

Liu, H., Bidigare, R. R., Laws, E., Landry, M. R. & Campbell, L. (1999). Cell cycle and physiological characteristics of *Synechococcus* (WH7803) in chemostat culture. *Mar. Ecol. Prog. Ser. 189*:17-25.

Liu, Y., Tsinoremas, N. F., Johnson, C. H., Lebedeva, N. V., Golden, S. S., Ishiura, M. & Kondo, T. (1995). Circadian orchestration of gene expression in cyanobacteria. *Genes Dev. 9*:1469-1478.

Lorenzen, H. (1957). Synchrone Zellteilungen von *Chlorella* bei verschiedenen Lict-Dunkel-Wechseln. *Flora (Jena), 144*:473-496.

Lorenzen, H. & Hesse, M. (1974). Synchronous cultures. In: Stewart, W. D. P. (Ed.) *Algal Physiology and Biochemistry,* Blackwells, Oxford, pp.894-908.

Lorenzen, H. & Schleif, J. (1966). Zur Bedeutung der kürzest möglichen Generationsdauer in Synchronkulturen von *Chlorella. Flora 156*:673-683.

McAteer, M., Donnan, L. & John, P. C. L. (1985). The timing of division in *Chlamydomonas. New Phytol. 99*:41-56.

Mann, N. & Carr, N. G. (1974). Control of macromolecular composition and cell division in the blue-green alga *Anacystis nidulans. J. Gen. Microbiol. 83:*399-405.

Marcus, Y., Schuster, G., Michaels, A. & Kaplan, A. (1986). Adaptation to CO_2 level and changes in the phosphorylation of thylakoid proteins during the cell cycle of *Chlamydomonas reinhardtii. Pl. Physiol. 80*:604-607.

Martin-Jėzėquel, V., Hildebrand, M. & Brzezinski, M. A. (2000). Silicon metabolism in diatoms: implications for growth. *J. Phycol. 36*:821-840.

Matsuda, Y., Koseki, M., Shimada, T. & Saito, T. (1995). Purification and characterization of a vegetative lytic enzyme responsible for liberation of daughter cells during the proliferation of *Chlamydomonas reinhardtii. Plant & Cell Physiol. 36*:681-689.

Matusiak-Mikulin, K., Tukaj, C. & Tukaj, Z. (2006). Relationships between growth, development and photosynthetic activity during the cell cycle of *Desmodesmus armatus* (Chlorophyta) in synchronous cultures. *Eur. J. Phycol. 41*:29-38.

Mihara, S. & Hase, E. (1971). Studies on the vegetative life cycle of *Chlamydomonas reinhardtii* Dangeard in synchronous culture. I. Some characteristics of the cell cycle. *Plant Cell Physiol. 12*:225-236.

Minami, S. A. & Goodenough, U. W. (1978). Novel glycopolypeptide synthesis induced bygametic cell fusion in *Chlamydomonas reinhardtii. J. Cell. Biol. 77*:165-181.

Mitsui, A. & Suda, S. (1995). Alternative and cyclic appearance of H_2 and O_2 photoproduction activities under nongrowing conditions in an aerobic nitrogen-fixing unicellular cyanobacterium *Synechococcus* sp. *Curr. Microbiol. 30*:1-6.

Mitsui, A., Suda, S. & Hanagata, N. (1993). Cell cycle events at different temperatures in aerobic nitrogen-fixing marine unicellular cyanobacterium *Synechococcus* sp. Miami BG 043511. *J. Marine Biotechnol. 1:*89-91.

Miyagishima, S., Takahara, M., Mori, T., Kuroiwa, H., Higashiyama, T. & Kuroiwa, T. (2001). Plastid division is driven by a complex mechanism that involves differential transition of the bacterial and eukaryotic division rings. *Plant Cell 13*:2257-2268.

Moore, L. R., Goericke, R. & Chisholm, S. W. (1995). Comparative physiology of *Synechococcus* and *Prochlorococcus*: influence of light and temperature on growth, pigments, fluorescence and absorptive properties. *Mar. Ecol. Prog. Ser. 116*:259-275.

Mori, T., Binder, B. & Johnson, C. H. (1996). Circadian gating of cell division in cyanobacteria growing with average doubling times of less than 24 hours. *Proc. Natl. Acad. Sci. USA. 93*:10183-10188.

Münzner, P & Voigt, J. (1992). Blue light regulation of cell division in *Chlamydomonas reinhardtii. Plant Physiol. 99*:1370-1375.

Murray, A. W., Solomon, M. J. & Kieschner, M. W. (1989). The role of cyclin synthesis and degradation in the control of maturation promoting factor activity. *Nature 339*:280-286.

Naumann, E. (1921). Notiz zur Systematik der Süßwasseralgen V. Über *Nannochloris,* eine neue Chlorophyceengattung. *Ark. Bot. 16*:1-19.

Okamoto, O. K. & Hastings, J. W. (2003). Novel dinoflagellate clock-related genes identified through microarray analysis. *J. Phycol. 39*:519-526.

Okita, T. W. & Volcani, B. E. (1978). Role of silicon in diatom metabolism. IX. Differential synthesis of DNA polymerases and DNA-binding proteins during silicate starvation and recovery in *Cylindrotheca fusiformis. Biochem. Biophys. Acta, 519*:76-86.

Okita, T. W. & Volcani, B. E. (1980). Role of silicon in diatom metabolism. X. Polypeptide labelling patterns during the cell cycle, silicate starvation and recovery in *Cylindrotheca fusiformis. Exp. Cell Res. 125*:471-481.

Oldenhof, H., Zachleder, V. & van den Ende, H. (2004). Blue light delays commitment to cell division in *Chlamydomonas reinhardtii. Plant Biol. 6*:689-695.

Oldenhof, H., Zachleder, V. & van den Ende, H. (2006). Blue- and red-light regulation of the cell cycle in *Chlamydomonas reinhardtii* (Chlorophyta). *Eur. J. Phycol.41*:313-320.

Oldenhof, H., Zachleder, V. & van den Ende, H. (2007). The cell cycle of *Chlamydomonas reinhardtii*: the role of the commitment point. *Folia Microbiol. 52*:53-60.

Pan, J. M., Haring, M. A. & Beck, C. F. (1997). Characterization of blue light signal transduction chains that control development and maintenance of sexual competence in *Chlamydomonas reinhardtii. Plant Physiol. 115*:1241-1249.

Paul, J. S. & Volcani, B. E. (1976). Photorespiration in diatoms IV: Two pathways of glycolate metabolism in synchronized cultures of *Cylindrotheca fusiformis. Arch. Microbiol. 110*:247-252.

Pirson, H. & Lorenzen, H. (1958). Ein endogener Zeitfaktor bei der Teilung van *Chlorella. Z. Botan. 46*:53-67.

Pirson, A. & Lorenzen, H. (1966). Synchronized dividing algae. *Ann. Rev. Pl. Physiol. 17*:439-458.

Reeves, C. D. & Volcani, B. E. (1985). Messenger RNA and polypeptide accumulation patterns in synchronized cultures of *Cylindrotheca fusiformis. J. Gen. Microbiol. 131*:1735-1744.

Rizzo, P. J. (1987). Biochemistry of the dinoflagellate nucleus. In: Taylor, F. J. R. (Ed.) *The Biology of the Dinoflagellates.* Blackwell Sci. Publ., Palo Alto, California, pp.143-173.

Rodriguez, M., Cho, J. W., Sauer, H. W. & Rizzo, P. J. (1994). Evidence for the presence of a cdc2-like protein kinase in the dinoflagellate *Crypthecodinium cohnii. J. Eukary. Microbiol. 40*:91-96.

Round, F. E. (1972a). The problem of size reduction in diatoms. *Nova Hedwigia, 23*:291-303.

Round, F. E. (1972b). The formation of girdle, intercalary bands and septa in diatoms. *Nova Hedwigia, 23*:449-463.

Rynearson, T. A. & Armbrust, E.V. (2000). DNA fingerprinting reveals extensive genetic diversity in a field population of the centric diatom *Ditylum brighrtwellii. Limnol. Oceanogr.45*:1329-1340.

Saito, T., Inoue, M., Yamada, M. & Matsuda, Y. (1998). Control of gametic differentiation and activity by light in *Chlamydomonas reinhardtii. Plant Cell Physiol. 39*:8-15.

Schlösser, U. G. (1981). Algal wall-degrading enzymes – autolysins. In: Tanner, W. & Loewus, F. A. (Eds.) *Encyclopedia of Plant Physiology, New Series.* Pp. 333-351. Springer-Verlag, Berlin.

Senger, H. (1961). Untersuchungen zur Synchronisierung von *Chlorella*-Kulturen. *Arch. Mikrobiol. 40*:47-72.

Šetlík, I., Berková,E., Doucha, J., Kubín, Š., Vendlová, J. & Zachleder, V. (1972). The coupling of synthetic and reproduction processes in *Scenedesmus quadricauda. Arch. Hydrobiol.(Suppl. 41) Algol. Stud. 7*:172-217.

Šetlík, I., Šetlíkova, E., Masojidek, J., Zachleder, V., Kalina, T. & Mader, P. (1981). The effect of translation and transcription inhibitors on the development of the photosynthetic apparatus in cell cycles of *Scenedesmus quadricauda.* In: Akoyunoglou, G. (Ed.) *Photosynthesis Vol 5: Chloroplast development.* Balaban International Science Services, Philadelphia, PA, pp. 481-490.

Šetlík, I. & Zachleder, V. (1984). The multiple fission cell reproductive patterns in algae. In: Nurse, P. & Streiblová (Eds.) *The Microbial Cell Cycle* pp. 253-279. Boca Raton, Florida; CRC Press.

Shalapyonok, A., Olson, R. J. & Shalapyonok, L. S. (1998). Ultradian growth in *Prochlorococcus* spp. *Appl.Environ. Microbiol. 64*:1066-1069.

Simanis, V. & Nurse, P. (1986). The cell cycle control gene *cdc2*$^+$ of fission yeast encodes a protein kinase potentially regulated by phosphorylation. *Cell 45*:261-268.

Sorokin, C. (1959). Tabular comparative data for the low- and high-temperature strains of *Chlorella. Nature 184*:613-614.

Sorokin, C. & Krauss, R. W. (1959). Maximum growth rates of *Chlorella* in steady-state and in synchronized cultures. *Proc. Natl. Acad. Sci. U. S., 45*:1740-1744.

Sorokin, C. & Krauss, R. W. (1965). The dependence of cell division in *Chlorella* on temperature and light intensity. *Am. J. Botan. 52*:331-338.

Sullivan, C. W. (1977). Diatom mineralisation of silicic acid. II. Regulation of $Si(OH)_4$ transport rates during the cell cycle of *Naviculosa pelliculosa. J. Phycol. 13*:86-91.

Sumiya, N., Hirata, A. Kawano, S. (2008). Multiple FtsZ ring formation and reduplicated chloroplast DNA in *Nannochloris bacillaris* (Chlorophyta, Trebouxiophyceae) under phosphate-enriched culture. *J. Phycol. 44*:1476-1489.

Suzuki, L. & Johnson, C. H. (2002). Photoperiodic control of germination in the unicell *Chlamydomonas. Naturwissenschaften, 89*:214-220.

Suzuki, L., Woessner, J. P., Uchida, H., Kuroiwa, H., Yuasa, Y., Waffenschmidt, S., Goodenough, U. W. & Kuroiwa, T. (2000). A zygote-specific protein with hydroxyproline-rich glycoprotein domains and lectin-like domains involved in the assembly of the cell wall of *Chlamydomonas reinhardtii* (Chlorophyta). *J. Phycol. 36*:571-583.

Sweeney, B. M. (1982). Interaction of the circadian clock with the cell cycle in *Pyrocystis fusiformis. Pl. Physiol. 70*:272-276.

Sweeney, B. M. (1984). Circadian rhythmicity in dinoflagellates. In: Spector, D. L. (Ed.) *Dinoflagellates.* Academic Press, Orlando, Fl., pp.343-365.

Sweeney, B. M. & Borgese, M. B. (1989). A circadian rhythm in cell division in a prokaryote, the cyanobacterium *Synechococcus* WH7803. *J. Phycol. 25*:183-186.

Sweeney, M. B. & Hastings, J. W. (1958). Rhythmic cell divisions in populations of *Gonyaulax polydra. J. Protozool. 5*:217-224.

Szurkowski, J., Baścik-Remisiewicz, A., Matusiak, K. & Tukaj, Z. (2001). Oxygen evolution and photosynthetic energy storage during the cell cycle of green alga *Scenedesmus armatus* characterized by photoacoustic spectroscopy. *J. Plant Physiol. 158*:1061- 1067.

Tamiya, H. (1966). Synchronous cultures of algae. *Ann. Rev. Pl. Physiol. 17*:1-26.

Thinh, L. V. & Griffiths, D. J. (1970). Physiological changes accompanying the recovery of cell division in "giant" cells of *Chlorella vulgaris* (Emerson strain). *Plant Cell Physiol.11*:621-629.

Thinh, L. V. & Griffiths, D. J. (1971). The contrasting effects of chloramphenicol and cycloheximide on the recovery of cell division and chlorophyll synthesis in "giant" cells of *Chlorella vulgaris* (Emerson strain). *Plant Cell Physiol. 12*:171-179.

Thinh, L. V. & Griffiths, D. J. (1972a). Ultrastructural changes accompanying the recovery of "giant" cells of the Emerson strain of *Chlorella.* I. Autospore production. *Arch. Mikrobiol. 87*:47-60.

Thinh, L. V. & Griffiths, D. J. (1972b). Ultrastructural changes accompanying the recover of "giant" cells of the Emerson strain of *Chlorella.* II. Chloroplast development. *Arch. Mikrobiol. 87*:61-75.

Thinh, L. V. & Griffiths, D. J. (1973a). The role of photosynthesis and respiration in the light-induced recovery of "giant" cells of the Emerson strain of *Chlorella. Plant Cell Physiol. 14*:1177-1185.

Thinh, L. V. & Griffiths, D. J. (1973b). Nucleic acid synthesis accompanying the recovery of cell division and chloroplast development in "giant" cells of the Emerson strain of *Chlorella. Plant Cell Physiol. 14*:497-504.

Thinh, L. V. & Griffiths, D. J. (1974). The effect of L-arginine on the growth of heterotrophic cultures of the Emerson strain of *Chlorella.* I. Effects on cell growth, cell division, chloroplast development and nucleic acid synthesis. *New Phytol. 73*:1087-1095.

Thinh, L. V. & Griffiths, D. J. (1977). The effect of L-arginine on the growth of heterotrophic cultures of the Emerson strain of *Chlorella.* II. Reversal of the arginine affect by chloramphenicol. *New Phytol. 78*:121-129.

Thinh, L. V. & Griffiths, D. J. (1980). The effect of L-arginine on the growth of heterotrophic cultures of the Emerson strain of *Chlorella.* III. Studies with the structural analogue Canavanine. *New Phytol. 84*:429-437.

Tischner, R. (1976). Zur induction der Nitrat-und Nitritreduktase in vollsynchronen *Chlorella*-Kulturen. *Planta, 132*:285.

Tischner, R. & Hüttermann, A. (1978). Light-mediated activation of nitrate reductase in synchronous *Chlorella. Pl. Physiol. 62*:284.

Trainor, F. R., Cain, J. R. & Shubert, L. E. (1976). Morphology and nutrition of the colonial green alga *Scenedesmus*: 80 years later. *Bot. Rev. 42*:5-25.

Treier, U., Fuchs, S., Weber, M., Warkarchuk, W. W. & Beck, C. F. (1989). Gametic differentiation in *Chlamydomonas reinhardtii*: light dependence and gene expression patterns. *Arch. Microbiol. 152*:572-577.

Tukaj, Z., Kubínová, A. & Zachleder, V. (1996). Effect of irradiance on growth and reproductive processes during the cell cycle in *Scenedesmus armatus* (Chlorophyta). *J. Phycol. 32*:624-631.

Umen, J. G. & Goodenough, U. W. (2001). Control of cell division by a retinoblastoma protein homolog in *Chlamydomonas. Gen. Dev. 15*:1652-1661.

Van Dolah, F. M. & Leighfield, T. A. (1999). Diel phasing of the cell cycle in the Florida red tide dinoflagellate, *Gymnodinium Breve. J. Phycol.35*: 1404-1411.

Van Dolah, F. M., Leighfield, T. A., Sandel, H. D. & Hsu, C. K. (1995). Cell division in the dinoflagellate *Gambierdiscus toxicus* is phased to the diurnal cycle and accompanied by activation of the cell-cycle regulatory protein cdc2kinase. *J. Phycol. 31*: 395-400.

Vaulot, D., LeBot, N., Marie, D. & Fukai, E. (1996). Effect of phosphorus on the *Synechococcus* cell cycle in surface Mediterranean waters during summer. *Appl. Environ. Microbiol. 62*:2527-2533.

Vaulot, D., Marie, D., Olson, R. J. & Chisholm, S. W. (1995). Growth of *Prochlorococcus,* a photosynthetic prokaryote, in the equatorial Pacific Ocean. *Science 268*:1480-1482.

Vergara, J. J., Berges, J. A. & Falkowski, P. G. (1998). Diel periodicity of nitrate reductase activity and protein levels in the marine diatom *Thalassiosira weissflogii* (Bacillariophyceae). *J. Phycol. 34*:952-961.

Vitová, M., Hendrychova, J., Čížková, M, Cepák, V., Umen, J. G., Zachleder, V., Bišová, K., Vítová, M., Kubín, Š. & Chová, J. H. (2002). Variety of cell cycle patterns in the alga *Scenedesmus quadricauda* (Chlorophyta) as revealed by application of illumination regimes and inhibitors. *Eur. J. Phycol. 37*: 361-371.

Waffenschmidt, S., Woessner, J. P., Beer, K. & Goodenough,, U. W. (1993). Isodityrosine cross-linking mediates insolubilization of cell walls in *Chlamydomonas. Plant Cell 5*:809-820.

Wang, D., Kong, D., Wang, Y., Hu, Y., He, Y. & Sun, J. (2003). Isolation of two plastid division ftsZ genes from Chlamydomonas and its evolutionary implications for the role of FtsZ in plastid division. *J. Exp. Bot. 54*:1115-1116.

Weissig, H. & Beck,, C. F. (1991). Action spectrum for the light-dependent step in gametic differentiation of *Chlamydomonas reinhardtii. Plant Physiol. 97*:118-121.

Werner, D. (1966). Die Kieselsäure im Stoffwechsel von *Cyclotella cryptica* Reimann, Lewin and Guillard. *Arch. Mikrobiol. 55*:278-308.

Wu J-T., Tischner, R. & Lorenzen, H. (1986). A circadian rhythm in the number of daughter cells in synchronous *Chlorella fusca* var *vacuolate. Pl. Physiol. 80*:20-22.

Yamaguchi, M. (1992). DNA synthesis and the cell cycle in the noxious red-tide dinoflagellate *Gymnodinium nagasakiense. Mar. Biol. (Berl.) 112*:191-198.

Yamamoto, M., Nozaki, H. & Kawano, S. (2001). Evolutionary relationships among multiple modes of cell division in the genus *Nannochloris* (Chlorophyta) revealed by genome size, actin gene multiplicity, and phylogeny. *J. Phycol. 37*:106-120.

Yamamoto, M., Nozaki, H. & Miyazawa, Y. (2003). Relationship between presence of a mother cell wall and speciation in the unicellular microalga *Nannochloris* (Chlorophyta). *J. Phycol. 39*: 172-184.

Zachleder, V. & Bišová, K. (2008). Accumulation, activity and localization of cell cycle regulatory proteins and the chloroplast division protein FtsZ in the alga *Scenedesmus quadricauda* under inhibition of nuclear DNA replication. *Plant Cell Physiol. 49*:1805-1817.

Zachleder, V. & Cepák, V. (1987). Variations in chloroplast nucleoid number during the cell cycle in the alga *Scenedesmus quadricauda* grown under different light conditions. *Protoplasma 141*:74-82.

Zachleder, V., Doucha, J., Berková, E. & Šetlík, I. (1975). The effect of synchronizing dark period on populations of *Scenedesmus quadricauda. Biol. Plant. 17*:416-433.

Zachleder, V., Kawano, S., Kepák, V. & Kuroiwa, T. (2004). The effect of nalidixic acid on growth and reproductive events in nucleocytosolic and chloroplast compartments in the alga *Scenedesmus quadricauda, Folia Microbiol. 49*:441-451.

Zachleder, V., Schläfli, O. & Boschetti, A. (1997). Growth-controlled oscillation in activity of histone H1 kinase during the cell cycle of *Chlamydomonas reinhardtii* (Chlorophyta). *J. Phycol. 33*:673-681.

Zachleder, V. & Šetlík, I. (1990). Timing of events in overlapping cell reproductive sequences and their mutual interactions in the alga *Scenedesmus quadricauda. J. Cell Science, 97*:631-638.

Zachleder, V. & Van den Ende, H. (1992). Cell cycle events in the green alga *Chlamydomonas eugametos* and their control by environmental factors. *J. Cell Science 102*:469-474.

Zuethen, E. (Ed.), *Synchrony in Cell Division and Growth.* Interscience, New York, 1964.

In: Daughter Cells
Editors: A. Hitomi et al., pp. 51-128

ISBN: 978-1-60876-790-8

Chapter II

STEM CELLS AND MORPHOGENETIC DEVELOPMENTAL PROGRAMS IN PLANTS

Tatyana Batygina
Komarov Botanical Institute of RAS, Saint Petersburg, Russia

"Nothing is more practical than a good theory"
"Plasticity is the main rule of all the living, it is life itself.
It is the possibility of many lives, many movements of life.
These movements follow one another. Marvelous harmony…"
Auguste Rodin

"Nothing is more practical than a good theory"
Gustav Kirhgoff

ABSTRACT

- Today's actual question is *creation of the science on stem cells* and development of its theoretical bases, taking into account comparative evolutionary cell biology and the *cardinal problem of the developmental biology – the problem of wholeness* for representatives of various kingdoms. This allows revealing correlations and studying correlative dependences of various structures at different levels of biohierarchy. The creation of every science is impossible without the application of methodology.
- In current work the system of nontraditional ideas about the nature and role of stem cells in ontogenesis, reproduction and evolution of plants is proposed.
- The main properties of plant stem cells have been developed by us, which has shown the integrity of morphogenous and reproductive processes at all stages of plant's life cycle. A special attention should be paid to the dynamics of the stem cells' pool and its correlative dependence with niche.

- System approach and analysis of data on male and female generative structures – egg cell at different stages of development, double fertilization in model objects: syngamy, triple fusion, plasmogamy and karyogamy taking into consideration the cell cycle, zygote, embryogenesis of model objects in representatives of a lot of flowering plants taxa allowed us to express nontraditional idea, that from the embryological point of view zygote is proximate stem cell and gives a rise for all other stem cells in plant ontogenesis, particularly those of shoot and root apices.
- Zygote contains information about architectonics and development of organization of an individual, which is connected with gradients and correlations between its parts, and consequently with heterogeneity of different axes of a future organism.
- The data of developmental biology and especially comparative embryology of representatives of various taxa of flowering plants enable to consider that formation of stem cells derived from zygote, are probably typical for all organs – flower, stem, leaf, root – and at all stages of the life cycle – sporophyte, gametophyte. Their localization is depending, first of all, on their functioning.
- Not every cell possesses the totipotency and not every totipotent cell is able to become the stem cell, and not every stem cell can produce an individual.
- One has to pay espesial attention to the role of somatic cells in organization of embryo's body, in reproduction, and particularly, in changing of morphogenetic programs in ontogenesis and evolution. This view of the role of somatic cells also corresponds to the data of discoveries made in the middle of the XXth century: stem cells, growing plants out of somatic cells, haploidy, parasexual hybridization, and embryoidogeny. These categories are for the first rime considered by us from the position of stem cells.
- A new phenomenon, embryoidogeny, was for the first time revealed by us in 1977 – a new type of vegetative propagation, where elementary structural unit is an embryoid (somatic embryo) forming out of somatic cells.
- Thus, not only zygote, but somatic cell as well is able to be the initial cell – stem cell of a plant, playing a big role in ontogenesis and evolution.
- The diversity of morphogenetic processes comprehends two components: potentially infinite somatic cells activity that is conditioned by the functioning of dormant centres' cells, being in intermediate differentiated state, and loss of stemness – senescence and apoptosis.
- The plasticity and polyvariety of these processes are conditioned by the diversity of the morphogenesis and origin of vegetative and generative structures in which the peculiar reserve of cells, possessing the stemness property, is situated. Hence in plant morphogenetic processes even third component is present – restoring of stemness in many respects depending on stemness degree.
- The arsenal of the data received in the XX-XXI centuries on biology of development including stem cells biology has allowed us to propose and discuss the three main perspective trends in study of stem cells – traditional and nontraditional.

- Nontraditional trends:

- gametes – zygote – embryo – individual – population – coenosis;

- different types of somatic cells considering their localization and function in the organism; possibility of homologization of some somatic cells with the zygote;
- Traditional trend:
- apical meristem of shoot apex and root apex in which organization somatic cells take part.

INTRODUCTION

Considering the study of stem cells, ***today's actual question is creation of the science on*** *stem cells* and development of its theoretical bases, taking into account comparative evolutionary cell biology for representatives of various kingdoms. In our opinion, while *developing the theory* we should begin with the elaboration of system complex and integrating approach enabling to study the *cardinal problem of the stem cell biology – the problem of wholeness*. This allows revealing correlations and studying correlative dependences of various structures at different levels of biohierarchy – the above-organism level, individual, organ, cell and molecular levels. A special attention should be paid to the **dynamics of the stem cells' pool and its correlative dependence with niche's. Comparative** evolutionary researches of the biology of stem cells of both plants and animals, their location and properties should be studied taking into account *developmental biology* of each taxon and also peculiarities of *the centre of origin, ethology and ecology*. These properties determine reproduction strategy, plasticity, reserves and the system of reliability of the organism. The research of stem cells in culture *in vitro* should be based on *the regularities and laws of developmental biology revealed in natural conditions*. Comparative study of the morphoprocesses in natural conditions and in culture in vitro is one of the principal positions in our experimental work – **"Let the nature be your only goddess" (Auguste Rodin).**

In current work the system of nontraditional ideas about the nature and role of stem cells in ontogenesis, reproduction and evolution of plants is proposed, based on the original results of long-term investigations and analysis of literature in area of developmental biology with use of system integrating approach.

The creation of every science is impossible without the application of methodology. According to the words of the great science theoreticist and zoologist, V.N. Beklemishev (1964, 1994), methodology is part of logic, one of the sciences of the study of the human mind, the spirit in its highest manifestations, and the technology of thought.

Besides that, a number of presuppositions, axioms and postulates underlie all conclusions of each branch of science, as V.N. Beklemishev supposes. For complete clarity sake it is necessary to find out, what is their accuracy, what is their obviousness and what are their admissibility and usefulness?! Especial attention the author paid to accuracy of notions.

The creation of the unified terminology for plants, animals and human taking into account their species specificity, is the serious problem, but each notion must be clear and correct (Van der Pijl, 1969; Batygina, 1987). This statement can be illustrated by the fact of misuse of such term as *«somatic embryogenesis»*, which in fact means formation of the embryo of uncertain origin – *somatic or sexual* !!! First of all, the term *«embryogenesis»* has been always used to designate formation of *the embryo* as result of fertilization (*syngamy*). *The notion «somatic»* refers to embryo's origin – out of *somatic cell of various origin,* while *the notion «embryogenesis»* is used for the *sexual embryo.*

We suppose, that it is necessary to use more correct term *"embryoid"*, which is the structural unit of a *new phenomenon* of vegetative propagation – the *embryoidogeny*, and the process of embryoid formation should be referred to as *embryoidogenesis*. This phenomen has a big significance in ontogenesis and evolution of plants (for details see: Batygina, 1977–2009).

Probably, the term *"somatic embryogenesis"* has been long used in literature not only because of *misunderstanding of the essence of the notion*, but also due to *conservatism* in using terms.

As it was truly mentioned by *world-wide famous* scientist Timofeev-Resovsky (1980) *creation of a new trend in science demands of mind to be prepared.*

Genetic and molecular analyses that have been undertaken during the last 50 years were mainly directed to revealing of the mechanisms of various structures initiation, cell proliferation induction, systems of cellular signalization, etc. The important role of phytohormones in morphogenesis was confirmed by these studies, especially that of auxin, cytokinin and their ratios. A lot of information available today on plant stem cell regulation and on what transcription factors control cell proliferation and maintenance of stemness, as well as how this ultimately leads to differentiation in morphogenesis (Bosh et al., 2008).

In my opinion and as it was truly regarded by the great Russian thinker-evolutionist I.I. Shmalgausen, when studying all these processes there should be paid more attention to correlative interdependences of separate morphoprocesses and structures which, from the point of view of general biology, are considered one of the key moments in morphogenesis of plants – proliferation, differentiation, dedifferentiation, etc. (I.I. Shmalgausen, 1968).

Many investigators suppose, that plant cells retain their ability to be a truly totipotent stem cell even when they are fully differentiated. (Gamborg et al., 1968; Reinert, Bajaj, 1977; Bosch, 2008). This question remains debatable; it should be clarified how fair the statement is, that all various types of differentiated cells are able to give rise to a new organism.

In our opinion, based on original data baked up with literature and stem cells properties revealed by us (Batygina et al., 2004; Batygina, Rudskiy, 2006; Batygina, 2009), the following statement is true:

Not every cell possesses the totipotency and not every totipotent cell is able to become the stem cell, and not every stem cell can produce an individual.

Already in the beginning of the 20th century, 1935, the great Russian biologist N.K. Koltzov believed that «… combination of … sciences – *genetics and embryology …*, as well as, *cytology and biochemistry*, will create a united science, which will allow solving general biological problems». Later the scientist announced appearing of a new science – *biology of development*, which was expected to study reasons and analysis of individual development. In 1967, in Moscow B.L. Astaurov had established the Institute of Biology of Development named after N.K. Koltzov, which studies various aspects of the life cycle, ontogenesis and evolution of individual representatives of plants, animals and of human beings.

Molecular genetics is one of the components of genetic-embryological researches. In the late 20th century the well-known geneticist M.E. Lobashov claimed that *«... in order not to be in isolation from embryology genetics should seek for the efficient equipment for its researches which lies in discoveries of new genetic methods increasing the resolving capacity of the genetics analysis. This requires «intellectual equipment» – synthesis of biological knowledge and creativity»* (M.E. Lobashov, 1969).

For many decades there existed the opinion that there is a *gap between sexual and somatic cells,* and consequently, between *asexual, vegetative propagation* – without meiosis and fertilization – where mainly the somatic cell is the structural unit, and *sexual one* with the zygote being the structural unit.

I would like to attract your attention to discoveries of 20th century, which show the importance of somatic cell in morphogenesis, ontogenesis and evolution: stem cells (Maksimov, 1909), investigations of "*dormant meristem*" in leaf (Naylor, 1932; Yarbrough, 1932; McVeigh, 1938; Batygina et al., 1995, 1996, 2006, 2009) and ovule (Batygina, 1979, 1991; Batygina, Vinogradova, 2007), *«quiescent centre»* in root apex (Clowes, 1954), *«méristème d'attente»* in shoot apex (Buvat, 1955), *growing plants out of somatic cells, haploidy, parasexual hybridization,* which has stimulated works in the field of morphogenesis of reproductive structures.

This has encouraged revealing of a new phenomenon, *embryoidogeny (formation of embryoid – somatic embryo) – a new type of vegetative propagation, where elementary structural unit is an embryoid (somatic embryo).*

Besides that, *universality of pathways of morphogenesis* (embryogenesis, embryoidogenesis, gemmorhyzogenesis) upon *individual formation* in *natural conditions* and in *culture in vitro* was discovered (Batygina, 1977-2009).

These discoveries have also favoured further elaboration of the theory of reproduction and of regularities of morphoprocesses and their correlation dependences in the field of morphogenesis of reproductive and vegetative structures in plants. *Especial attention was paid to working out of stem cell theory* (the main properties of stem cells revealed, there has been developed the program of the study of cell lines and their participation in organogenesis with changing of stemness character and degree, etc. – Batygina, 1979 – 2009).

It should be noted that due to achievements in the field of comparative and experimental embryology the XXth century has seen such discoveries as types of fertilization – pre- and postmitotic, new types of embryogenesis - Paeoniad- and Graminad-, tubifloral type of endosperm development, three types of karyogamy for seed plants and the new forth type (M-type) of karyogamy for flowering plants. The progress of the embryology in the XXth century has been connected with the use of the latest approaches (the system approach) and methods (electron microscopy, autoradiography, microsurgery, modeling), resulting in revealing of specific peculiarities of meiosis with the help of endoenzyme markers, dimorphism of the sperm, and in executing of fertilization *in vitro*, and in obtaining of unique data on cytoskeleton of reproductive and vegetative structures, etc.

Today developing the stem cells theory is within the science of developmental biology and should be considered from the point of view of the integration of various sciences. It should be noted that for many years there have been studied various aspects of *descriptive and comparative embryology* taking into consideration classical notions, in particular, *the laws of embryogeny: of origin, numbers, position, destination and economy* (Soueges, 1937; Johansen, 1950). This together with the revealing of *formation of the epyphisis and hypophisis* in the embryo of most of flowering plants at the early stages of development have become a considerable contribution to the study of *morphogenesis* of plants (Hanstein, 1870; Soueges, 1934; Raghavan, 1990). The *essence* of these structures lies in *organization of apical meristems, shoot and root of a future plant.*

Despite variety in composition of a shoot, root and other similar organs, it is characteristic of all higher plants to have the quiescent centre or domain of non-differentiated cells, which can even consist of one cell, in the apical meristem. Also, the time of their laying down is taxon specific and normally it happens in the fifth or sixth cell generation and is realized through asymmetric divisions. As a result, the whole structural domain of non-differentiated cells, as the initial of all following processes of organogenesis, by definition can be formed by cells of independent lines. *Diversity of known types of embryogenesis also demonstrates independence of arising of the shoot and root system organs. For example, in Graminad-type of development of an embryo both root tissues and shoot tissues can appear in very different cell lines* (Batygina, Barlow, 2005, personal communication). However, it is the cells of quiescent centre which are considered "stem cells" in traditional narrow sense of the term.

The term *«stem cell»* in the botanical literature, as well as in the zoological and medical ones, is used only as a *functional notion.* There are no genetic and epigenetic markers known yet to identify stem cells. The known complex WUS/CLV for stem cells of the shoot apex of *Arabidopsis* is only to some extent similar to QHB of root's stem cells. Apart form general spatial location stem cells are also *joined by processes proceeding in the domain.* Relatively *symmetric divisions* are connected with self-maintenance of stem cells pool, while *asymmetric ones* – with *transition* of daughter cell to a certain way of *differentiation and specialization.* Stem cells derivatives – initials of cell lines – create the *niche of stem cells.* The surroundings of the domain of differentiating cells differs principally by presence of sister cells.

Study of the phenomenon of "stem cells" occurs to be one of the perspective trends in developmental biology. Laboratory of Embryology and Reproductive Biology of Komarov Botanical Institute, St-Petersburg, Russia, has been for many years working on general biological problem of origin and role of stem cells in ontogenesis of plant organisms (Batygina 1979-2009; Batygina, Freiberg, 1979; Batygina, Titova, Shamrov, Bragina, Vasilyeva, Rudskiy, 2004; Rudskiy et al., 2005; Batygina, Rudskiy, 2006).

The main properties of plant stem cells have been developed by researchers of the laboratory, which has shown the integrity of morphogenous and reproductive processes at all stages of plant's life cycle (see Chapter).

Not every cell possesses the totipotency and not every totipotent cell is able to become the stem cell. Totipotent stem cell is able to form an individual by different pathways of morphogenesis in plants – embryogenesis, embryoidogenesis and gemmorhizogenesis (Batygina, 1987-2009).

System approach and analysis of data on *male and female generative structures – egg cell at different stages of development, double fertilization in model objects: syngamy, triple fusion, plasmogamy and karyogamy taking into consideration the cell cycle, zygote, embryogenesis of model objects* in representatives of a *lot* of flowering *plants* taxa allowed us *to express nontraditional idea,* that from the *embryological point of view zygote is proximate stem cell (stem cell of 1st order) and gives a rise for all other stem cells (of the next orders) in plant ontogenesis, particularly those of shoot and root apices. Every plant cell is able to be in different morphophysiological states, the characteristics of one of them mostly correspond to the properties of stem cell, the others correspond to actively dividing meristematic cell. The transitions between these states are stochastic and in most cases histogenesis and organogenesis are connected with the loss of so called stemness, but any other transitions are also possible* (Barlow, 1997; Batygina et al, 2004; see Introductory chapter).

The study of descriptive, comparative, experimental developmental biology and especially comparative embryology of representatives of various taxa of *flowering* plants enables to consider that *formation of stem cell of orders derived from zygote, are probably typical for all organs – flower, stem, leaf, root – and at all stages of the life cycle – sporophyte, gametophyte, their functioning, first of all, depending on their localization and function.*

It is from the point of view of stem cells that *initial cells of somatic embryos (embryoids) of various origin in natural conditions and in culture in vitro* should be considered. In this connection, it is of a great importance to study *the phenomenon of embryoidogeny – a special category of vegetative propagation* (Batygina, 1977-2009).

For the first time it is grounded, that *the mature Zygote in plants – an initial cell of sexual embryo – is a unique differentiated totipotent stem cell, «progenitor»* of stem cells of other lines. *It contains information about architectonics and development of organization of an individual, which is connected with gradients and correlations between its parts, and consequently with heterogeneity of different axes of a future organism* (Batygina et al., 2004).

Usually totipotent stem cell, the zygote, which is a result of sexual process – fusion of two gametes, develops into multicellular organism. In some cases in plants and animals in the zygote upon the gametes fusion or, possibly, at later stages of organism development, complicated morphogenic changes occur defining formation of two or more organisms – the monozygotic twins (or triplets, etc.). These individuals form as result of asexual (vegetative) propagation and change of morphogenic program from sexual to asexual process. This phenomenon is a bright evidence of the fact, that the zygote – stem cell – possess different potential and cloned at earliest developmental stages giving various number of individuals with the same genetics (except mutations, polyploidy etc.). Numerous data of such kind are obtained in various plant species and referred to as monozygotic (cleavage) embryoidogeny, the kind of asexual propagation (orchids, etc.). Facts obtained in representatives of Paeonia confirm that the zygote and its derivatives maintain stemness, which manifests itself in the peony at the stage of coenocite-cell structure of the sexual embryo as a multitude of initials forming out of the epidermis cells of sexual embryo and giving embryoids (cloning). Their genetics is similar with the sexual embryo.

Stem cells in peony ontogenesis (zygote, epidermal cells of coenocytic-cellular structure – initials of embryoids) – play the key role in changing of developmental program (sexual ↔ asexual) and cause the formation of multiple ontogeneses in life cycle due to formation of sexual and somatic (embryoid) embryos.

Considering this fact in peony evolution it should be noted that switching of programs from sexual to asexual propagation in plants and animals is a compromise solution of the genetics and ecological problem in reproductive structures, where it was convincingly shown that there are no gap between sexual and somatic cell, as it used to be stated before.

The phenomenon of formation of monozygotic twins proves the fact that it is possible to talk about the role of sexual and somatic cell and their homology. Adaptive significance of this phenomenon lies in keeping genetic material quite homologous during many years.

Zygote – organism – is an open system, dependent not only on the very individuality, but on the structure of above-organism systems – population, coenosis, biogeocoenosis.

Not only zygote, but somatic cell as well is able to be the initial cell - stem all of a plant.

It is from the point of view of stem cells that initial cells of somatic embryos (embryoids) of different origin in natural conditions and in culture in vitro should be considered. In this connection the study of a special category of vegetative propagation – embryoidogeny – gains a great significance (Batygina, 1977-2009).

The diversity of morphogenetic processes comprehends two components: potentially infinite somatic cells activity that is conditioned by the functioning of dormant centers' cells, being in differentiated state, and loss of stemness – senescence and apoptosis.

Similar processes can take place at different ontogenesis stages and in all structures, e. g. the formation of embryoids from epidermal cells of sexual embryo in Paeonia (Batygina, 2009). The plasticity of leaf tissues – Bryophyllum, Gnetum provides the possibility of formation of embryoids *from a single or group of cells (dormant meristem=stem cells)*, in other words, the *repeating of early ontogenesis stages in the tissues of adult plant* is observed (Batygina et al., 1996, 2005). Programmed destruction of definite part of leaf primordium cells (*Anubias*) leads to formation of epidermis owing to the *transdifferentiation* of interior tissues (Rudsky et al., 2005). The *plasticity* of generative structures' cells of *maternal plant* (*Poa*) permits formation of *embryoids* from *nucellus and integument* (Batygina, 1991b). Thereby cells of *embryoids* during the development use *the niche of the sexual embryo*. The phenomenon of *gametophytic apomixis* widely spread among angiosperms consists of the initiation of somatic embryo (*embryoid* – T.B.) formation without fertilization from egg cell or synergid – cells of female gametophyte (Czapik, 1997). *Microspore* of *Triticum* is a *totipotent cell*, capable of switching of the developmental program from gametophytic to sporophytic pathway of development through direct and indirect secondary embryoidogenesis. Upon the experimental conditions microspore turns to equal divisions (in contrast to unequal ones upon the usual pathway of development) and further to formation of *embryoids*, so called *haploids*.

The *plasticity* and *polyvariety* of these processes are conditioned by the diversity of the morphogenesis and origin of vegetative and generative structures in which the peculiar *reserve of cells*, possessing the stemness property, is situated. Hence in plant morphogenetic processes even the third component is present – the *restoring of stemness* in many respects depending on stemness degree.

Development of the theory of reproduction has revealed *a considerable role of the somatic cell in ontogenesis and evolution*. For embryologists it is clear that upon the sexual and agamic reproduction (out of somatic cell) the organisms carry information about various aspects of organization being initiated at different structural basis. The information about the general architectonics, mainly, exists at the zygote or somatic cell level. *Independent of the individual's origin: out of the zygote or somatic cell, a new organism is born.*

Data of *classical descriptive and comparative embryology* provide evidence of the correlation between the most important acts of morphogenesis – *the type of divisions, pathway of morphogenesis, character of laying down of the morphological axis/axes and their changes in the process of their genesis, polarity in plants in the egg cell, zygote-embryo and at later stages of development.* The specific peculiarity of morphogenesis lies in correlative changeability of macro- and microstructural components of the developing organism: the general type of symmetry, gradients, architectonics, etc.. It should be mentioned that in ontogenesis the continuity of information about organization is observed at molecular, cellular and organism levels. There is the correlative dependence of development of different components (mitochondria, chloroplast, nucleus, dictyosomes, etc.) found at different levels of biohierarchy (Tokin, Korotkova, 1977; Korotkova, 1979).

Studying stem cells we should not underrate interconnections in development of stem cells and niches. Irregularities in these correlations can lead to abnormal morphogenesis of various structures – cells, apices, organs. There should be mentioned a special role of availability of *«critical mass»* of various components – organelles, cells, biochemical components – as the regularity of passing of any morphological pathway (T. Batygina, 1987). Differentiation of embryological structures and organism, on the whole, can be accompanied by formation of several critical masses. They determine «the threshold of factors» necessary for realization of one of morphogenesis pathways; it is also taxon-specific, see also the law of numbers by R. Souèges.

The phenomenon of embryoidogeny shows *the significance of the role of somatic cells in organization of embryo's body, in reproduction, and particularly, in changing of morphogenetic programs in ontogenesis and evolution. This view of the role of somatic cells also corresponds to the data of those five discoveries which were made in the middle of the XXth century: stem cells, growing plants out of somatic cells, haploidy, parasexual hybridization, and embryoidogeny. Besides there were revealed such structures as dormant meristem, méristème d'attente and quiescent center.*

Embryologists have shown the fact, that there is no gap between sexual and somatic cells, revealed the importance of somatic cells in ontogenesis and evolution and raised the question about possibility of homologization of some totipotent somatic cells and zygote.

The arsenal of the data received in the XX-XXI centuries on biology of development including stem cells biology has allowed us to propose and discuss the three main perspective trends in study of stem cells – traditional and nontraditional.

Nontraditional trends:

- *gametes – zygote – embryo – individual – population – coenosis;*
- *different types of somatic cells considering their localization and function in the organism; possibility of homologization of some somatic cells with the zygote;*

Traditional trend:

- *apical meristem of shoot apex and root apex in which organization somatic cells take part.*

Progress in the study of biology of stem cells will be connected with development of the theory of developmental biology and the theory of reproduction in ontogenesis and evolution. The *integration of all three trends* in research of stem cells from the point of view of wholeness at *various levels of biohierarchy* will enable developing a number of *hypotheses*, as well as basics of *the stem cells theory*, and formulating *new laws of development* taking into account the existing data of *classical works* and basing on *general regularities*, existing *laws of biology*, *morphogenetic correlations* considering *ethology* and *ecology*. The development of the theory is one of the bricks in the construction of the theory of developmental biology and the theory of reproduction.

Development of the theory of stem cells is one of the bricks in the construction of the theory of reproduction.

Propagation: Its Modes, Types and Forms

During global climate change, anthropogenic factors firstly damage the reproductive structures in plants (embryo, stamen and seed), resulting in impairment and often disappearance of vegetative cover. In this connection, the preservation of the natural plant gene pool, restoration of soil seed bank and finding new forms and sorts in the shortest terms require knowledge of the whole perspective of events that take place from the origin of the plant organism and determining the course of individual development up to the formation of fruits and seeds.

All processes resulting in the increase of biological units are referred to as propagation. It is one of the general properties of a matter residing in different structures, from molecules to populations.

The systems of plant reproduction are considered to be examined both in relation to the character of genetic information transferred to posterity and the modes of its realization in ontogenesis. Due to the factors of environmental influence, the adaptive changes occur not only in genetic information but also in the modes of its transferring.

The phenomenon of propagation in plants is of a great theoretical and practical significance. The discoveries of the twentieth and twenty-first centuries—plant breeding from somatic cells, parasexual hybridization, haploidy, stem cells, M-type of caryogamy at fertilization in angiosperms, the phenomenon of metamorphosis in the ontogenesis of highly specialized parasitic plants, etc.—have activated work in the sphere of developmental biology, morphogenesis of reproductive structures and ethology. The world's leading biologist repeatedly emphasized how important it is to study the phenomenon of plant reproduction for understanding the evolution and genetics of flowering plants (Grant, 1977, 1981).

There is a necessity of integration and creation of "...interdisciplinary disciplines that build bridges between traditional specialties and thereby open the pathway to the creation of universal science, which should serve as a sort of framework uniting splintered sciences into a single one" (Takhtajan, 1998: p. 11). The formation of "integrated embryology" (T.B.) is required, which will lead to understanding of genetics and reproductive biology as well as cytology and biochemistry in order to solve general biological problems.

Long-standing complex investigations of various phenomena and processes related to reproduction of flowering plants allow formulating the general theses of reproduction theory. They are:

- the principles of the formation of reproductive structures (alternation of haploid and diploid generations, universality of morphogenesis pathways, status and interaction of types, modes and forms of seed and vegetative propagation that determine the reproduction strategy of species);
- the principles of structural organization of embryonal structures (polyvariety of development, space-temporal coordination);
- the theory of critical periods, in the course of which the change in the developmental program, establishment of embryo autonomy, etc. take place.

The knowledge of reproduction theory permits making prognoses on the posterity genotype and opens new perspectives in the control of certain ontogenesis stages. Embryological information is somewhat important also in the creating of new selection programs and industrial technologies with untraditional modes of new plant form and sort breeding.

The Modes of Organism Formation, Propagation and Renewal

Propagation can take place at different levels—molecular, cellular, tissue, organ, organism and population. Cell division is the basis of propagation (cytogony). According to traditional classification, plants have *three types of propagation: sexual, asexual, and vegetative*. In spite of a large number of investigations devoted to different aspects of propagation phenomenon, the biological entity of definite types, modes and forms of reproduction and their interactions are considered to be not clear enough. The question on the volume of notions of *"sexual process"* and *"sexual propagation"* is also debatable. All these terms require definition and, as L. van der Pijl noted, "...differentiation of terms is *not rather the play of words* but an absolutely *essential condition* in order to understand the nature of things" (1969, p. 15). We tried to express our untraditional view of this problem.

On the example of the notion *"sexual process"* one could easily see how much the terminology, concerning propagation, is mixed up. The sexual process in plants in its typical form is understood as the fusion of two sexual cells – gametes and the formation of zygote. Interpreting the notion "sexual process" in flowering plants most investigators ignore one of its key periods – meiosis. From our point of view *the sexual process includes meiosis and fusion of gametes (from different meioses), which results in arising of a zygote, i.e., a new individual.* Thus during sexual process the increasing number of individuals does not occur, since a single new organism forms, as a rule, from the zygote, developing as a result of the fusion of male and female gametes. *The enlarging of sexual progeny* is provided just by *the multiplicity* of generative structures (ovules, pollen grains, gametes, zygotes) and so by the multiplicity of sexual processes. That is why, when *only one sexual process* takes place, *reproduction* should be mentioned.

The term *"sexual process"* is often replaced in literature by the notion *"sexual propagation"* which, according to the notion mentioned above, is not correct, and these terms could not be considered synonyms (see Batygina, Vasylieva, 2002).

The famous Italian botanist, Battaglia (1963), while examining the phenomenon of reproduction from the general biological view, formulated the notion that organisms are able to arise *in the same phase—sporophyte from sporophyte (homophasic increasing in number, "multiplication")—or, in the alternative (antithetic) phase, sporophyte from gametophyte (heterophasic increasing in number, "reproduction").*

It should be mentioned that one speaks about asexual type of propagation in different members of plant kingdom "in the narrow sense"—*multiplication by spores* and "in the broad sense"—*vegetative propagation.* Multiplication by spores is typical for algae, bryophyta, licopodiophyta, equisetophyta and ferns. Pisyaukova (1980) suggested using *for the flowering plants the term "vegetative multiplication" only*, considering the usage of the term *"asexual multiplication"* to be in general tautology. According to her opinion the spores of flowering plants *lost their main functions, connected with multiplication and dispersal.*

Thus, the analysis of literature and original data *allows decline the terms "sexual propagation" and "asexual propagation" related to the flowering plants, exclude them from the generally accepted classification of propagation types and use instead of them the notions "seed propagation" and "vegetative propagation".*

In the native and foreign literature ones often use the term "reproduction" (French *reproduction*, German *reproduktion*). It should be noted that only the detailed analysis of article context, where this term applied, permits understanding of the meaning utilized by the authors—reproduction, propagation or renewal.

All questions related to the problem of *propagation* appeared to be *pure rhetoric* at first sight. But unfortunately, as we satisfied, *interpretation of many notions does not reflect the essence of phenomena both from genetic and general biological positions.* The debatable character of the notions *"propagation"* and *"renewal"* is also considered to be the confirmation of this. It should be mentioned that the first applies to the individual and the second to the population. The term "propagation" is not applied to the systems of above organism level. At such distinguishing of the notions the "vegetative renewal" is understood as the renewal of population by vegetative propagation (Levina, 1981). Accordingly, the "seed renewal" is understood as the renewal of population by seed propagation.

Seed propagation, as well as vegetative renewal, depends on numerous different environmental factors: meteorological, edaphic, allelopathic, parasitic, competition and therefore occurs to be the *stochastic process.* However *sexual* and *asexual processes* of individual formation, *heterophasic and homophasic reproduction, seed and vegetative propagation* and *renewal* elapse conjugately, the certain correlations, which permit to remain *the homeostasis of species and population*, are revealed between them. But the question on *the ratio of different types, forms and modes of propagation in various taxa believe to be the most difficult and discussible.* The solution of this problem has a great importance for the study of plant adaptation, preserving of biological diversity and biological resources as well as for the elaboration of biotechnological approaches for the mass copying of value genotypes.

Seed Propagation

The flower is the most important reproductive organ of angiosperms; in its ovary, the processes of *micro- and megagametogenesis, pollination and fertilization* occur, resulting finally in *the formation of fruits and seeds containing a new sporophyte, the embryo.* The flowers are extremely diverse in structure and colour, their diameter varies from millimeters (for example, orchids) up to the metre and more (for example, *Rafflesia*).

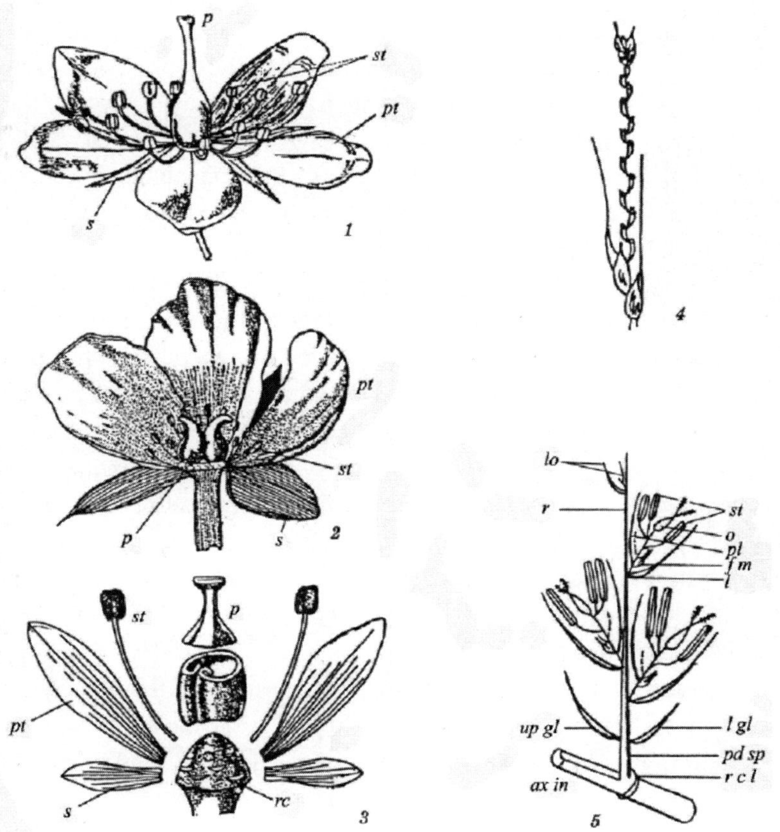

Figure 1. Structure of the flower and inflorescence: (1) general view of the flower; (2) the peony flower (in longitudinal section with a lot of stamens and two pistils; other have been removed); (3) separate parts of the flower; (4, 5) the structure of the ear (4) and triflorous spikelet (5) in cereals; *up gl* – upper glume; *pl* – plea; *o* – ovary with two stigmas;
lo – lodicule (flower glume at upper half-grown flower); *pt* – petals; *pd sp* – pedicel of the spikelet; *l gl* – lower glume; *l* –lemma; *r* – rachilla; *ax in* – common axis of inflorescence; *p* – pistil, *r c l* – rudiment of the covering leaf; *st* – stamen; *f m* – flowering membrane;
rc – receptacle; *s* – sepal.

The flower is the totality of sterile and fertile structures developed on the floral disk. The *sepals and petals* belong to the sterile flower structure; *stamens and carpels* belong to fertile ones (Figure 1). The borders between sterile and fertile structures are conventional in some plants.

In the classical concept, the members of different groups of plants (mosses, club-mosses, ferns, gymnosperms and angiosperms) undergo in their life cycle the alternation of generations: *sporophyte and gametophyte.* The term "alternation of generations" was introduced by Hofmeister (1851). Meiosis occupies the central place in reproductive cycle, at the alternation of *diploid and haploid phases.* From each cell with the diploid chromosome number the *haploid* sexual cells (gametes) are produced in the course of meiosis. Haploid stage of plant development is represented by *gametophyte* (beginning from the growing of micro- and megaspores up to zygote). An organism developed *from zygote up to the producing of spores is referred to as the sporophyte.* In the process of evolution towards angiosperms, the gradual reduction of gametophyte and domination of sporophyte occur. The sporophyte of angiosperms is represented by trees, bushes and herbs. The gametophytes are reduced and are formed as a result of several mitotic divisions. The development of an *embryo sac* (female gametophyte) is usually reduced to three divisions, and the development of *pollen grain* (male gametophyte) to two.

V. A. Poddubnaya-Arnoldi (1976) considered this phenomenon of "alternation of generations" from a different point of view. She assumed that sporophyte and gametophyte in angiosperms are so much connected with each other, that *they should not be examined in isolation with each other*, as far as *the plant is a total organism.* One could speak only about *the developmental phases in the life cycle of plants.* However, as R. E. Levina (1981) emphasized, the introduction of the notion "alternation of generations" has played a positive role in science, as it availed to reveal the unity of origin in different plant groups. We share the opinion of V. A. Poddubnaya-Arnoldi and R. E. Levina. *Theoretical elaborations on the research of polyembryony phenomenon from the positions of the reliability system of the organism (plasticity and tolerance)* allow us express the idea that *the dividing on sporophyte and gametophyte is conditional in flowering plants* (Batygina, Vinogradova, 2007).

While examining the question of seed propagation, we should once again mention that the peculiar features of reproduction life in plants are to a great extent conditioned by immobility of their adult individuals. The formation of *special structures* (*diaspora*) which make easier the species propagation and dispersal became one of the consequences of this. In the course of many millions years of plant world evolution the spore stay to be universal diaspora (algae and higher cryptogamous). Later, in seed plants, this function transferred to *the seeds* that help them to occupy the dominant position on the land.

The seed, one of the units of propagation and dispersal typical for seminal plants, contains *the initial of new plant — the embryo* — and specialized accumulating tissue (endosperm, perisperm, etc.) enclosed into the seed coat (testa, spermoderm) (Figure 2).

Seed propagation in plants is provided by the account of formation of enormous number of seeds with embryo. One should note that in the same seed two processes of new individual producing can take place: *sexual* and *asexual.* Owing to this, the seed can contain *several embryos of different origin* (*polyembryony*) and give the seedlings of different genetics (*matroclinal posterity*—from nucellus and integuments (2n=2n) or *the individuals* (*zygote with new genetics – n+n*).

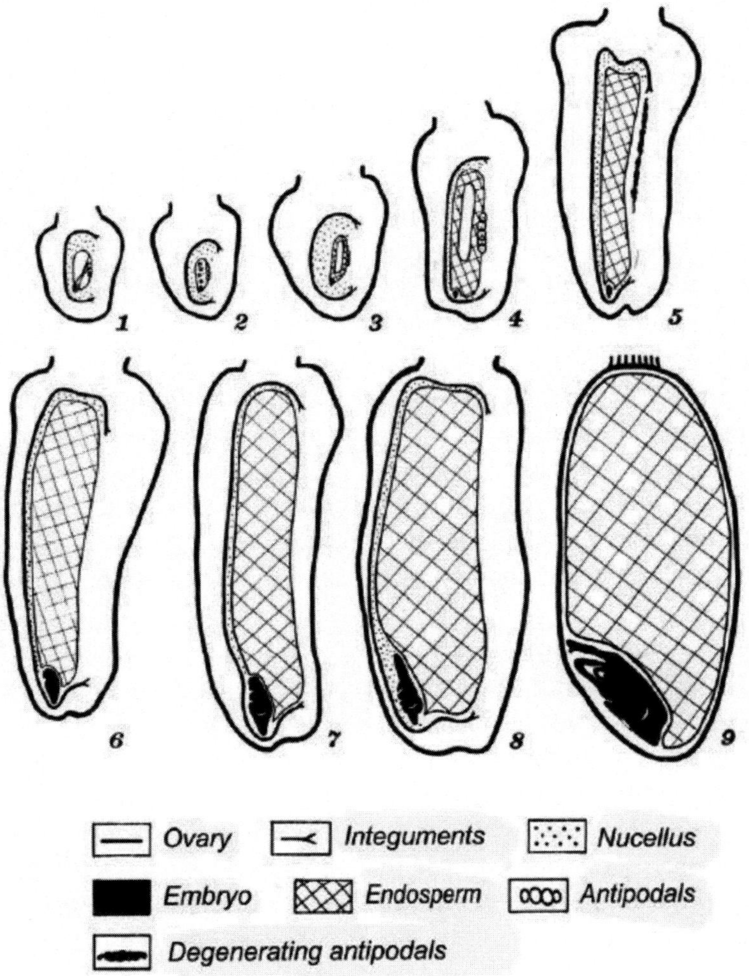

Figure 2. Successive phases of the development of the embryo and endosperm in the caryopsis of wheat, Diamant : (1) after 5–6 h; (2) 1 day; (3) 2; (4) 3–4; (5) 7–8; (6) 9–10; (7) 11–12; (8) 13–15; and (9) 25–30 days after pollination (from Batygina, 1987**a**).

Seed propagation of plants leads to increasing of species quantity. The final evidence of propagation intensity is considered to be the seed crop.

Vegetative Propagation

In the classical meaning, vegetative propagation is *the increasing of species individuals* in number as the result of separation *of the living available parts of the plant vegetative body* (buds, shoots, roots, etc.); in many cases, this process is accompanied by *regeneration*, i.e., *restoration* of missing organs in separated plant parts. This is essentially distinguished the vegetative propagation from the seminal one, in the course of which during *seed germination the individual of full value "is being born" from the embryo at once.*

Posterity, arising as a result of vegetative propagation, independently on the place of its formation (vegetative or generative structures), produces clones—totality of the genetically homogenous individuals, identical by genotype to the maternal organism. According to new data on vegetative propagation, the clone sometimes is supposed to consist of not rather uniform individuals as a result of mutational, etc., processes.

Vegetative propagation occuring *without the participation of a man* is referred to as *natural,* while the one happening with his participation is *termed artificial.* In natural conditions, the efficiency of vegetative propagation is defined by *the age* (or age condition) of maternal plant, *the longevity of its physiological contact* with vegetative posterity and by *the degree of remoteness and rejuvenation of the later.*

Certain types and modes of vegetative propagation were distinguished: *sarmentation and particulation* (Smirnova, 1974; Barykina, 2000a, b). The structural unit of sarmentation usually is a brood bud. In botanical literature the notion "sarmentation" should be used for identification of one of the natural vegetative propagation modes, at which the intensive capture of the territory observes and the highest degree of survival of daughter plants, that undergo quick rejuvenation, riches. By the depth of posterity rejuvenation such mode of vegetative propagation usually approaches to the seminal one (Levina, 1981; Barykina, 2000a).

The bud is a *monopolar structure* presented by initial shortened shoot or its part (Figure 3). In a broad sense, both the dormant buds and the tops of growing shoots are called buds. In a narrow sense, only the structures that stay in rather dormancy are considered to be buds. Moreover, by the type of laying down of initials the buds are divided into *vegetative and generative (flower)* with the initials of reproductive organs (flowers and inflorescences). Besides the pure *vegetative* buds there are also *vegetative-generative* ones, in which the series of vegetative metamers are laying down and the apical cone is transferred into initial flower or inflorescence. *The bud that contains the initial of a single flower is refered to as a flower bud.* By the *position on the plant and origin the terminal, lateral and adventive buds* are distinguished. The terminal buds are formed on tops of shoots; the lateral in the leaf axil.

We established for the first time that new individuals at *vegetative propagation* in flowering plants could be produced by means not only of the bud, but the somatic embryos as well (Figure 4). In this connection, I consider it possible to express *the untraditional view on the types and modes of the seed and vegetative propagation, their ratio in the system of flowering plant reproduction, morphogenesis pathways standing in their basis and mechanisms determining the change of developmental program in ontogenesis and phylogenesis.*

Despite of many studies, a biological sense of some modes, types and forms of propagation is still unclear, as well as their relationship and the evolutionary role. A *viviparity* (lat. *vivus,* living; *pario,* to give birth) takes a special place among them.

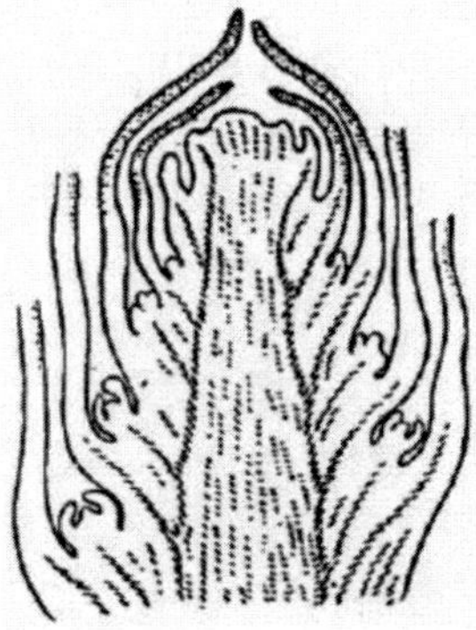

Figure 3. Longitudinal section of a bud.

However, the notion "viviparity" is still uncertain and debatable because of insufficient information available on its structural-functional bases. There is no common opinion what the term "viviparity" means: either germination of seeds on the maternal plant (in some *Rhizophora, Bruguiera, Kandelia, Ceriops* and *Avicenia*) or young plant development on the different parts (leaf, stem, inflorescence) of the mother plant (in *Poa viviparum, Polygonum viviparum, Cardamine, Bryophyllum*). However, most of the botanists classify both a germination of mangrove seeds as well as vegetative diaspors in other plant species under viviparity.

Studies of viviparity are fundamentally important for research of morphogenesis, relationships between maternal and daughter organisms, transport of nutrients and hormones, their influence on the daughter organism development, formation of its autonomy, and reproductive strategy of species. It is also important for solving such applied tasks as a selection of optimal conditions for seed storing and their germination, cultivation of the endangered species in vitro for further repatriation.

Animal viviparity has been already known to Aristotle (fourth century BC). He used it as a character to divide the animals in viviparous and ovoviviparous. Linnaeus also applied this character classifying species from the genera *Festuca* and *Poa*.

Scientific names of some plants (*Festuca ovina var, vivipara, Polygonum viviparum, Eriogonum viviparum, Poa viviparum, Remusatia vivipara, Asplenium viviparum*), as well as animals (*Xenopleura viviara, Viviparus viviparous, Zoarces viviparous, Barbus viviparous, Lacerta vivipara, Patriella vivipara, P. paravivipara*) were created being based on the term "viviparity".

Viviparity in plants is a special type of reproduction wherein an embryos in a seed, fruit or brood bud develop and form seedlings (propagules) without a dormancy period of the maternal plant. Genotype of the descendants depends on the propagule origin.

At the moment this phenomenon has been discovered in 281 species of the flowering plants belonging to 55 families (Nymphaeaceae, Ranunculaceae, Orchidaceae, Poaceae) (after Batygina, 2006). Besides, there are 197 species of viviparous ferns of 8 families. The viviparous species of *Bryophyllum* (Crassulaceae) is considered to be one of the most interesting objects of great practical significance.

Viviparity in animals is a type of sexual reproduction wherein either larva or juvenile develops inside the body of the maternal organism, and is released free of egg envelopes. Viviparous species are already known among ammonites.

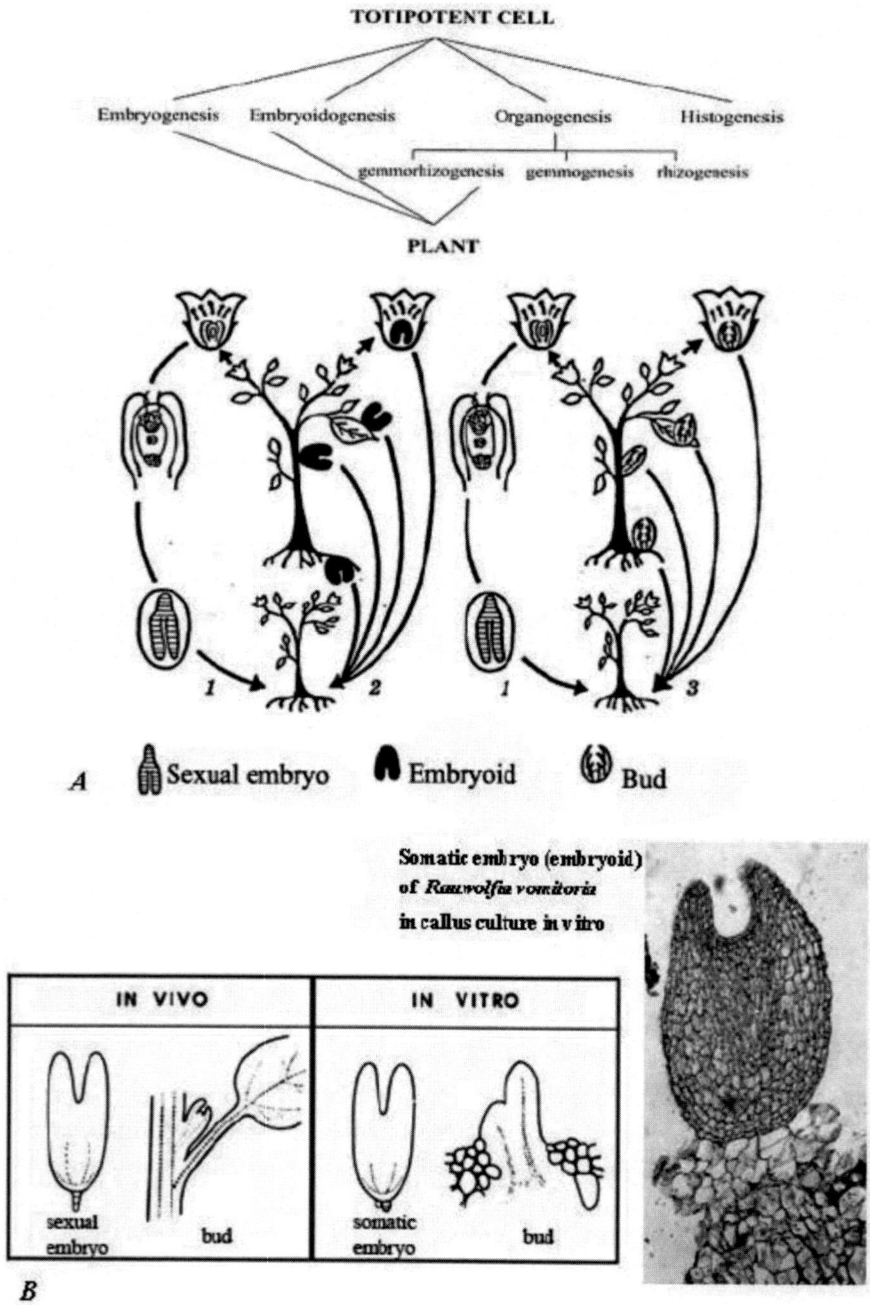

Figure 4. A. Pathways of morphogenesis in the system of reproduction of flowering plants: (1) embryogenesis, sexual mode of reproduction; (2) embryoidogenesis and (3) organogenesis, asexual modes of reproduction (from Batygina, 2000); B. Structural items of sexual and asexual reproduction *in vivo* and *in vitro* (from Haccius, 1978).

Genetic Grounds for Viviparity

With normal development of a seed, the phase of maturation involves the synthesis of reserved nutrition substances, ceasing of the embryo's growth and development of resistance to drying. It is known that abscisic acid (ABA) is the key regulator of gene expression in late embryogenesis, which correlates responses to various extreme conditions. The change in synthesis and sensitivity to ABA can be one of the reasons leading to viviparity.

Considering content and degree of sensitivity to ABA viviparous mutants of corn were divided into two classes. In mutants of the first class (*vp2, vp5, psl* (*vp7*), *vp8, vp9*) there can be observed the decreased level of ABA and change in synthesis of carotinoid (with exception of *vp8*).

In *vp1*, the mutant related to the second class the level of endogenous ABA in the embryo is not changed, though the mutant is not sensitive to exogenous hormone. Besides, in *vp1* mutant the synthesis of anthocyans in cells of aleurone layer is depressed. The product of *Viviparous-1* (*VP-1*) gene is included in regulatory proteins, stimulating maturation and dormancy of caryopsis. The product of *VP-1* gene is the factor of transcription and is required for the expression of *Cl, Globulin* и *Em* genes. Besides, the protein *VP1* acts as a repressor of genes of α-amilase, functioning at germination in cells of aleurone layer.

In seeds of viviparous mutants of corns, genes participating in synthesis and recognizing of ABA have been discovered. Embryos of one of mutants (*vp14*) have normal sensitivity to ABA. However content of ABA in mutant embryos was by 70% lower, than in embryos of a wild type, which point at the disorder in synthesis of ABA. Protein *VP14* catalyses the reaction of disintegration of carotinoids.

A lot of viviparous mutants of *Arabidopsis* have phenotypes similar to those of viviparous mutants of corn (for example, *abi3* mutants). Mutation of *abi3* occurs in the period of dormancy of the seed, gaining of spare proteins and fats, decay of chlorophyll, ability to respond to the action of ABA and resistance to drying. Characteristics of the existing mutant *abi3*-alleys have confirmed the hypothesis that the protein *AB13* participates in the cascade of reactions of perception and trasduction of ABA. Normal perception of ABA by the *AB13* gene is a necessary but not sufficient condition for many critical stages connected with embryogenesis in Arabidopsis. Similar successions and analogous phenotypes of *VP1* and *AB13* prompts the idea that they are functionally homologous genes.

In *Arabidopsis* mutants *leafy cotyledon* (*lec1*) and *fus3* early germinated seeds are rarely onserved. Comparison of these mutants provides the ground to believe that genes *LEC1* and *FUS3* can be responsible for connected but not identical functions during embryogenesis. They code regulatory factors which activate a wide range of embryogenetic programs starting from the heart stage. *FUS3* and *LEC1* positively regulate increased content of protein *AB13* in the seed.

Viviparity is determined by the genotype of the embryo and is not dependent of the genotype of the endosperm.

It is possible, that mechanisms of regulation of viviparity in the mangrove will be similar to those of in mutants of corn and *Arabidopsis*. Probably, it will be possible to find some similarity in regulatory processes taking place at vegetative viviparity.

Nontraditional Notions of the Types and Modes of Reproduction

Morphogenesis Pathways

The active application of the method of *in vitro* culture in experimental and biotechnological investigations constantly excites interest to the series of unsolved problems of morphogenesis repeatedly examined in classical morphology and gives a rise certain new questions. For instance, what are the pathways and modes of morphogenesis at the formation of a new individual in the culture *in vitro* and could they be compared with morphological events occurring during the formation of new individual in natural conditions? What are the initial cells giving somatic embryos in the culture *in vitro* and in natural conditions characterized by? Why do the isolated somatic cells in one taxon produce the buds, in others the roots and in some the somatic embryos? Is there parallelism in the development of sexual and somatic embryos, forming in natural conditions and in the culture *in vitro*? The question on the modes and mechanisms of switching over developmental program from one morphogenesis pathway to another in the process of ontogenesis is considered to be one of the actual.

In the 1970s, contradictory opinions concerning morphogenesis pathways in the culture *in vitro* existed. According to the generally accepted view, the formation of new individual in flowering plants realizes either by sexual mode— the sexual process preceding embryogenesis (heterophasic reproduction) or by asexual (vegetative) mode—the formation of buds and roots (homophasic reproduction), i.e., regeneration occurs (Battaglia, 1963; Batygina, 1999). However it is appeared that upon the homophasic reproduction new individual creates not only by regeneration (organogenesis—gemmorhizogenesis), but also by means of somatic embryo formation—embryoidogenesis.

In the light of data on flowering plant reproduction *in situ, in vivo* and *in vitro* we have revealed for the first time that *the vegetative propagation* is presented by two types: *embryoidogenous* (somatic embryos) and *gemmorhizogenous* (bud plus roots) but *not only a single gemmorhizogenous, as earlier considered.* At the formation of new individual not two (embryogenesis and organogenesis, as earlier considered) but three pathways of morphogenesis exist: *embryogenesis, embryoidogenesis* and *organogenesis.* The pathways and modes of new individual formation given are universal both in natural and experimental conditions (Batygina 1984) (Figure 4A). It allows us distinguish a new category of vegetative propagation – embryoidogeny and examine the role of this phenomenon in the system of flowering plant reproduction. Long standing elaboration of theoretical bases of plant reproduction permits to *establish the status and interaction of different types, modes and forms of seed and vegetative propagation*, realization of each in a rather extent determines the *reproductive strategy of the species.* The creation of new classification of types and modes of flowering plants reproduction and propagation considers being a result of this elaboration (Batygina, 1987a, b, 1989a, b, 1990a, b, 1991a, 1992, 1993a, 1994, 1996a, 1997, 2000, 2005, 2006a, 2009; Batygina Vasilyeva, 2002; Batygina et al., 1978).

The theoretical elaborations in the sphere of embryoidogeny already have permitted to obtain the certain plant forms and sorts, including that on the basis of androcliny phenomenon, this is what the present monograph devoted to (see also "Embryological bases of wheat androcliny", 2005; Batygina et al., 2007).

Embryoidogeny: A New Category of Vegetative Propagation

Embryoidogeny (Greek *embryon,* embryo; *oidos,* view; *genus,* origin) is one of the two types of homophasic reproduction of flowering plants *in situ, in vivo* and *in vitro*, the elementary structural unit of which is an embryoid. *Embryoid* is the incipient individual forming asexually *in situ, in vivo* and *in vitro*. Synonyms are *somatic embryo* and *embryo-like structure*.

The main thesis of the concept of embryoidogeny is the belief in the universality of the morphogenesis of embryoids as well as sexual embryos, developing in natural conditions and in experimental ones, in in vitro culture (Batygina, 1996a, b, 1998, 2006a, b, 2009; Batygina, Zakharova, 1997a, b).

Embryoids, like embryos, are characterized by the formation of its own *new axis* (on relation to the maternal organism), joining polar developing shoot and root apices. As a rule embryoid does not obtain the vascular system common with maternal organism (close radicular pole). Genesis of embryoid, as well as its shape and sizes are taxon specific. *The main features established for sexual embryos (polarity, cellular and histogenic differentiation, autonomy, etc.) are typical for embryoid.* During embryoid germination as at embryo germination a new individual just "births" (Batygina, 1987b, 1989a, b, 1993a, b, 1997, 2000, 2006a, b, 2009).

Together with this, *embryoid possesses the principal (by the origin) resemblance with the bud*, that is firstly in *the mode of formation – homophasic reproduction* (sporophyte → sporophyte), that *unites them as elementary structural units of vegetative propagation.*

The main difference of embryoid (somatic embryo) from the bud is that *the latter from the very beginning* of its development represents only *the part of a whole organism – "monopolar" structure (has only shoot apex and the open raducular pole, the cotyledons are not formed)*, connected with the maternal plant by means of vascular system (Figure 4B). Just in the process of further development its connection with maternal organism interrupts and as far as the root (roots) develops it becomes bipolar. However in embryoids of some species as in the buds the short-time connection with the maternal organism is observed; moreover, the longevity of contact and the stage of embryoid development on which this contact is realized are differed in various plant species (see below). *It allows speaking about the existence of transitional, intermediate forms between embryoid and bud.* The research of such intermediate forms and the revealing of their *structural-functional characteristics* have a great significance regarding their utilization *in the work on cloning plants in biotechnological practices* (for example, the definition of monopolarity or bipolarity of above mentioned structures).

Thus, while *distinguishing the embryoidogeny into the special type of reproduction and propagation, we used two criteria: ontogenetic (homophasic reproduction, accompanied by neither meiosis nor the process of gamete fusion, i.e., asexual mode of new generation formation; uniparental heredity) and morphological (bipolar organization of the structure with shoot and root apices and new polar axis). As to embryogeny (heterophasic reproduction), the formation of new individual occurs to be a result of sexual process, i.e., meiosis and gamete fusion (biparental heredity).*

The tendency of the somatic embryo formation realizes *at all stages of flowering plant ontogenesis, beginning with the zygote*. Embryoids can arise on *the different structures and organs of plants exogenously or endogenously*, usually from one somatic cell, rare from embryonal cellular complex. The phenomenon of embryoidogeny is observed in species, growing in various ecological zones.

Depending on the origin and location of somatic embryos on maternal plant, *two main forms* of embryoidogeny can be distinguished: *floral* (reproductive) and *vegetative* (Figure 5).

<table>
<tr><td colspan="3">With a alternation of generations</td><td colspan="11">Without a alternation of generations</td></tr>
<tr><td colspan="3">Sexual</td><td colspan="11">Asexual</td></tr>
<tr><td colspan="5">Seed propagation</td><td colspan="9">Vegetative propagation</td></tr>
<tr><td colspan="2">Gamospermy</td><td colspan="3">Agamospermy</td><td colspan="9">Aspermy</td></tr>
<tr><td colspan="2">Amphimixis</td><td>Apomixis (gameto-phytic)</td><td colspan="11">Amixis</td></tr>
<tr><td colspan="3">Embryogeny</td><td colspan="5">Embryoidogeny</td><td colspan="6">Gemmorhyzogeny</td></tr>
<tr><td rowspan="3">Capsella, Pisum</td><td rowspan="3">Avicennia, Rhizophora</td><td rowspan="3">Antennaria, Hieracium, Taraxacum</td><td colspan="2">Floral</td><td colspan="3">Vegetative</td><td>Floral</td><td colspan="3">Vegetative</td><td rowspan="2">Sarmentation</td><td rowspan="2">Particulation</td></tr>
<tr><td>Embryonal (cleavage)</td><td>Ovular (nucellar, integumentary)</td><td>Foliar</td><td>Cauligenic</td><td>Rhizo-genic</td><td></td><td>Foliar</td><td>Caulige-nic</td><td>Rhizoge-nic</td></tr>
<tr><td>Erythronium, Orchis, Paeonia</td><td>Allium, Citrus</td><td colspan="3">Bryophyllum, Crassula, Ranunculus</td><td>Allium, Festuca, Poa</td><td colspan="3">Hammarbya, Lilium</td><td>Ajuga, Paris, Stachys</td><td>Aconitum, Carum, Plantago</td></tr>
<tr><td></td><td>Vivipary</td><td></td><td colspan="2"></td><td colspan="3"></td><td colspan="4">Vivipary</td><td colspan="2"></td></tr>
<tr><td colspan="3">Embryogenic</td><td colspan="5">Embryoidogenic</td><td colspan="6">Gemmorhizogenic</td></tr>
<tr><td colspan="14">TYPES OF REPRODUCTION</td></tr>
</table>

Figure 5. Modes and types of reproduction and propagation in flowering plants; • • **gametophytic** apomixis, androcliny (from Batygina, 2000 with alterations).

Floral (reproductive) *embryoidogeny* is the formation of embryoids in the flower and seed structures. It comprises in its turn embryonal[1] **monozygotic "cleavage"** (*monozygotic* splitting) on the basis of zygote or embryo; ovular, from integumentary and nucellar cells in ovule; and also gametophytic embryoidogeny, from the cells of micro- and megaspores, male and female gametophytes.

It is interesting to examine the phenomenon of *embryonal monozygotic embryoidogeny*, the formation of monozygotic twins, as one of the modes of homophasic reproduction.

The formation of monozygotic twins is the result of producing of the somatic embryos (one or more) from a single initial cell, zygote or the cell of multicellular embryo by means of different modes (equal, unequal division etc.); it is likely conditioned by the character of division, the time and the place of the formation of these twins. Per se, it is considered to be asexual propagation and the cloning of the organism at different ontogenesis stages.

The term embryonal monozygotic embryoidogeny emphasizes that the somatic embryos arise from the somatic cells of the sexual embryos (for instance, from suspensor cells) at different, as a rule, first stages of their development. . So a single zygote gives rise to the two monozygotic twins; in this case, one of them occurs to be a sexual embryo; another one, an embryoid. At equal zygote division, the two monozygotic twins derive by asexual propagation, i.e., such twins by their origin are considered to be somatic embryos

[1] Introduction of the terms "embryonal", "ovular" and "gametophytic" embryoidogeny as well as "foliar", "cauligenous", "rhizogenous" embryoidogeny (see later) permits correct identification not only of embryoid origin (from somatic cells of embryo or ovule), but also the embryogenesis stages at which they form.

(embryoids), probably, of the same genetics because of meiosis process lack before their formation.

The notion "monozygotic twins" occurs to be rather capacious and comprehends all cases of the formation of genetically identical clones from a single zygote at different stages of embryogenesis. From the other hand, the notion of *monozygotic embryoidogeny* is likely to be considered not only in a broad sense (as a synonym of the monozygotic twins formation), but also in a narrow sense, limiting by the cases when the clones are derived immediately by the account of zygote division (splitting).

The phenomenon of monozygotic embryoidogeny was observed in different members of flowering plants: *Loranthus* (Loranthaceae), *Erythronium, Tulipa* (Liliaceae), *Limnocharis* (Orchidaceae) (Ernst, 1918), *Lobelia* (Lobeliaceae) (Crete, 1938), Papaveraceae (Ilyina, 1968), Poaceae (Erdelská, 1996) and in many other species of flowering plants.

Jeffrey (1895) in details described this phenomenon in *Erythronium americanum.* After fertilization the zygote divides with the formation of irregular group of cells ("embryogenic mass") at the lower end of which the protrusions develop (as a rule in 2–3 or 4 in number) becoming independent embryos (embryoids).

In *Limnocharis emarginata* (Hall, 1902) zygote divides transverse into two cells, thereby the largest basal cell in certain cases is able to produce embryogenic mass of cells developing similar to that in *Erythronium.*

In the members of the Orchidaceae family (*Eulophia epidendrae,* Swamy, 1943) *several somatic embryos (embryoids,* T.B.), *derived from the somatic cells of the sexual embryo could arise in the same embryo sac* (Figure 6). According to the author, "budding" is possible. The figures he gave and the discussion of data Fagerlind fulfilled (1946) can demonstrate that "the bud" and "protrusions" could be considered as embryos (embryoids, T.B.).

In *Eulophia epidendrae*, three different varieties of the appearance of such embryos (embryoids, T.B.) were discovered:

- after fertilization, the zygote divides with the formation of embryogenic mass, from the chalazal cells of which numerous embryos (embryoids, T.B.) arise;
- the zygote results in filiform proembryo that "branches", and the branches transform into a new embryo (embryoid, T.B.);
- shoots that "bud" from the proembryo develop into new embryos.

The special type of "cleavage" embryoidogeny is observed during embryo development in the members of the *Paeonia* genus (Figure 7). The formation of sexual embryo in all representatives of peony is considered to be unique as at the first stage of its development the coenocyte produces (from zygote) without cell wall formation (Yakovlev, Yoffe, 1961). At the later period of sexual embryo development (heterophasic reproduction) numerous embryo-like structures produce at the coenocytic-cellular and cellular stages, *one of which modifies into the sexual embryo*, according to Yakovlev (1983) concept. *According to our notions, these structures represent the somatic embryos, arising from epidermis cells of sexual embryo.* This occurs to be one of the forms of monozygotic embryoidogeny (Batygina, 1987b, 1989a, b, 1992; Brukhin, Batygina, 1994).

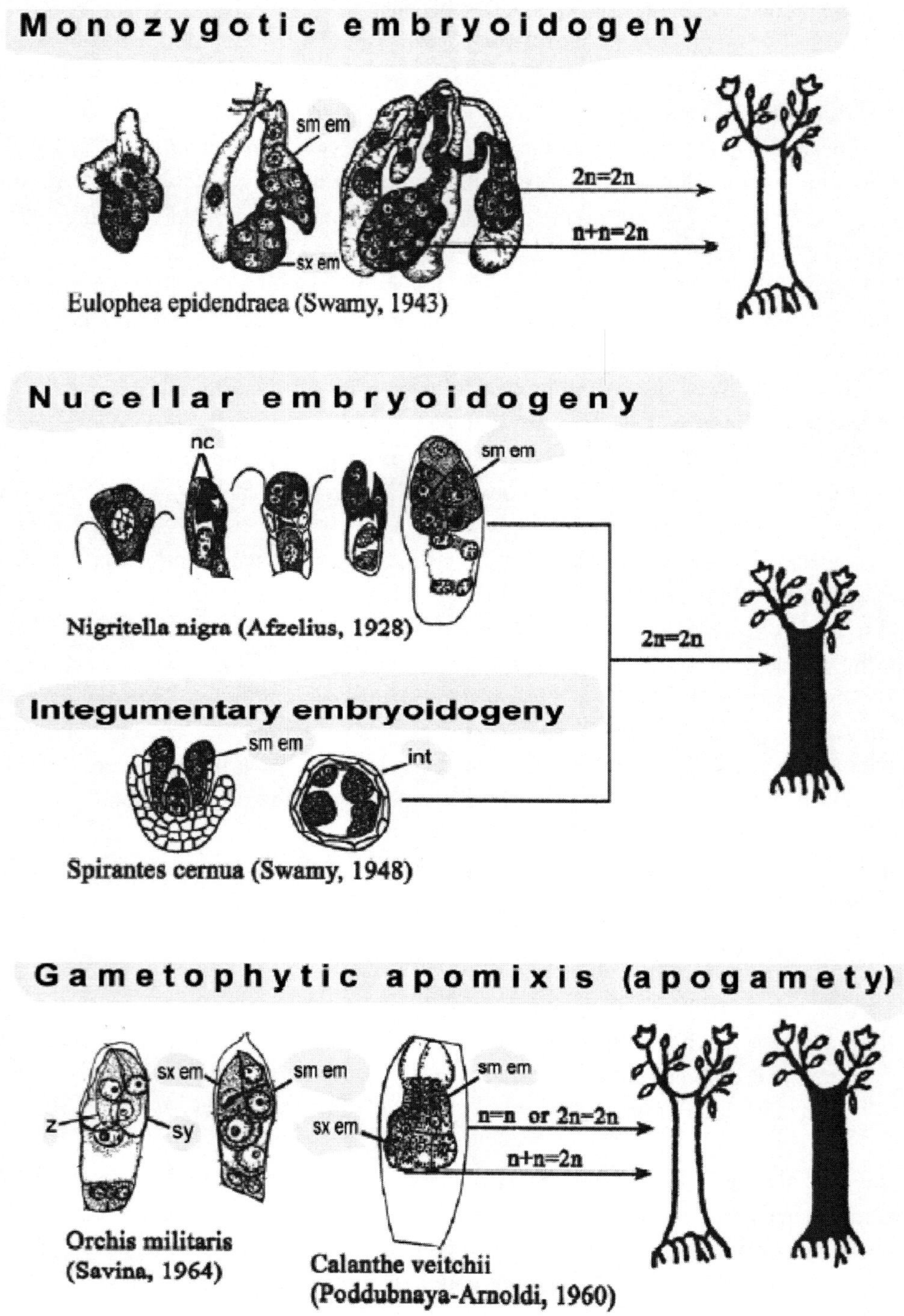

Figure 6. Pathways of formation of the somatic embryo with polyembryony in orchids: *z* – zygote; *int* – integument; *nc* – nucellus; *sx em* – sexual embryo; *sy* – synergid; *sm em* – somatic embryo; ■ – uniparental inheritance; □ – biparental inheritance (from Batygina, 2000).

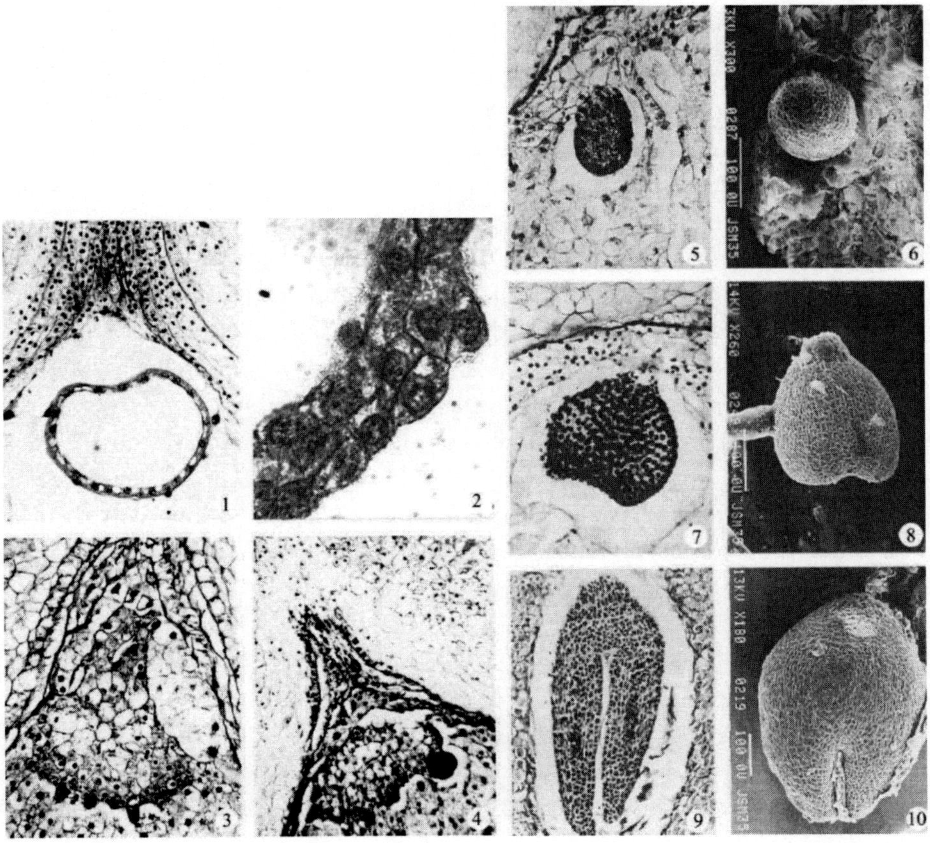

Figure 7. Monozygotic embryoidogeny in representatives of genus *Paeonia.* (1) multi-nuclear coenocyte; (2–4) coenocyte-cellular structure with somatic embryos; (5–10) globular (5, 6), heart-shaped (7, 8) and torpedo-shaped (9, 10) stages of the development of one of embryoids at coenocyte-cellular proembryo (the others degenerate), endosperm cellular, inner integument is practically destroyed; (1–5, 7, 9) LM and (6, 8, 10) SEM (from Brukchin, Batygina, 1994).

So, *in the mature seed, the only somatic embryo (embryoid)* remains as a result of competing development; the rest die (*apoptosis*) at the different developmental stages. *The sexual embryo also degenerates*; the somatic embryos produce on it, herewith its destruction takes place at the different stages of seed development in various peony species. In this case *the cloning of organism happens at the first stages of its development* (embryo) and in the mature peony seed there is, as a rule, *a single somatic embryo* (embryoid), represented the *clone* of *sexual embryo* but not *a proper sexual embryo*, as it was expected earlier (Yakovlev, Yoffe, 1961; Yakovlev, 1983). This is believed to be a striking example of *transition from one mode of sporophyte formation—sexual (n+n)—to another—asexual (embryoidogenesis—2n=2n).* In the process of *evolution* the representatives of one taxon obtained the ability to homophase reproduction (*sporophyte→sporophyte*) at the very early ontogenesis stages, owing to that the vegetative propagation by cloning takes place (uniparental heredity) (Figure 8).

The detailed cyto-embryological investigations revealed in the developing peony seed *two types of successivelyt forming embryos, differing in origin: sexual embryo with non differentiated organs, which stops developing and is replaced by the formation of another, somatic embryo, differentiated in all general organs* (shoot and root apices, cotyledons).

While examining the reproduction in different peony representatives, its important peculiarity should be mentioned. It is in the fact that the vegetative propagation, presented by two types, embryoidogenesis and gemmorhizogenesis, becomes the main mode of peony reproduction. At embryoidogenesis the formation of somatic embryo (embryoid) takes place in the seed, the stem cell of epidermis of sexual embryo occurs to be the initial of it. At gemmorhizogenesis, playing the general role during the formation of peony shrub, the laying down of monopolar structures, the buds and the shoots arising from them, occur. I would like focus attention to the fact that in the process of bud laying down during the formation of both shrub and flowers the stem cells of apex play a significant role as well as at embryoidogenesis. Thereby it should be noted once again that two reproductive modes in the peony represent per se vegetative propagation. Although in the peony the sexual process occurs (gamete fusion resulting in zygote producing), owing to "falling down" of the final stages of sexual embryo development, the sexual mode of individual formation in general sense is absent.

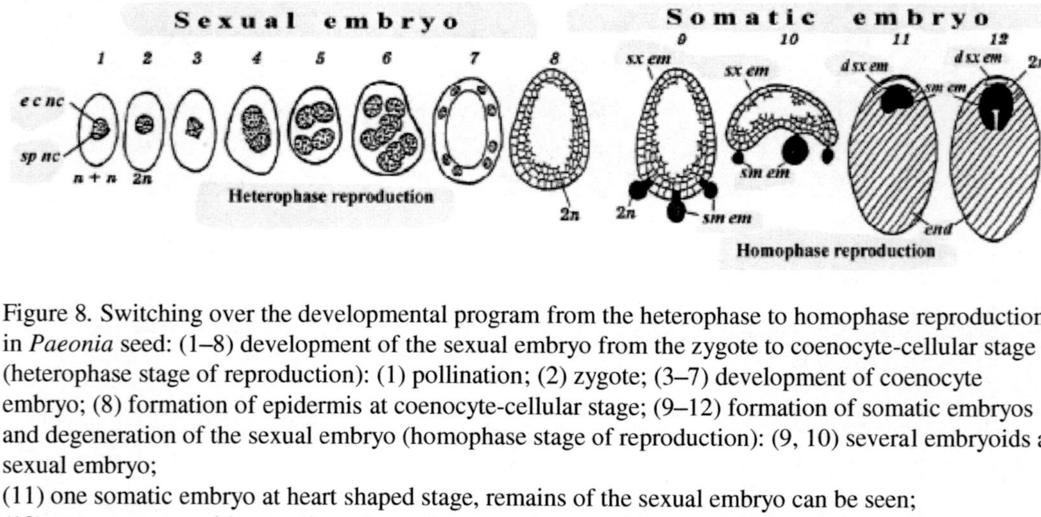

Figure 8. Switching over the developmental program from the heterophase to homophase reproduction in *Paeonia* seed: (1–8) development of the sexual embryo from the zygote to coenocyte-cellular stage (heterophase stage of reproduction): (1) pollination; (2) zygote; (3–7) development of coenocyte embryo; (8) formation of epidermis at coenocyte-cellular stage; (9–12) formation of somatic embryos and degeneration of the sexual embryo (homophase stage of reproduction): (9, 10) several embryoids at sexual embryo;
(11) one somatic embryo at heart shaped stage, remains of the sexual embryo can be seen;
(12) mature sperm with somatic embryo; *d sx em* – degenerating sexual embryo; *sx em* – sexual embryo; *sm em* – somatic embryo; *end* – endosperm; *sp nc* – sperm cell nucleus;
e c nc – egg cell nucleus (from Batygina, 1997).

All of the factors mentioned above allow taking into consideration the importance of reproduction aspects listed above in the research of the peony population structure and coenosis formation with its participation.

Complete multistage morphogenetic program of peony life cycle is related with subprograms — activity of stem cells, cloning and terminal differentiation at different biohierarchy stages.

Some experimental data concerning the influence of different factors (radiation, morphactins, morphoregulators, etc.) on the development of sexual embryo of *Eranthis* and other plants in culture *in vitro* (Haccius, 1965, 1978, Figure 9) are comparable to the phenomenon described for peony (Batygina, 1992). It permits to assume that the stress situations in nature likely induced the formation of somatic embryos on the base of dieing sexual embryo. In some cases (*Paeonia*) in nature this process turned to be fixed in the course of evolution; in others (*Eranthis*), the realization of embryoidogenic pathway of morphogenesis fulfils sporadically, only in stress situations.

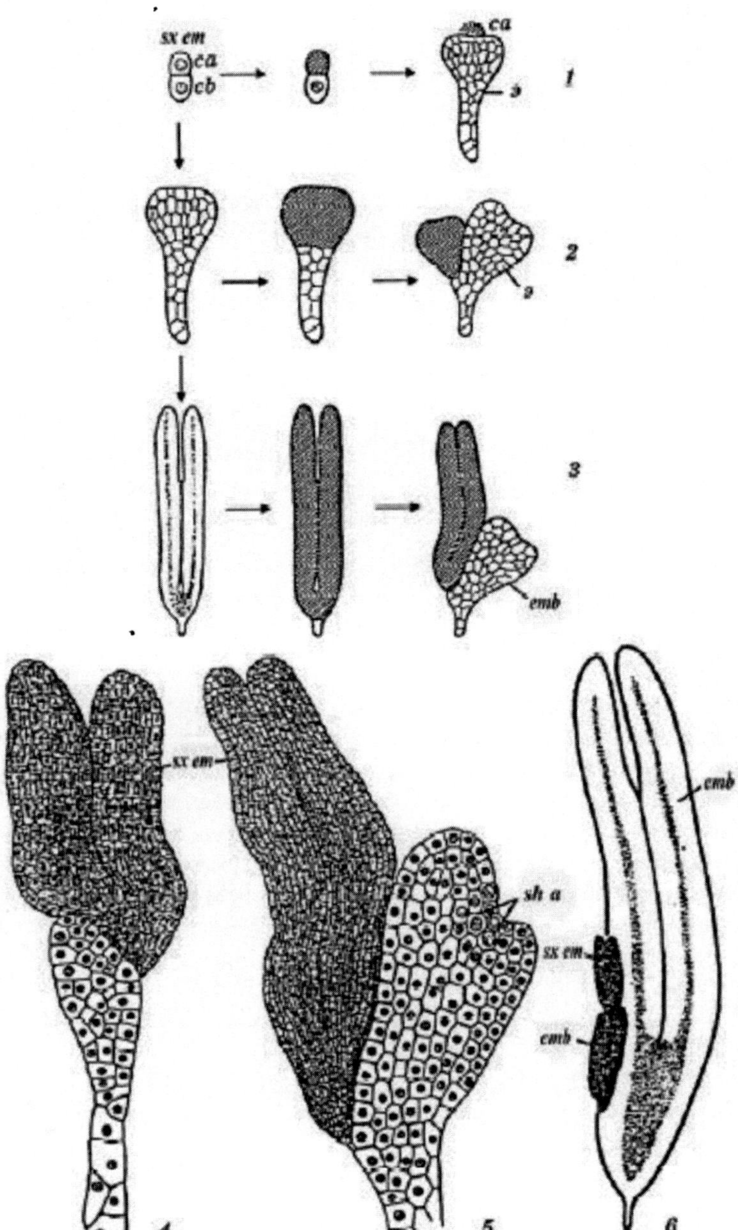

Figure 9. Formation of the somatic embryo (embryoid) from suspensor cells of the sexual embryo in *Eranthis hiemalis* in culture *in vitro*: *sh a* – shoot apex; *sx em* – sexual embryo; *emb* – embryoid (from Haccius, 1965).

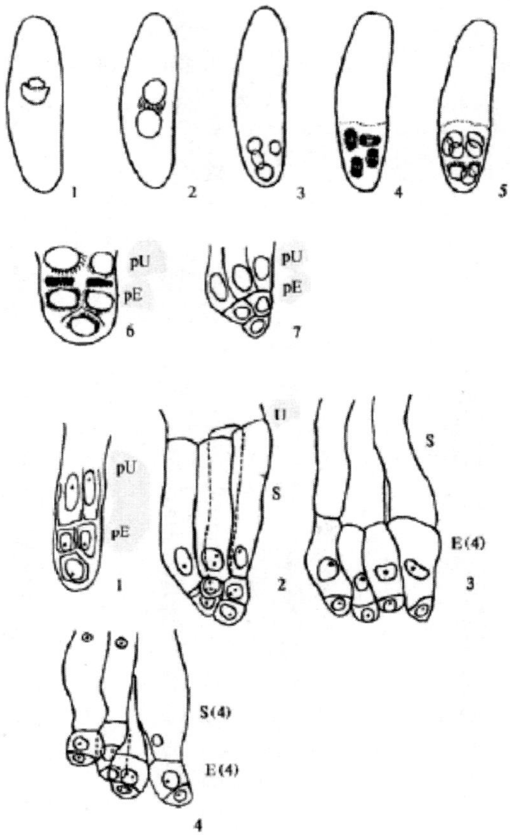

Figure 10. Two-stage embrygenesis of the gymnosperms: A. General scheme of embryogenesis in Cupressaceae and Taxodiaceae families (from [Dogra, 1984]):
(1) pollination; (2–5) free-nuclear development of proembryo; (6) formation of cell walls; (7) cellular proembryo; B. Embryogenesis *Thuja orientalis* (from Singh, Oberoi, 1962): (1) primary proembryo; (2) formation of the suspensor; (3) secondary proembryo, cleavage; (4) development of secondary embryos, formed as a result of cleavage; **pU** – primary upper (open) tier; **pE** – primary embryonic group; U – upper and open tier at later stages of development; S – suspensor tier of secondary proembryos;
E – embryonic group of secondary proembryos (the terminology according to Dogra, 1984) (from Teryokhin, 1996).

In all cases of morphogenetic changes of the complete program of life cycle both in natural conditions (*Paeonia*) and in culture *in vitro* (*Eranthis*), stem cells determine the mechanism of critical stages in the ontogenesis of different angiosperms, which is taxon specific.

A special case of monozygotic embryoidogeny is the embryogenesis of coniferous and other gymnosperms. After fertilization in the zygote of gymnosperms, several free-nuclear divisions (from 2 up to 8) usually occur. Many investigators refer to the coenocytic zygote as a free-nuclear proembryo (Dogra, 1984, a.o.). After divisions and direct transitions, nuclei complete the process of cell formation on the basis of secondary spindle, resulting in the formation of the primary cellular proembryo. After this, as a result of cell divisions, *the secondary proembryo is created in which longitudinal splitting or, correctly, dividing into cells (cleavage) takes place, realized owing to the differences in the rate of their growth.* The cleavage in coniferous embryos is the regular process, leading to the formation of separate embryos from the "embryonal units" on the each end of the suspensor "tube".

Regarding the remaining terminology of Dogra, on the whole, for the description of coniferous embryology, E. S. Teryokhin (1996) suggests substituting the terms "primary embryo" and "secondary embryo" for the terms "primary zygote" and "secondary zygote" accordingly. In confirmation to this, he gave the examples of the cleavage in *Actinostrobus pyramidalis* and *Callitris rhomboidea,* whereby it is realized just at the transition from free-nuclear coenocytic primary zygote to cytokinesis, or in *Ephedra*, in which the transformation of "the primary zygote" into "the secondary zygotes" is fulfilled by another means but also at the unicellular state of the former.

Thus, two-phase of embryogenesis (formation of "primary" and "secondary" zygotes) and the process of cleavage, i.e., separation of secondary zygotes and proembryo leading to polyembryony and embryo competition are considered to be the important peculiarity of embryogenesis in most taxa of the present gymnosperms (Teryokhin, 1996). In this connection it should be mentioned that the species with cleavage type of polyembryony in gymnosperms believe to be ancestors with reference to the taxa with unexpressed polyembryony (Johansen, 1950). Two phases of gymnosperm embryogenesis, as Teryokhin proposed (1991), allow considering zygote of angiosperms to be the homolog only to the "primary" zygotes of gymnosperms and the embryos of angiosperms are not essentially homologous to the secondary embryos of gymnosperms except *Gnetum*.

Speaking about monozygotic and polyzygotic embryoidogeny, the phenomenon of fused "Siamese" embryos *in vitro* should be examined. So under the influence of inhibitors of polar auxin transport (NPA, quercetin) on the isolated sexual embryos of wheat at the globular stage or at the early stage of transition to organogenesis, two phenotypes of "Siamese embryos" form "back to back" with numerous scutellums, turned to each other by dorsal sides and situated terminally in relation to the apical shoot meristems; "heat-shaped Siamese embryos" with the scutellums turned to each other at the ventral sides and situated laterally in relation to the apical shoot meristems (Fischer et al., 1997). Similar phenotypes with numerous scutellums and apices were observed in embryoids forming in the culture *in vitro* of anthers in spring soft wheat at higher concentration of synthetic auxin 2,4-D (Seldimirova, Titova, 2007). These data demonstrate the generality of morphogenetic process laying in the base of embryogenesis in seed and embryogenesis *in vitro* and general mechanisms resulting in the both cases monozygotic embryoidogeny.

The phenomena similar to monozygotic embryoidogenesis in plants and leading to the polyembryony were noted for the embryogenesis of animal and human (Kanaev, 1968), when the initiating of monozygotic twins appeared to occur at different stages — from the stage of two blastomers until the stage of gastrulation, i.e., that stage when initiation of intestines happens. In most animals it is a rare phenomenon; however, in some mammalia (two species of armadillo) at the rather late developmental stage normally several embryos arise as a rule instead of a single one.

Thus, both in plants and in animals at the formation of monozygotic twins, the cloning of zygote and embryo takes place.

Monozygotic and polyzygotic twins are a fine illustration of the displaying of the *polyembryony* phenomenon. In the case of seed plants, polyembryony (polyembryoidogeny, T.B.) in a broad sense (sensu lato) is the formation of several embryos both in a single seed and on the separate structures of a plant (flower, leaf, stem and root). In animals (for example in insects, mammalia etc.), analogous phenomenon manifests in the formation of several embryos in reproductive organs. This phenomenon is considered to be universal and occurs both in plant and animal organisms.

Because of the debatability of some aspects of the extremely difficult problem in developmental biology (polyembryony, embryoidogeny) concerning the time of formation, trigger mechanisms, a change of developmental programs, mono-, polyzygotic and Siamese twins both in animals (human) and plants (*in situ* and *in vitro*) is likely required in the discussing of certain notions used by different investigators.

The analysis of literature has revealed that usage of the term "cleavage" in different languages—Russian, *кливаж*; English, *cleavage*; French, *clivage*—infers different meanings, and the various dictionaries give various semantics of these terms. Probably, this is the reason for ambiguity of the determination of this notion. For instance, during investigating of the first ontogenesis stages in animals the notion "splitting" uses the most often, in western literature the "cleavage stage" is distinguished in the embryo development characterized by the fact that the larger zygote and its derivatives undergo a series of cell divisions without compensator growth of the cells formed. Russian zoologists as a rule do not use the term "cleavage stage", because they apply to this phenomenon the Russian term "splitting". However in rare cases this term is also used in plants (Yakovlev et al., 1957). The notion "cleavage" concerning the plants is used for the characteristics of the phenomenon of "splitting", "disintegration" of zygote and embryo wall. These notions need to be corrected with special accent that these terms could be simultaneously used both for plants and animals. Perhaps, the lack of nuances in behaviour of certain structures at embryogenesis, in particular the structure of walls at different stages of cell development, requires fixed attention and correction in their using.

In conclusion, I would like to join the opinion of famous embryologists Brien (1956) and O.M. Ivanova-Kazas (1977) working a lot with the problem of the asexual propagation in animals, who consider that all facts on the asexual propagation given often brings the dissonance in the concert of classical biology because they do not always conform with generally accepted views. Our nontraditional approach to the study of vegetative reproduction in plants and the discovery of a new category of vegetative propagation in plants (embryoidogeny) illustrates this state.

Nucellar and integumentary embryoidogeny (the formation of somatic embryos from the cells of nucellus and integuments—ovule covers) was noted in more than 250 plant species. There is large information on nucellar embryoidogeny in fruit trees, such as *Citrus, Mangifera, Eugenia,* etc. The number of polyembryonic seeds can vary in different species. Thus, in *Citrus microcarpa,* the seeds usually produce 21 seedlings, whereas in *C. unshiu,* about 40.

Nucellar and integumentary embryoidogeny is also observed in some orchid species (*Nigritella nigra, Spiranthes cernua*) at different stages of ovule development, often just before embryo sac formation (Afzelius, 1928; Swamy, 1948).

Embryoids are able to arise from somatic cells of nucellus and integuments by different modes and at different stages of ovule development (Modilewsky, 1931; Button et al., 1974; Batygina, Mametyeva, 1979; Batygina, Freiberg, 1979; Batygina, 1991). Thus, in *Poa pratensis* the embryoids form either from a single cell of nucellus (dormant meristem), or through embryonal cellular complex as a result of proliferation of nucellus cells (Figure 12). Independently on the mode of embryoid formation, in the mature seed at nucellar and integumentary embryoidogeny one is dealing with the clone of maternal plant presented by somatic embryos. However, in this *Poa* genus together with agamospermous seeds, the seeds with sexual and nucellar embryos are observed.

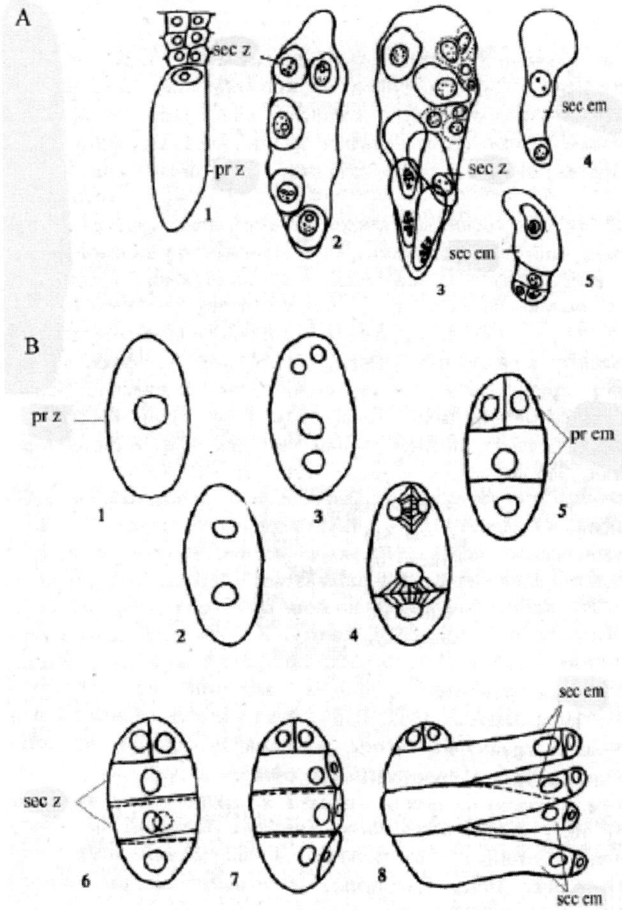

Figure 11. Different types of two-stage embryogenesis in gymnospermous plants:
(A) two-stage embryogenesis in *Ephedra* with original (coenocyte) type of monozygotic polyembryony: (1) primary zygote *Ephedra trifurcata* (from Land, 1904); (2–5) formation of secondary zygotes and development of secondary embryos in *E. foliata* (from Khan, 1943); (B) two-stage embryogenesis in *Actinostrobus piramidalis* with cleavage type of monozygotic polyembryony: (1) primary zygote; (2–4) coenocyte zygote; (5, 6) formation of secondary zygotes; (7) cleavage; (8) development of secondary embryos; sec em – secondary embryos, sec z – secondary zygote, pr em – primary embryo, pr z – primary zygote (from Saxton, 1913 and Teryokhin, 1996).

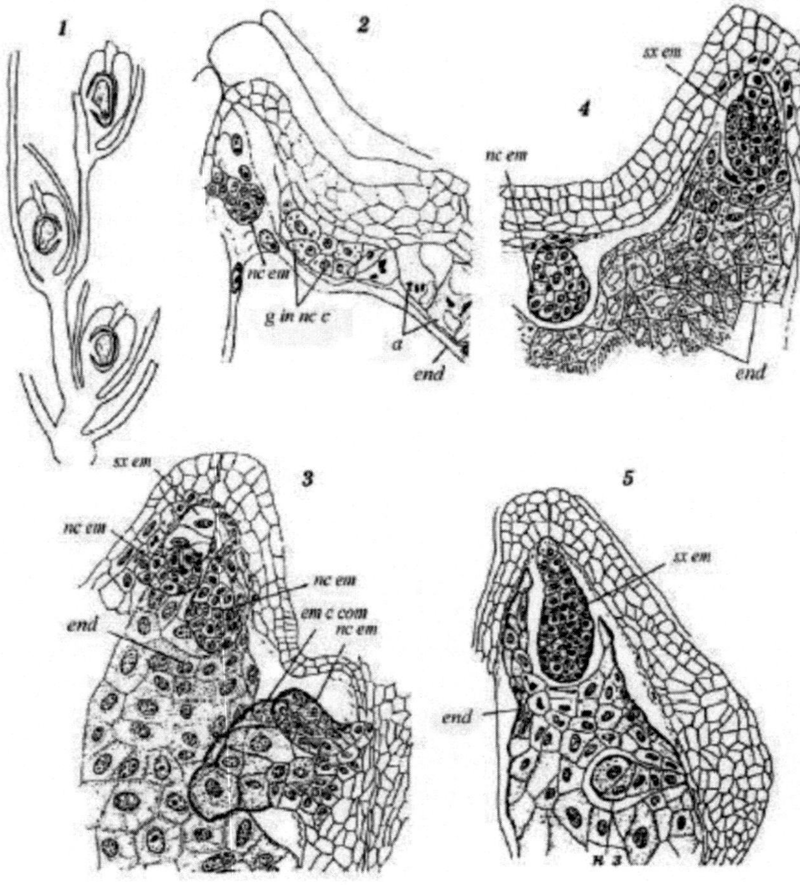

Figure 12. Various modes of formation of embryoids in the embryo sac in *Poa pratensis:* (1) spikelet; (2) micropylar part of the embryo sac with nucellar embryoid and group of initial nucellar cells, out of which embryoids are formed; (3) micropylar part of the embryo sac, sexual and nucellar embryos can be seen, embryonic cellular complex; (4, 5) sexual embryos at different stages of development and nucellar embryos, being formed out of singular cells of dormant meristem; ***a*** – antipodals, *g in nc c* – group of initial nucellar cells, *nc em* – nucellar embryo, *sx em* – sexual embryo, *em c com* – embryonic cellular complex, *end* – endosperm (from Batygina, Freiberg, 1979).

Nucellar embryoidogeny is given attention because of various causes. One of them is the capacity of obtaining of plants without viruses. At usual vegetative propagation the cuttings are known often to be infected by different patogenes whereas nucellar embryos and seedlings planted from them are lack viruses. Some authors explain this by that the nucellus and neighbored tissues do not contact with each other by vascular system.

Geneticists and plant breeders use the high capacity of some plant species (*Mangifera, Citrus*) to form nucellar embryos as the main criterion of genotype selection for cross-breeding, ensuring thereby the diversity in generation. According to the data of many authors, **nucellar embryoids derivate "rejuvenated" seedlings**, which resemble ones that have arisen from sexual embryos. Maheshwary (1950) emphasized that seedlings of *Citrus,* forming *from vegetative buds on sporophyte, develop without spines*, whereas seedlings from *nucellar embryos develop into the plants with spines*. Clones of *Citrus*, always being vegetatively propagated (by cuttings), finally become weaken and sterile (Cook, 1938; Hodgson,

Cameron, 1938; Frost, 1938). It is remarkable that gardeners renew the *Citrus* clones using *nucellar seedlings, developing better then plants, obtained from cuttings.*

There are cases in which, *in the same seed, the formation of embryoids precedes parallel to embryogenesis* (e.g., *Citrus, Poa*). We use the term "clone" for embryoids and seedlings arisen from them. It should be noted that *in some representatives of angiosperms* (Citroideae, Orchidaceae, etc.), *two different clones—two generations—can form in the same seed*: the clone of maternal organism (by means of nucellar and integumentary embryoidogeny) and the clone of daughter individual (*by means of cleavage monozygotic embryoidogeny*).

In that case when in the same seed there are *embryoids of different* origin: *maternal (ovular) and new daughter (embryonal)* the plants of two *genotypes develop in population*. It is the reason of the appearance of genetic heterogeneity of seeds which together with vegetative embryoidogeny finally conditions the *genetic heterogeneity of populations.*

Gametophytic Embryoidogeny

According to our theoretical ideas, under certain stress situations in nature, in certain plant species *the anther and the ovule with all of their structural elements* are considered to be utilized as *the units of propagation and dispersion*. It is confirmed by a great number of data on *gametophytic apomixis* at which the embryos—somatic and parthenogenetic—are able to arise in seeds from haploid, diploid, etc. cells of megaspores and their derivatives (Figure 6). A good example of this state is the phenomenon of haploidy ("androcliny"[2])—the changing of the program of development of microspore cells and pollen grain from gametophytic to sporophytic one under the influence of definite environmental factors and the obtaining of new individuals without fertilization by means of different morphogenesis pathways: embryoidogenesis (direct or indirect through callusogenesis) and gemmorhizogenesis (homophasic reproduction—vegetative propagation) (Figure 13). This phenomenon was found in very different angiosperm species and presents a subject of the given monograph.

The data on gametophytic apomixis and androcliny, in our opinion, continuously testifies to the preservation in evolution of morphogenetic potentials (reserves) of reproductive structures, realization of which, however, is displayed only under certain conditions. This is the reason for proposing that *in some species of flowering plants in nature the spores fail in their main functions connected with reproduction and dispersal*, as Pisyaukova assumed (1980). In this connection, it is necessary to remind once again all investigators who work in approaches and methods of generative and reproductive structure cultivation that in culture *in vitro* only those potencies and reserves that are typical of all organism cells in natural conditions are capable of being realized (Batygina, 1987a, 1994b, 2005; Batygina et al., 1978).

Vegetative embryoidogeny is *the formation of embryoids on vegetative organs*; it includes *foliar embryoidogeny (on leaf), cauligenous embryoidogeny (on stem) and rhizogenous one (on root).* The arrangement of embryoid formation to the definite vegetative organ is taxon-specific.

2 From our point of view, the term "androcliny" that Khokhlov introduced in 1976 (see also Tyrnov, 1998) should be discussed.

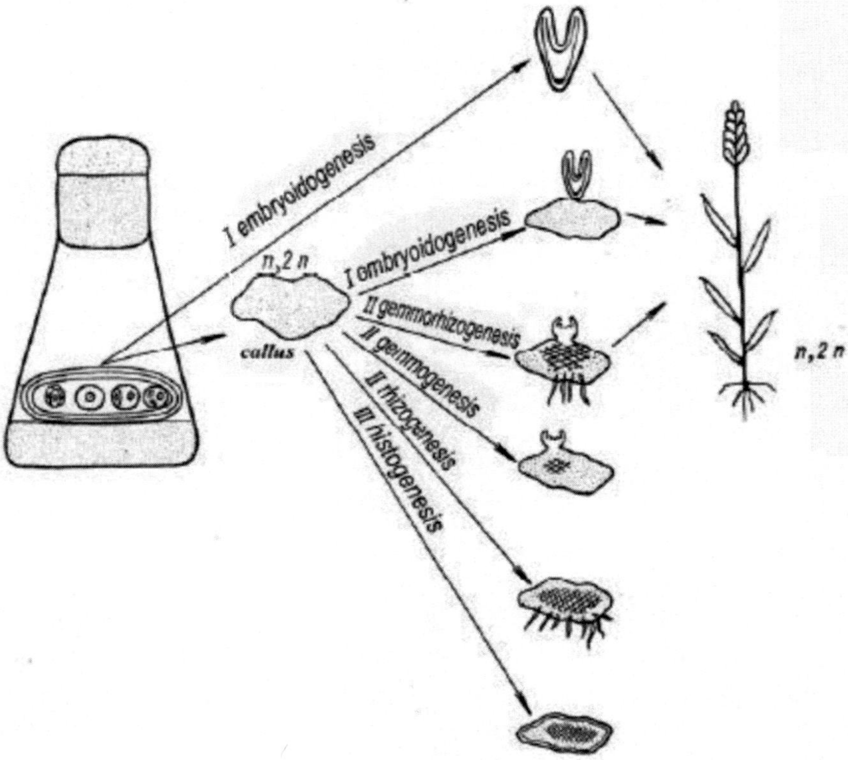

Figure 13. Possible pathways of morphogenesis in culture *in vitro* of anthers of flowering plants (from Batygina, 1984).

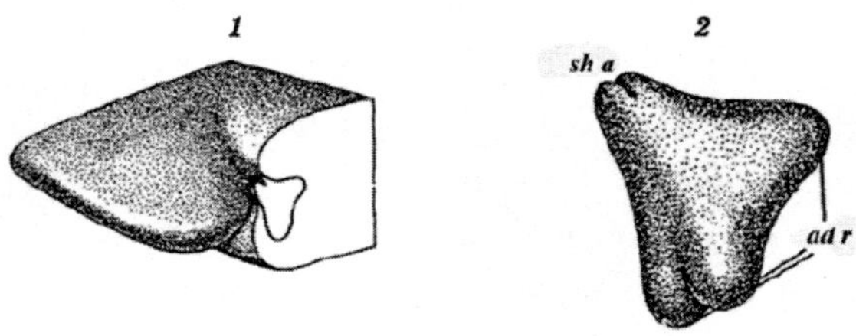

Figure 14. Leaf embryo *Bryophyllum pinnatum*: (1) somatic embryo inside the leaf tissue; (2) isolated somatic embryo with shoot apex (*sh a*) and adventive roots (*ad r*) (from Yarbrough, 1932).

Foliar Embryoidogeny

Two representatives of Crassulaceae family, *Bryophyllum calycinum* and *B. daigremontianum*, were studied in detail. Uncompleted data concerning the morphogenesis of structures arising from vegetative organs results so that there is no unit terminology for their identification. Thus, for example, the structures formed on the leaf of different *Bryophyllum* species are referred to as "buds" (Howe, 1931), "embryos" or "leaf embryos" (Naylor, 1932; Yarbrough, 1932; Warden, 1968), "leaf pseudobulblets" (Johnson, 1934; Resende, 1954), "embryoids" (Batygina, 1987, 2006a).

In one of the latest works concerning the research of reproduction in the genus *Bryophyllum* (=*Kalanchoë*), the authors, while discussing the process of the structure formation on the leaf in different species, speak about *embryogenesis* which is, of course, not correct because this term is "occupied" for the development of sexual embryo to which the meiosis and sexual process precede. The somatic embryo on leaf derives from the leaf somatic cells. Therefore, they should be correctly referred to as embryoids, and the phenomenon of their formation is termed the phenomenon of embryoidogeny.

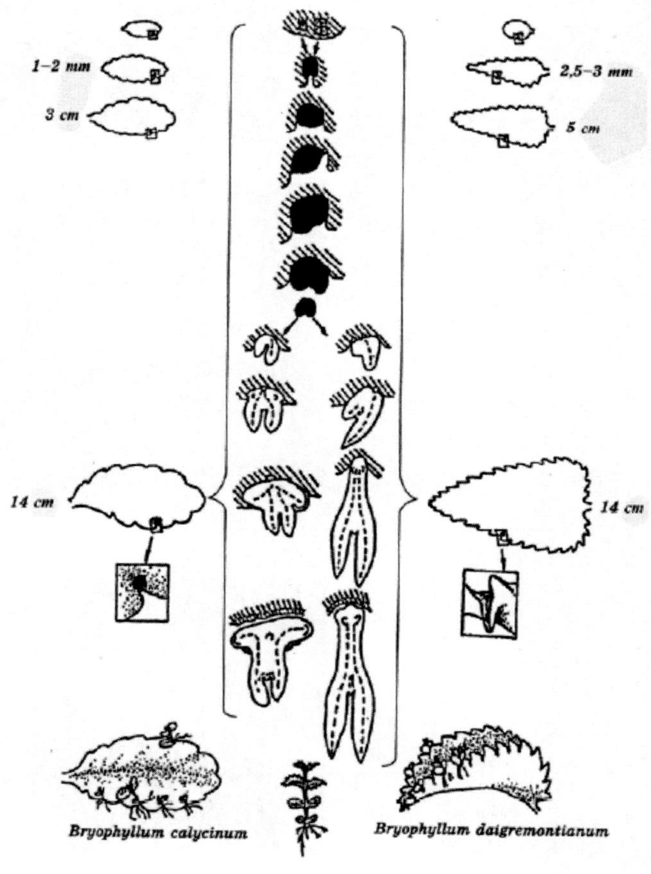

Fig. 15. Development of the vegetative propagule on a leaf. (1) scheme; (2) the edge of a young leaf of 10 mm length with an indent on it, a marginal meristem and conducting bundle can be seen; (3) the edge of a mature leaf with an indent on it, meristematic zone, out of which the propagule and conducting bundle are formed; (4, 6) vegetative propagule at the heart-shaped stage of development; (5, 7, 8) vegetative propagule at the torpedo-shaped stage with two unequal cotyledons and two adventive roots (in the cavity of the section at [5] only one of them can be seen), the shoot apex has a poor morphologic manifestation; (7) view from above; (8) bottom view; (9) the edge of a young leaf of 15 mm length with the meristematic zone, out of which the propagule is formed; (10) vegetative propagule at globular stage of development; (11–13) vegetative propagule at the torpedo-shaped stage, two unequal cotyledons, two adventive roots and shoot apex can be seen; (14) prominence of the edge of a leaf, where the zone left after separating of the propagule can be seen; (2–8) *Bryophyllum pinnatum*; (9–14) *B. daigremontianum*; (2–5, 9–11) LM; (6–8, 12–14) SEM; *ad r* – adventive root, *sh a* – shoot apex, *in* – indent, *m z* – meristematic zone, *c b* – conducting bundle, *ct* – cotyledon (from Batygina et al., 1996).

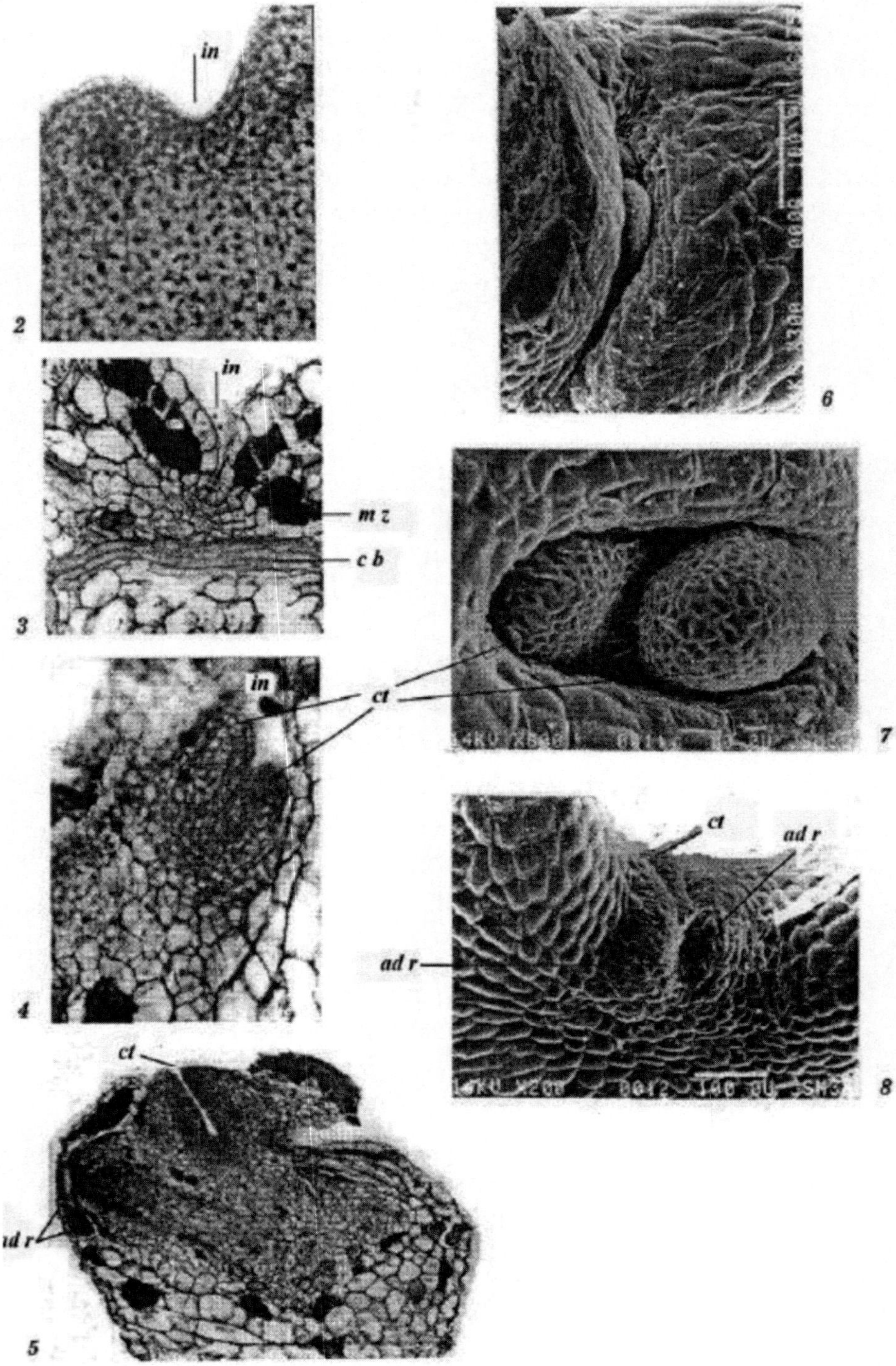

Figure 15. (Continued 1).

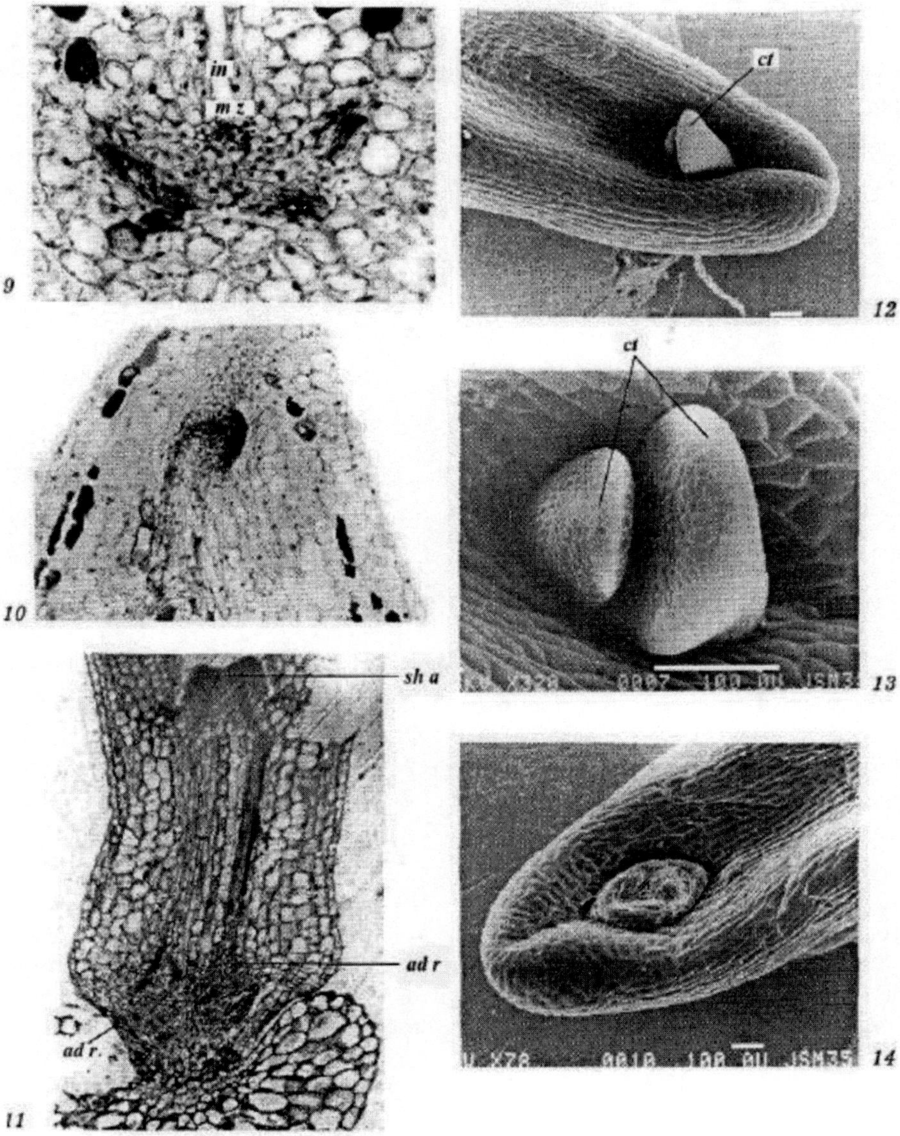

Figure 15. (Continued 2).

The results of investigations of *Bryophyllum calycinum* and *B. daigremontianum*, different in developmental biology, are the evidence in favour of the last term. In *B. daigremontianum*, the constant producing of somatic embryos from the leaf stem cells is observed, and in *B. calycinum* only in stress situations (Bragina et al., 1995; Batygina et al., 1996). In the species studied the leaf structures are formed endogenously, from the meristem, which Yarbrough (1932) called “dormant meristem”. Temporal investigation we fulfilled differs from the works of other authors by that we observed for the first time the cells of “dormant meristem” (stem cells) when the leaf reaches only one millimetre. This zone in *B. calycinum* is located *in leaf excision*, whereas in *B. daigremontianum*, *in excision projection* (Figure 15). It conditions the *endogenous formation of embryoids*. At early stage of leaf development the cells of meristematic zone do not contact with the conductive system. This is likely one of the reasons that it stays “dormant meristem”. The contact between them *is established at the later stage of leaf development, after formation of 7–10 layers of cells with dense plasma in this zone.*

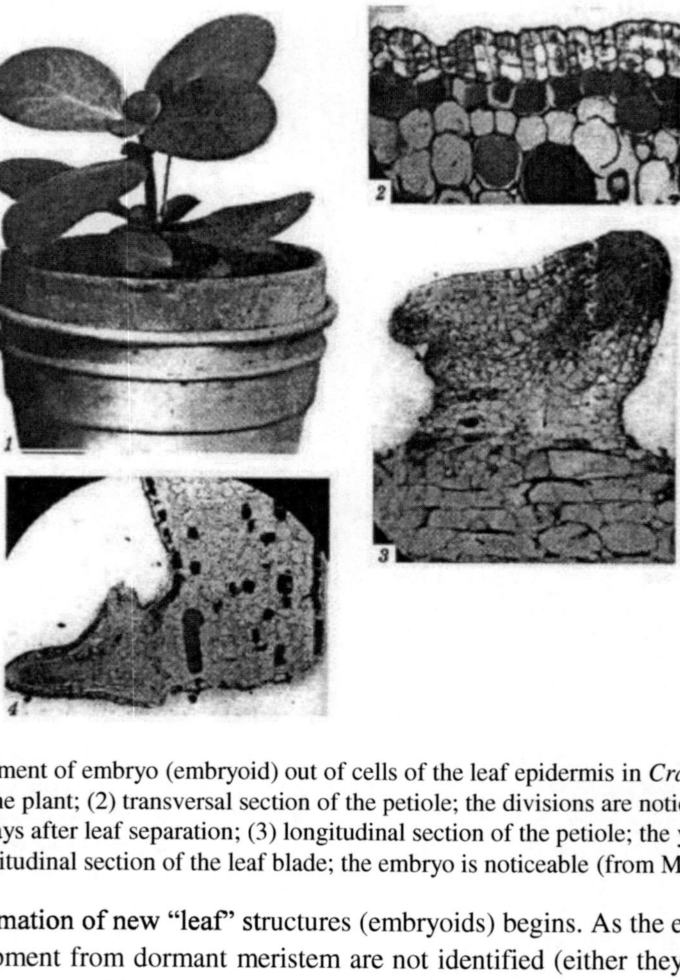

Figure 16. Development of embryo (embryoid) out of cells of the leaf epidermis in *Crassula multicava*. (1) appearance of the plant; (2) transversal section of the petiole; the divisions are noticeable in the epidermis cells 4 days after leaf separation; (3) longitudinal section of the petiole; the young embryo is noticeable; (4) longitudinal section of the leaf blade; the embryo is noticeable (from McVeigh, 1938).

Later, the formation of new “leaf” structures (embryoids) begins. As the earliest stages of embryoid development from dormant meristem are not identified (either they arise from one or several cells), one could speak about the resemblance of these structures with sexual embryo only at later developmental stages. The sexual embryo of dicots is known to undergo globular, heart-shaped, torpedo-shaped stages and the stage of maturation. *The same stages can be distinguished in the development of leaf structures in two species of Bryophyllum, but "the globular" stage differs by the morphology from classical globular stage of sexual embryo*. The formation of “cotyledons” in these structures occurs in common with sexual embryos of dicots. In both species one of “cotyledons” is usually smaller then another at early developmental stages. Later they became actually of identical sizes, the conductive system differentiates in them, and starch is accumulated. The plumule with two leaf primordia forms between cotyledons.

The vegetative structures of *B. calycinum* and *B. daigremontianum* are characterized by the lack of apical meristem of the main root, instead of which the system of adventive roots is formed. The laying down of adventive roots in *B. daigremontianum* is observed in the area of "hypocotyl" at the stage of late torpedo (well developed "cotyledons" with formed conductive system and the starch in their cells and the plumule with two leaf primordia). The laying down of adventive roots in the vegetative structure of *B. calycinum* is observed at the stage when "cotyledons" are well developed, the conductive system begins to form in them, but the starch is even lack and the plumule is even not differentiated. Adventive roots are connected with leaf conductive system.

In the genesis of the vegetative structure in species investigated, the differences in the organization of its basal part, the time and the place of adventive root initiation were revealed.

Subsequently, the roots break through leaf tissues and come to opening excision. The separation of seedlings from the leaf happens due to the formation of separating layer of cells, the part of them is filled with tannins.

Vegetative structures can be found on the maternal leaf up to the stage when 6 pairs of leaves are formed in them. Vegetative structures of *B. daigremontianum* in condition of high humidity in long day are able to produce new vegetative structures on the third pair of leaves.

Comparable morphological and histogenic analysis of vegetative structures allows ones to consider these structures in the both *Bryophyllum* species to be similar. Such features as: the origin from dormant meristem; the proceeding of main developmental stages, inherent in sexual embryo (globular, heart-shaped, torpedo-shaped); the presence of cotyledons and well developed plumule; bipolarity of their development could be regarded as argument. The formation of adventive roots on the base of dormant meristem derivatives is the specifics of these foliar embryoids. Short-term connection of embryoid with the conductive system of leaf *does not mean yet, that this structure is a bud*, but indicates its *transition form (state) between embryoid and bud*. These data confirm our opinion (Batygina et al., 2006).

The diversity of modes of somatic embryo formation in different members of *Bryophyllum*, differed by constitution or induction of such formation, already speaks about the definite specifics of their reproduction biology and strategy of reproduction related with their ecology and centre of origin. The work of last times should be mentioned concerning molecular-genetic aspects of investigation of various *Bryophyllum* species, characterized by different peculiarities of their development. The genes *KdSTM, KdECI* and *KdFUS3* were separated for the first time in *B. daigremontianum*, the products of which in some species express at embryogenesis and organogenesis and by which in rather extent the blocking of the formation of seeds with normal developed embryo is determined (Garcês et al., 2007).

According to the new data of Garcês et al. (2007) on evolution of asexual reproduction in *Bryophyllum* genus, in the morphogenesis process in *B. daigremontianum* two programs work—organogenesis and embryogenesis (embryoidogenesis – T.B.). These authors consider that the induced formation of plants and viability of seeds in some species are the ancestral features; according to their opinion, the asexual reproduction is initiated as a process of leaf organogenesis and then as an embryogenesis (embryoidogenesis – T.B.) process in leaves in response to the loss of sexual reproduction in this genus.

This question is believed to be discussible and, in our opinion, the phenomenon of embryoidogeny (vegetative propagation) and formation of somatic embryos on leaves is the primary in evolution and the seeds, produced as a result of fertilization, the secondary one. It is also confirms by the evolution of asexual propagation in animal kingdom (Ivanova-Kazas,

1977). Because of this we could not agree with the view of Garcês et al. (2007) that the seed reproduction (double fertilization) is the primary (see also Takhtajan, 1964).

In the other representative of Crassulaceae family—*Crassula multicava*—the formation of embryo-like structures from the derivatives of epidermis cells in leaf or petiole (i.e., exogenously) was described (McVeigh, 1938). These bipolar structures as in the case of *B. daigremontianum* and *B. calycinum* for a long time maintain contact with maternal organism.

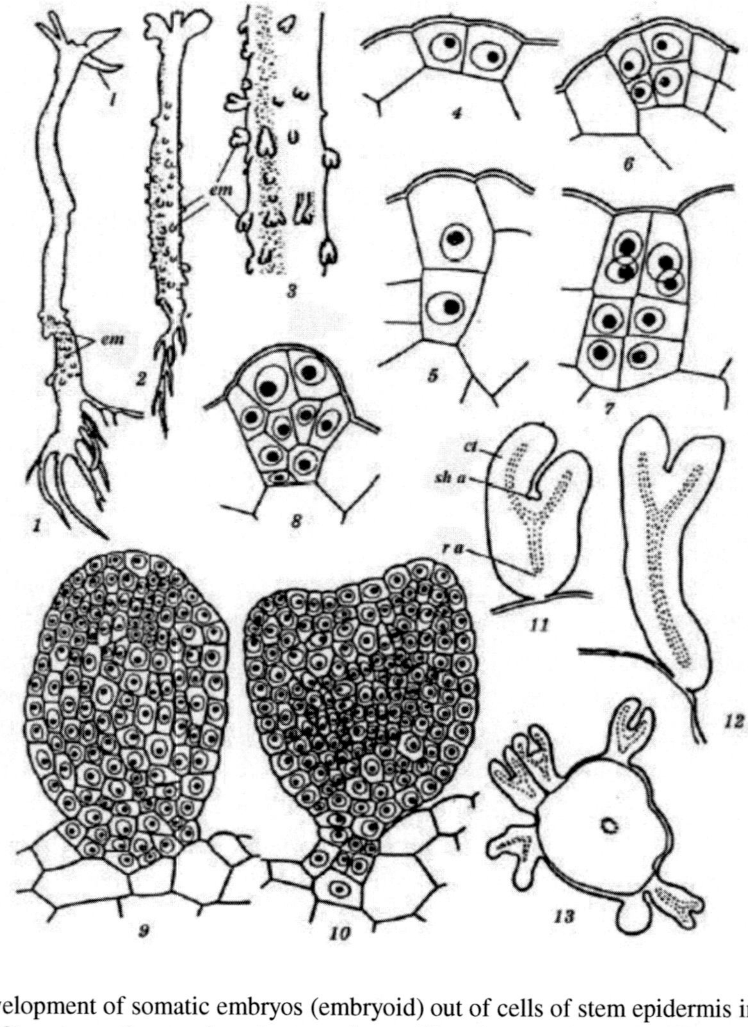

Figure 17. Development of somatic embryos (embryoid) out of cells of stem epidermis in *Ranunculus sceleratus*: (1, 2) one-month somatic embryos at lower (1) and upper (2) parts of the stem; (3) the part of the stem with embryoids; (4) two epidermis cells, which will produce embryoids; (5, 6, 7) two-, four-, and eight-celled embryoids; (8) more advanced stage of development of embryoid; (9) multi-celled embryoid; (10) somatic embryo with germs of cotyledons; (11) somatic embryo with cotyledons and root apex; (12) the same, more advanced stage of development of embryoid; (13) transverse section of the stem with developing embryoids on it; *r a* – root apex;*sh a* – shoot apex; *ct* – cotyledon; *em* – embryoids; *l* – leaves; (1–13) schemes;
(from Konar, Nataraja, 1965).

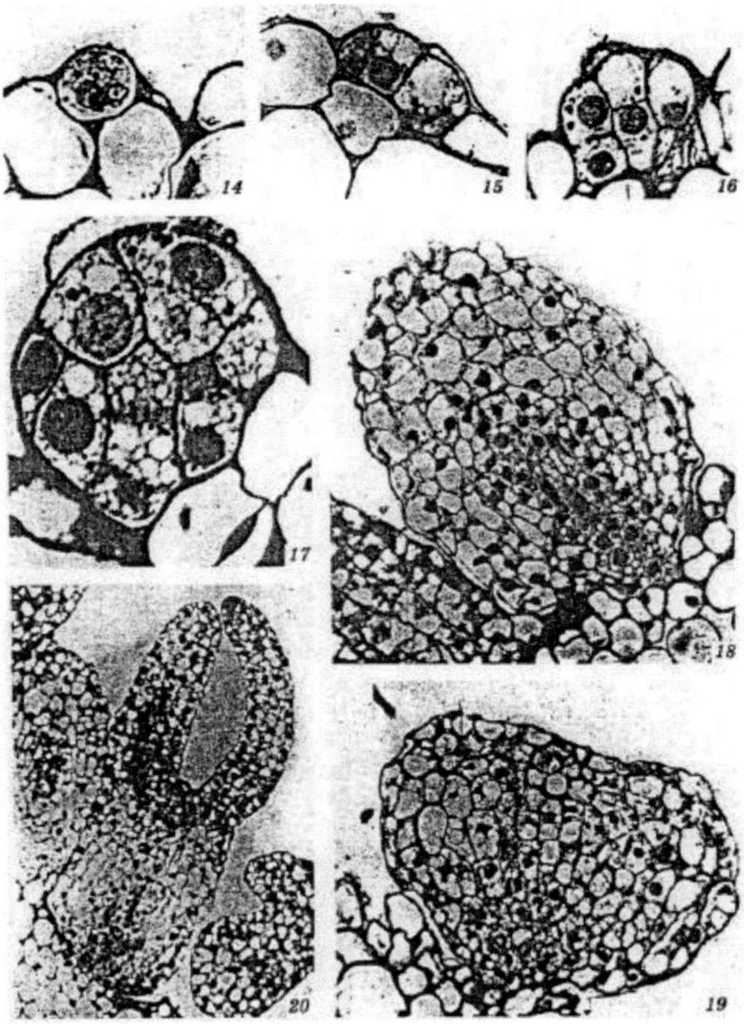

Figure 18. (14–20) Successive stages of development of embryoid from cells of stem epidermis in *Ranunculus sceleratus*; absence of its connection with the conductive system of the maternal organism, microphotos (from Konar, Nataraja, 1965).

Comparative analysis of the morphogenesis of several sexual embryos, somatic structures on leaves as well as embryoids formed in *in vitro* culture permits us to consider the using of notion "embryoid" for vegetative structures in different *Bryophyllum* species and in *C. multicava* to be more correct then using of other terms. Thus, we are of the same opinion as Naylor (1932), Yarbrough (1932) and McVeigh (1938), who suggested refer new vegetative structures formed on leaves to as "leaf embryos" or "embryos", but not as buds! *The term "embryoid" applied to Bryophyllum and in C. multicava we introduced in literature later.*

Most investigators compare leaf embryos with seeds (Yarbrough, 1934; Haccius, 1978). Dispersal of vegetative diaspores on different distances from maternal plant depends on the mode of their formation and the character of development. Seedlings arisen from these leaf embryos look like "tumbleweed".

Cauligenous embryoidogeny was firstly described in details in *Ranunculus sceleratus in situ* and *in vitro* (Konar, Nataraja, 1965; Konar et al., 1972). The embryoid of this plant as in *C. multicava* forms exogenously – from the derivatives of epidermis cells, its development proceeds similar to sexual embryo. At no stage of their development do cauligenous embryoids connect with the conductive system of the maternal organism (Figures 17 and 18), unlike the foliar embryos of *Bryophyllum* and *Crassula*.

Rhizogenous Embryoidogeny

In the literature, there are only fragmentary data, requiring additional investigations (see: Batygina, 1989a).

Parallelism in the Development of Sexual and Somatic Embryos

The resemblance and differences in the development of sexual and somatic embryos as well as the bud—elementary structural units of propagation — is the general problem of morphogenesis. For the first time the problem on the possibility of comparable study of sexual and somatic embryos obtained *in vivo* and *in vitro* and also the examination of these structures from gemmorhizogenesis positions was considered (Batygina, 1996b; Batygina, Zakharova, 1997). In this relation, we attracted the materials from the fundamental investigations: *to what extent does the morphogenesis of the somatic embryo submit to the laws of development of sexual embryo—to the laws of origin, numbers, position and purpose and the law of economy as well as the laws of cell divisions* (Souèges, 1939; Johansen, 1950). Unfortunately, insufficient attention is paid to the role of somatic embryo (embryoid) as an elementary structural unit in plant reproduction system that is more economic and adaptive, especially for the problems of biotechnology and selection.

The terminology and classification of embryoids (adventive embryos) and "embryo-like structures" for a long time remains the subject of discussion.

The opinion arose that the embryoid lacks the main features typical of a sexual embryo, namely: the initial cell of somatic embryo has no polarity axis; the regular model of cell divisions is not observed in embryoid (irregular course of divisions); the laws of cell divisions are destroyed; lack of suspensor; the embryoderm initiation is destroyed; lack of initials and typical centres of polar organization—epiphysis and hypophysis; (often) lack of hypocotyle initials; the normal organization of main root is disturbed; lack of pleroma, periblema, etc. On the bases of these the conclusion was done that the embryoid is closer by the morphology to the adventive but, but not to the sexual embryo.

However numerous literature and original data on the development of sexual and somatic embryos in natural conditions and in the culture in vitro testify that the features mentioned above not always could be used for the determination of morphological status both of the sexual embryo and embryoid.

The system approach to the investigation of reproductive systems allowed comparison of the formation types both of sexual and somatic embryos *in situ, in vivo* and *in vitro* and revealing the parallelism in their development. Let us examine from the positions of such an approach: initial cells of sexual and somatic embryos; sexual embryos of plants of different taxa *with various types of embryogenesis and seedling formation*; somatic embryos *differing on the place of arising* (seed, leaf, stem, root) and development on the plant *in natural conditions* as well as on the mode of the formation and development *in the culture in vitro*.

Initial Cells of Sexual and Somatic Embryos

The character of differentiation of initial cells in both sexual and somatic embryos to a great extent conditions the course of embryo- and embryoidogenesis. Identification of such cells continues to be one of the main problems of morphogenesis both of sexual and somatic embryos. In this relation, the solution to the following questions is considered immediate:

- What cell in the embryo sac could be considered an analog of somatic embryo's initial cell: egg cell (gamete), zygote (sporophyte) or probably another gametophyte cell (synergids, antipodal cells—at gametophytic apomixis)?
- What period of the mitotic cycle do these cells undergo, and what is the temp of the cycle proceeding (in particular, interphase)?

The egg cell of some plant species in natural conditions is known to produce the parthenogenetic embryos—embryoids possessing their own morphogenetic algorithm. The questions regarding the developmental stages—differentiation, specialization, etc. in which the egg cell is able to produce the parthenogenetic embryo and what is the trigger—also remain debatable.

In this relation, the initial cells of somatic embryo *in vitro* are worthwhile compare not only with egg cell but also with zygote as well as with initial cells that in natural conditions produce somatic embryos in the seed (cleavage, nucellar, integumental), on the vegetative (foliar, cauligenous and rhizogenous) and generative (androclinic) organs.

Initial Cells of Sexual Embryo

In flowering plants of all investigated taxa, during embryo sac development according to Polygonum-type, the first stages of egg cell (gamete) formation are rather universal. The initials of egg cells are morphologically similar (size and shape can slightly vary) and totipotent (Figure 19). They have the nucleus centrally situated, typical vacuolization and a wall. In the course of development of "the initial of egg cell" into "the formed" and then in "the mature" egg cell, its sizes are increasing, the form is changing, and the polarity and symmetry are establishing. Thereby the property of totipotency is partially being lost. The "mature" egg cell is characterized by the specific vacuolization of cytoplasm, nucleus transition and wall modification in its apical part.

At the time of fusion of female and male sexual nuclei in the formed "zygote initial", the process of differentiation begins. It continues during zygote formation; its cytoplasm, wall and nucleus undergo a series of modifications resulting, on one hand, in the formation of a taxon-specific appearance of "mature zygote", and, on the other hand, in the obtaining again of meristematic state (totipotency) and capability to proliferation. So in the system "egg cell-zygote" the multistage processes occur: meristematization, differentiation, specialization, dedifferentiation and meristematization again. The egg cell and zygote are comparable with the initial of the somatic embryo only at the stages of its development characterized by its capability to division (proliferation). However, up to now it is not reliably established whether such processes happen in the cells producing somatic embryos *in situ, in vivo* and *in vitro*. It should be taken into consideration that environmental conditions have a great influence on the development of initial cells. The initial of egg cell as well as immature and mature egg cell, and also as zygote form in the complex system of embryo sac and gradually in the course of development become isolated and rather autonomic.

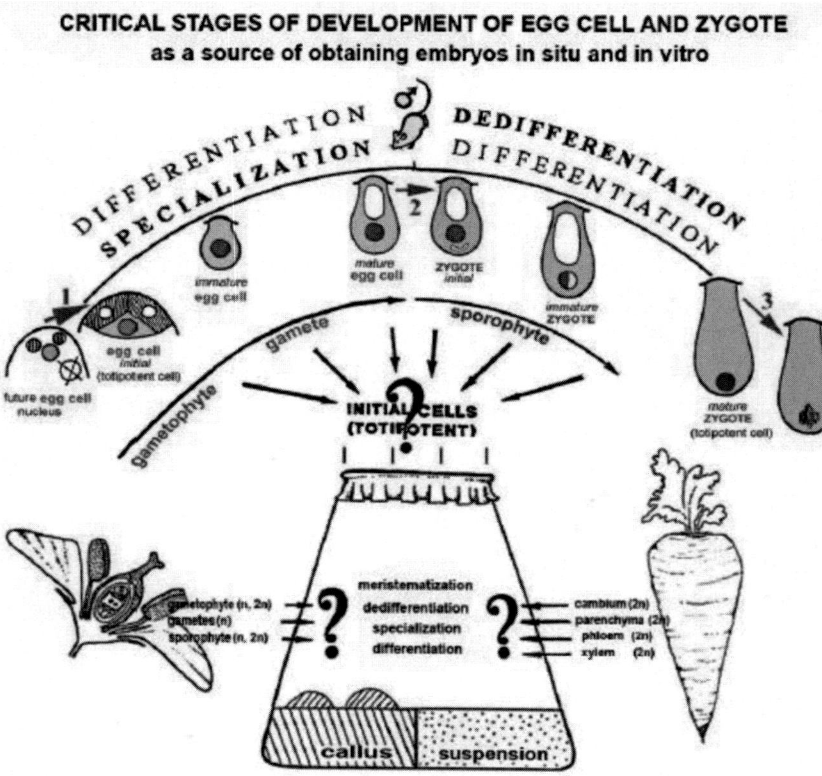

Figure 19. Possible pathways of formation of initials cells of sexual and somatic embryos in natural conditions and in culture *in vitro* (from Batygina, Zakharova, 1997 with alterations).

Initial Cells of Somatic Embryos (Monozygotic, Nucellar, Integumentary, Foliar, Cauligenous and Rhizogenous) in Natural Conditions

In various species of angiosperms, these cells are quite uniform by morphology and structure. They possess a series of common features with quickly dividing meristematic cells: little sizes, dense cytoplasm, large nucleus with noticeable enlarged nucleolus and many small vacuoles, and are characterized by high metabolic activity, intensive synthesis of RNA (Figure 20). The initial cells of somatic embryo before the first division usually obtain the polarity (the vacuole forms, and the nucleus is often situated eccentrically). In somatic embryos as in sexual ones initial cells are genetically determined, submit to the general laws of cell division and their further development occurs under the system of a whole individual.

It is likely provides the normal formation of somatic embryo, seedling and the plant in natural conditions.

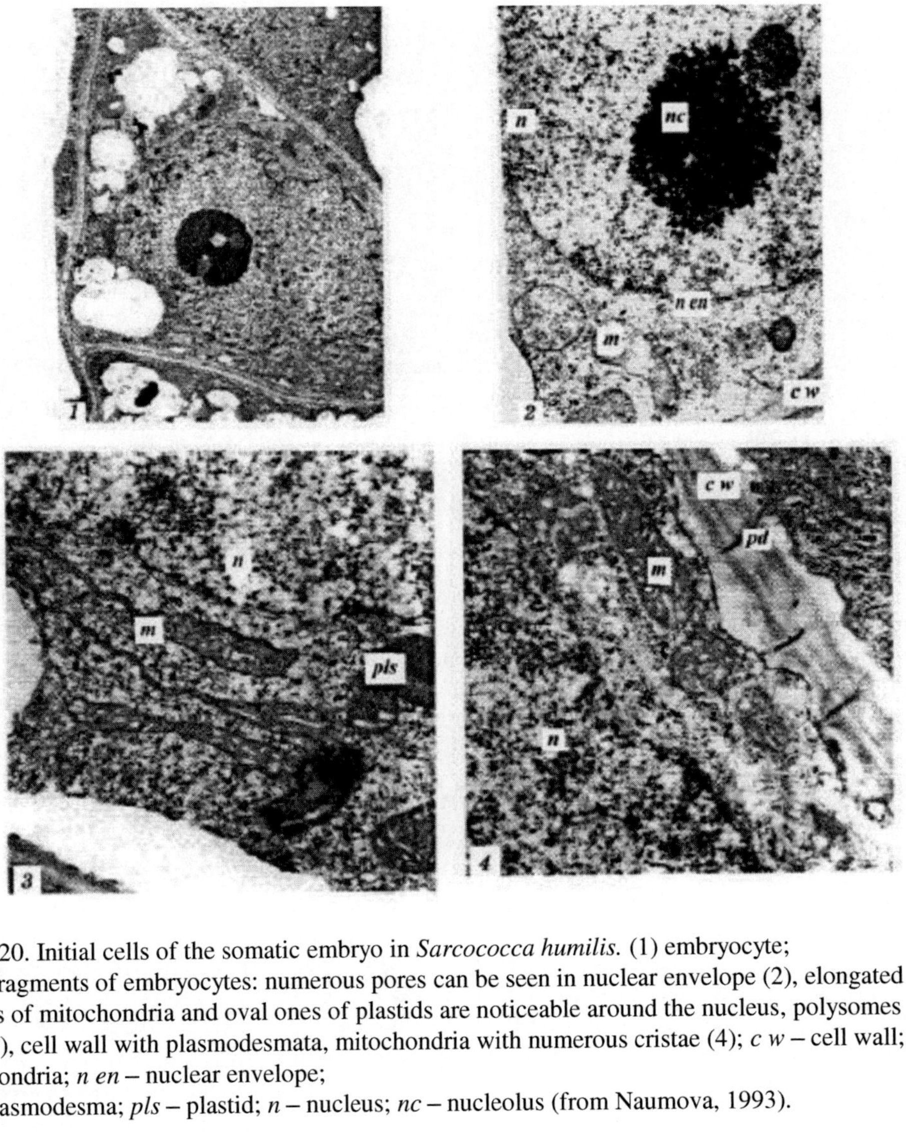

Figure 20. Initial cells of the somatic embryo in *Sarcococca humilis.* (1) embryocyte; (2–4) fragments of embryocytes: numerous pores can be seen in nuclear envelope (2), elongated profiles of mitochondria and oval ones of plastids are noticeable around the nucleus, polysomes can be seen (3), cell wall with plasmodesmata, mitochondria with numerous cristae (4); *c w* – cell wall; *m* – mitochondria; *n en* – nuclear envelope;
pd – plasmodesma; *pls* – plastid; *n* – nucleus; *nc* – nucleolus (from Naumova, 1993).

Initial Cells of Somatic Embryos in In Vitro *Culture*

Identification of such cells is one of the difficult problems of morphogenesis because, in most cases, the explant is heterogeneous in structure. The formation of somatic embryo in the culture *in vitro* is known to proceed by the pathway of direct or indirect embryogenesis. The direct embryogenesis is typical for the cells that have been determined even before their explantation; only trigger is required stimulating this process. Indirect embryogenesis occurs in the culture of cells capable to proliferation and callus formation.

There are two model systems, different in character of cell interaction, in tissue culture: callus and suspension cultures. In both heterogeneous model systems triggering of the processes of proliferation, differentiation and specialization of the initial cells and the embryo cells is to a great extent determined by the critical mass of cells and structures (Batygina, 1984) (Figures 21, 22).

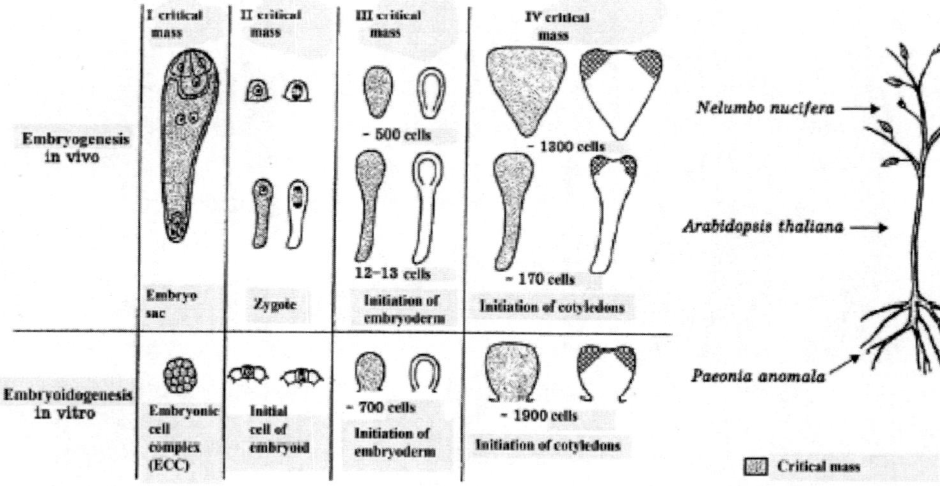

Figure 21. Critical mass of cells during the development of sexual and somatic embryos (from Batygina, 1984).

Callus is the heterogeneous structure (system), arising as a result of cell proliferation (mutations are possible) both on the wounded surface of certain structures of plant organism (exogenously) and in tissue depth (endogenously). More often it consists of unequal cells (embryonal cell complexes, ECC), totipotency of which could be different, that conditions their various morphogenesis pathways (embryoidogenesis, organogenesis and histogenesis). The notion on callus as the amorphous structure or tissue arisen by irregular growth considers being not quite correct. The callus represents a heterogeneous integrated system (not obligatory tissue), because it derives as a rule from initially different cells of generative or vegetative organs (Figure 19). Its morphogenetic potencies are taxon-specific and can change in the process of genesis depending to rather extent on different factors (temperature, moisture, time of cultivation etc.). Some facts permit stating that morphogenetic potencies both of proper callus and its constituent elements—group of cells—and their regulatory capacity as well as morphogenesis pathways could be different (organogenesis, embryoidogenesis or histogenesis). All this is to a great extent defined by the conditions of callus growing and also by the character of interactions between cell groups in callus, that, in its turn, is conditioned by their form, sizes (critical mass), ratio of different ECC, etc.

It is likely that because of this there is no need to use so many various terms (*morphogenetic callus*, *centre of differentiation*, *zone of secondary differentiation*, *compact embryogenous callus*, *green callus*, etc.) for the identification cell groups in callus possessing regenerating potencies realized by different morphogenesis pathways (embryoidogenesis, organogenesis, histogenesis). The notion "callus" is likely should be discussed, and one should call these structures *embryonal cell complex* as most scientists do (see Batygina, 1987a). Moreover, I consider to be expedient the addition to ECC the various definitions emphasizing its specifics, e.g., *ECC white* or *green*, *dense*, *compact*, etc. .

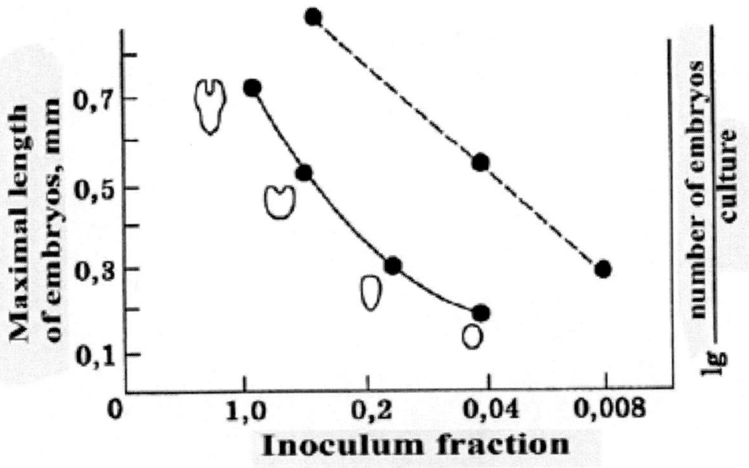

Figure 22. Influence of density of suspension culture on differentiation of embryoids in *Daucus carota* (from Batygina, Vasilyeva, 2002).

In callus culture, the interaction between cells could be different. The polarity and division in these cells are usually defined by their location in relation to callus surface. So, for example, in callus culture of wheat (*Triticum aestivum*), the endogenous laying down of meristematic zone consisting from the layers of cells directly oriented in relation to callus surface was revealed. In the future, the cells of this zone become the initials of somatic embryos (Titova, Batygina, Nikiforova, orig. data, Figure 22). However, at cultivating of wheat microspores, exogenous embryoid formation was observed (initials, the cells of callus epidermis) (Figure 24). In suspension culture (e.g., *Daucus carota*) the initial cells are polar before division, possess large vacuole and the nucleus driven to the cell wall.

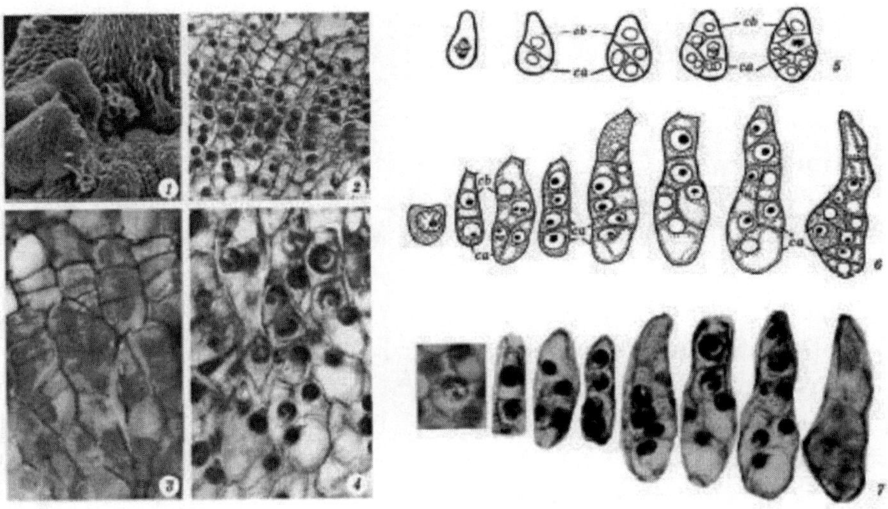

Figure 23. Endogenous formation of embryoids in *Triticum aestivum* in culture *in vitro*.
(1, 2) callus tissue with meristematic zone, where the initiation of embryoids take place;
(3, 4) initial stages of development of embryoids differently oriented in relation to the callus surface;
(5–7) deviations in the development of somatic embryo (6, 7) in comparison with sexual embryo (5);
(1) SEM; (2–4, 7) LM; (5, 6) scheme (Titova, Nikiforova, Batygina, orig. data).

The initial egg cell and the forming egg cell in the case of parthenogenesis or the mature totipotent zygote undergoing the complex pathway of morphogenetic modifications and capable of proliferation are likely to be considered the analog of the initial cell of diploid somatic embryo *in vitro*.

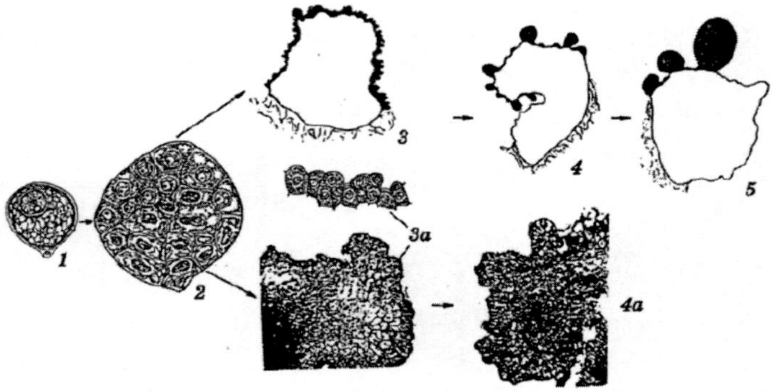

Figure 24. Exogenous laying down of initial cells of secondary embryoids in wheat callus. (1) microspore; (2) irregular primary embryoid; (3, 3**a**) callus, formed out of primary embryoid and exogenous laying down of initial cells of secondary embryoids; (4, 4**a**, 5) formation of secondary embryoids (from Shamrov et al., 1988).

At present, because of a lack of factual data on the morphogenetic processes in explant cells, it is very difficult to compare the development of the early somatic embryo *in vitro* and egg cell–zygote in natural conditions. Nevertheless, on the basis of existing information, the definite parallelism in the development of initial cells of sexual and somatic embryos could be distinguished. The high level of totipotency occurs to be the main property of all initial cells.

The obtaining of a great number of somatic embryos and regenerates in different taxa in *in vitro* culture depends on the completeness of our knowledge of the genesis of initial cells and their main characteristics (period of mitotic cycle, wall structure, state of cytoplasm, quantity and dispersal of organelles, etc.).

Sexual Embryo

The high degree of totipotency typical for mature zygote preserves in the cells *ca* and *cb*. While the embryo develops, in the process of histogenous differentiation and integration of certain cell complexes, it keeps only in the definite cell clasters (morphogenetic fields) of the embryo, seedling and plant. The significant diversity of *ca* and *cb* cells contribution in the embryo body construction should be mentioned. Using this feature, R. Souèges (1939) created the classification of embryogenesis types distinguishing six main types (megarchetypes). *He revealed in the embryo of some angiosperm species the epiphysis and hypophysis areas and their initials, the derivatives of which produce the apices of shoot and main root* (Figure 25). The formation of these regions as well as their initials usually takes place at early stages of embryo development, but the time of their differentiation is taxon-specific.

An onto-phylogenetic approach permits us (Batygina, 1997, 2006) to distinguish at least five embryo groups differing in this feature:

1) the typical initials of epiphysis and hypophysis and their centres are present (e.g., *Geum rivale*);
2) only the typical initial of hypophysis and its centre are present (e.g., *Capsella bursa-pastoris, Arabidopsis thaliana*);
3) only the typical initial of epiphysis and its centre are present (e.g., *Polemonium caeruleum*);
4) the typical initials of epiphysis and hypophysis and their centres are absent (e.g., members of Poaceae family);
5) the question of the presence of typical initials and centres of epiphysis and hypophysis in orchids (e.g., *Gymnadenia conopsea*) is debatable.

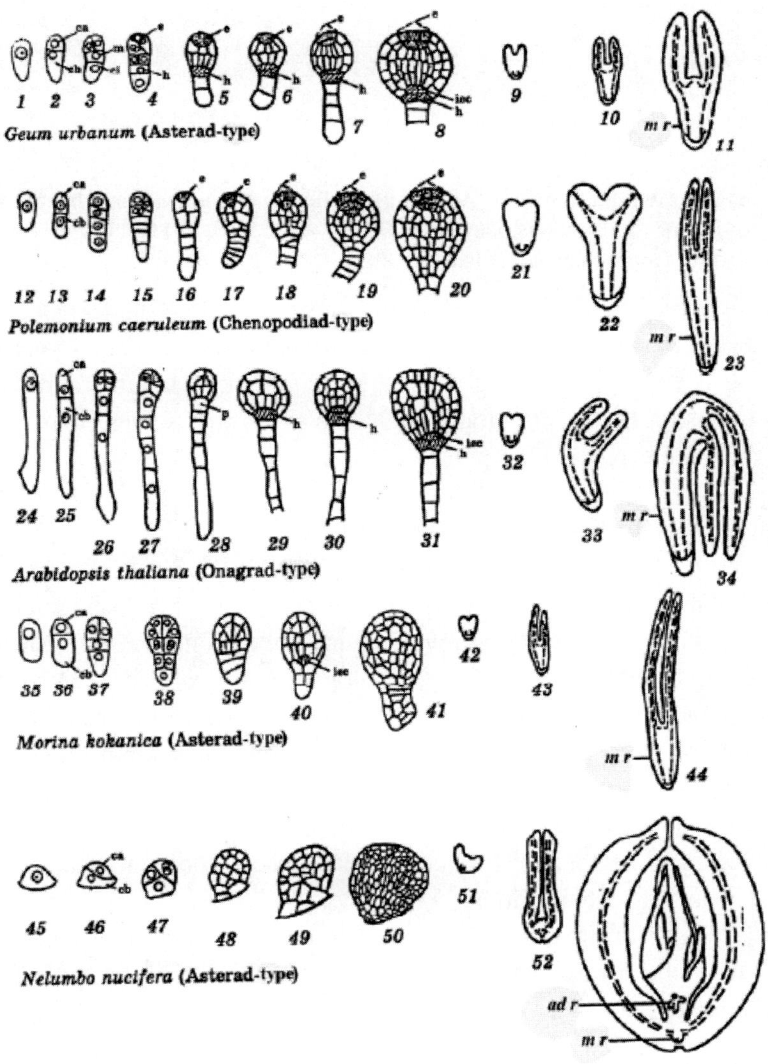

Figure 25. (Continued).

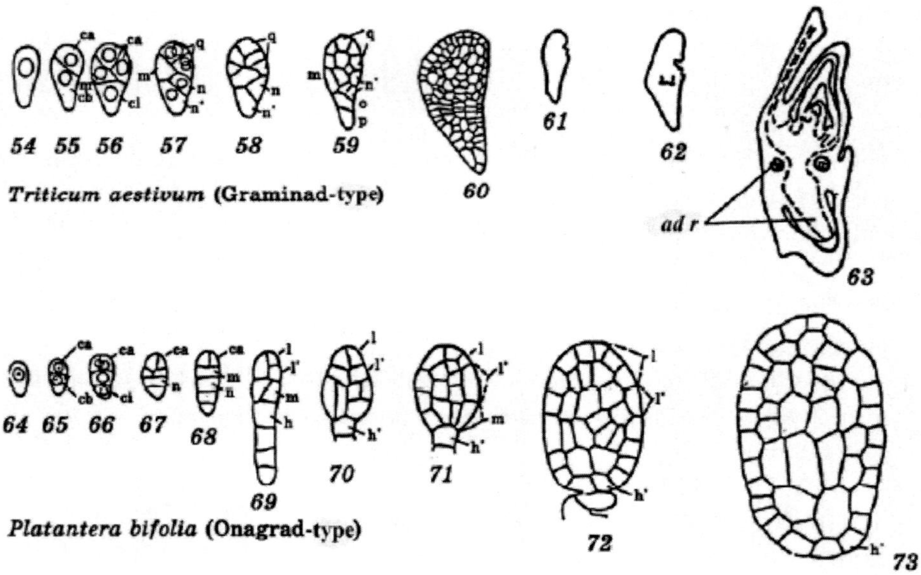

Figure 25. Polymorhism of sexual embryos in natural conditions. (1–73) successive stages of development of embryos with different degree of differentiation of epiphysis and hypophysis in dicotyledonous and monocotyledonous plants; (11, 23, 34, 44, 53, 63, 73) embryo in mature seed; points show epiphysis, epiphysial centre; shading—hypophysis, hypophysial centre; *ad r* – adventive root, *m r* – main root (from Batygina, Zakharova, 1997).

In the embryos of the first four groups, the morphological axis and polarity form, the shoot and root apices (in the first three groups, the main root, and in the fourth, the adventive root) develop normally. In the fifth group the mature embryo possesses polarity but the shoot and root apices are not expressed morphologically.

Transitional Forms between Embryo and Bud

The analysis of literature and original data on the development of sexual and somatic embryos in the natural conditions has revealed that the structures exist differing from the typical sexual embryo and the bud by the morphology. Because of this we introduced the term "transition form" under which we understand the structure possessing the features of embryo (e.g., the presence of globular, heart-shaped, torpedo-shaped developmental stages) and bud (the formation of adventive roots in the process of regeneration) (Batygina, 1993a, b). The embryos of such plants as *Nelumbo nucifera* (Nelumbonaceae), *Ceratophyllum demersum* (Ceratophyllaceae), the members of Poaceae, Orchidaceae, Orobanchaceae families illustrate the possible pathways of such transition.

In the embryos of *Nelumbo* and *Ceratophyllum* epiphysis and hypophysis are absent. In the embryo of *Nelumbo* (hydrophyte) while germinating the replacing of a main root by the system of adventive roots that initiate in the bases of bud leaves even at the late embryogenesis stages happens. But *Nelumbo* embryo from the very beginning of its existence occurs to be bipolar (Snegirevskaya, 1964; Kolesova, Batygina, 1988).

The embryo of *Ceratophyllum* (hydatophyte) is likely also bipolar at the first stages of its development owing to the group of cells which could be considered as the initials of the main root. In mature embryo the only well developed bud is present and the main root is absent, at the same time the adventive roots do not form both in seminal and postseminal development. In the seedling of *Ceratophyllum* the typical bipolarity is absent.

In most representatives of Poaceae family (xerophytes) the formed bud and well developed adventive roots (their number varies depending on plant species, place and time of growing, e.g., frost and spring) are present in the mature embryo. Some authors assume the main root exists but in evolution process it had been modified in coleorhiza (see Batygina, 1974, 1987a).

In members of Orchidaceae family the mature embryo lacks the typical shoot and root apices although it is bipolar (difference of two zones). Later in the process of seedling (protocorm) formation together with the bud the adventive root appears. As a result the secondary polarity establishes (Teryokhin, 1977; Batygina, Vasilyeva, 1983).

In parasitic plants (Orobanchaceae, etc.) the bipolarity of embryo with different reducing degree is likely observed from the very early stages of its development but later it loses. The secondary polarity appears only at seed germination (in contact with plant-host) thereby the true roots do not produce. *The embryo of parasitic plants* is represented the fine model for the studying of the reduction of *the typical initials of epiphysis, hypophysis as well as shoot and root apices* (Teryokhin, 1977). The great variability of the first developmental stages which displays in the different degree of *ca* and *cb* cell derivative participation in embryo body construction believes to be the peculiarity of its development. Often the first divisions are irregular but the development of this embryo does not allow applying it to any type of embryogenesis.

The series of sexual embryos of flowering plants given shows that the morphogenesis of sexual embryo is conditioned by the origin of its initials (zygote or egg cell in the case of parthenogenesis), type of embryo sac, ovule type and also by the peculiarities of taxon developmental biology. *The plasticity of embryo cells is rather high that is likely conditioned by the high degree of their totipotency.* This property makes possible *the deep morphogenetic modifications at different stages of embryo and seedling development with the preserving of their general organization.* These modifications in the process of adaptive evolution took place perhaps as a result of the disturbance of definite morphogenetic correlations at various biohierarchy levels. The consequences of these modifications could manifest at the very first (irregularity of cell divisions, different participation of *ca* and *cb* derivatives in embryo body construction etc.), medial (lack of hypophysis and epiphysis) and late (lack of the main root and adventive roots not only in embryo but also in seedling) stages of embryo development as well as in the general reducing of embryo differentiation level, beginning with early stages of its development. In the course of these reorganizations in the process of sexual embryo development the time of the appearance of primary polarity changes and also the replacing of it by the secondary polarity in course of seedling formation process becomes possible.

It is possible to interpret the entire complex of features listed above as evidence of potential evolutionary trends towards the transition from classical sexual embryo to the bud or from the bud to the sexual embryo (Batygina, 1990, 1993).

Embryoid (Somatic Embryo)

Morphogenesis of somatic embryo (occasionally and a seedling) in natural conditions occurs in the system of the maternal organism, under its influence. Perhaps it conditions the normal proceeding of the whole embryogenesis and a high percentage of seedling yield, unlike the system "the culture of somatic embryo *in vitro*" where the percentage of normal plant-regenerants is not high.

The Development of Embryoid in Natural Conditions (See Natural Embryoidogeny)

The somatic embryo of *Paeonia* members formed as the result of cloning of epidermal stem cells of sexual embryo occurs to be the bright example of natural embryoidogeny (monozygotic cleavage) (Figures 7, 8, 26). Peony somatic embryo by morphogenesis character is similar to the typical sexual embryos of dicots (parallelism in development). The first stages of embryoid development proceed analogously to Asterad-type of embryogenesis (Figures 26, *1-21*). At initial cell stage the establishment of polar axis begins. Later the normal development of all histogens of root and shoot apices happens. The organization of somatic embryo in peony seed resembles the classical model of sexual embryo. In somatic embryos at the nucellular and integumentary embryoidogeny the initials of epiphysis and hypophysis and their centres are absent but the development of endosperm, shoot and root apices, the formation of a main root with all its elements proceed normally as in sexual embryos do, however the question on the character of the first embryo divisions stays debatable (regular or irregular).

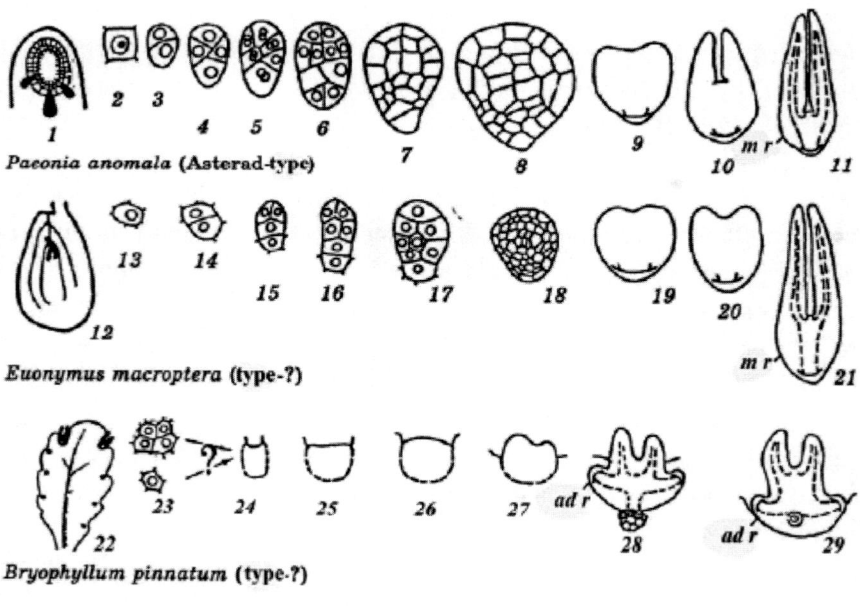

Figure 26. Polymorphism of somatic embryos in natural conditions: (1–29) successive stages of development of embryoids (1–11, cleavage; 12–21, integumental; 22–29, foliar); *ad r* – adventive root, *m r* – main root (from Batygina, Zakharova, 1997 with alterations).

Another mode of somatic embryo formation in Crassula multicava, where the somatic embryo derives from epidermal leaf cells, is considered interesting. In this case the typical initials of epiphysis and hypophysis and their centres are not revealed, the development of shoot and root apices in these embryos proceeds similarly with sexual ones (Figure 16).

In *Bryophyllum* species the structures formed on the leaf have no initials of epiphysis and hypophysis and their centres. But during their development, as was noted above, “globular”, “heart-shaped” and “torpedo-shaped” stages could be observed that speaks most probably about their belonging to somatic embryos or to the transitional form between embryo and bud. Morphogenesis of these structures is in rather extent similar to that of sexual embryo in grains (formation of shoot apex and adventive roots) (Figures 26, 22-29).

Thus, the structure and genesis of somatic embryos are taxon-specific and are likely in a great extent conditioned by the place of initial cell formation and that conditions at which the development of individual and all population in a whole occur. Morphogenetic resemblance of some embryoid types with the adventive bud confirms the existence of transitional forms between these structures.

The Development of Embryoid in In Vitro *Culture (Artificial or Experimental Embryoidogeny)*

As it was mentioned, *in vitro* culture presents to the investigators two different model systems for somatic embryo obtaining—suspension and callus. Let us examine the morphogenesis of such embryos on the example of different species of flowering plants contrasted by series of features. At the endogenous laying down of somatic embryos in the callus of *Triticum aestivum* the first divisions proceed similarly as in the sexual ones (Graminad-type of embryogenesis). However, the disturbance of the first pattern of cell division of their initial cell (equal, symmetrical instead of unequal,asymmetrical) and accordant further lying down of cell walls were observed more often that leads to abnormal histogenesis and the death of embryoids formed (Figure 23). The same picture of developmental disturbances is seen in the sexual plants of grains at distant hybridization.

In the development of somatic embryo in *Aconitum heterophyllum* in callus culture *in vitro*, the clear resemblance with Onagrad-type of embryogenesis, typical for many others members of Ranunculaceae family, was discovered (Figure 27). In the same time the different abnormalities, e.g., early differentiation of xylem elements in embryoids at heart-shaped stages could be observed. One of the causes of this is probably in the disturbance of differentiation at proembryo stage. The definite parallelism also was discovered in the formation of somatic embryo in callus and suspension cultures and sexual embryo *in vivo* in *Daucus carota*.

On the basis of information obtained, the following conclusion could be made. The resemblance in development of sexual and somatic embryos manifests in the general regularities of morphogenesis: polarity (formation of a new axis), symmetry (radial and bilateral), cellular and histogenous differentiation, morphogenetic and morphophysiological correlations, and capacity for proliferation. There are critical stages in the genesis of both embryos: laying down of the first partitions, protoderm, differentiation of organs, autonomy, etc. In sexual and somatic embryo the polymorphism, the transitional forms between the embryo and the bud and some abnormalities are noted. Moreover, the development of sexual embryo is known to proceed according to the laws of cell division and embryogeny. However, the variability of the first zygote division and the specifics of proceeding with further embryogenesis stages (8 types and more than 50 variations of embryogenesis), and the presence of transitional forms testify that these rules and laws fail to be absolute. A similar situation is observed also at the development of somatic embryos both in natural conditions and in culture *in vitro* (Batygina, 1996).

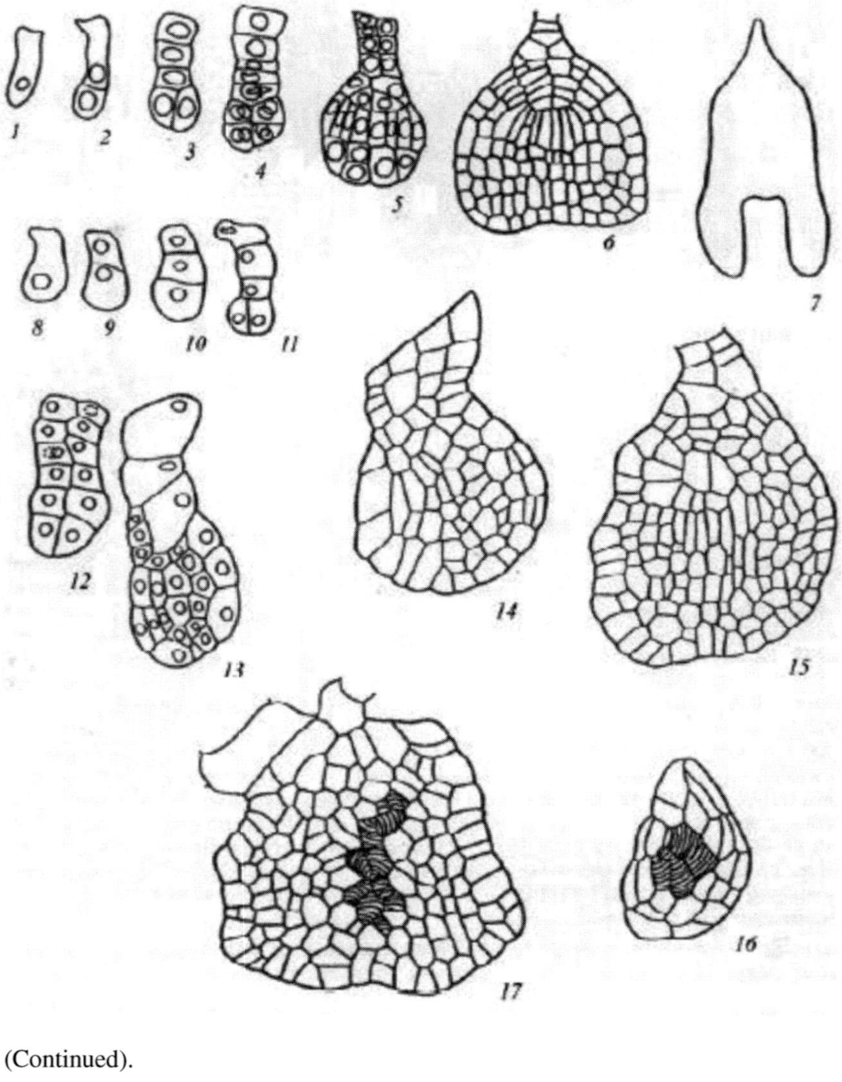

Figure 27. (Continued).

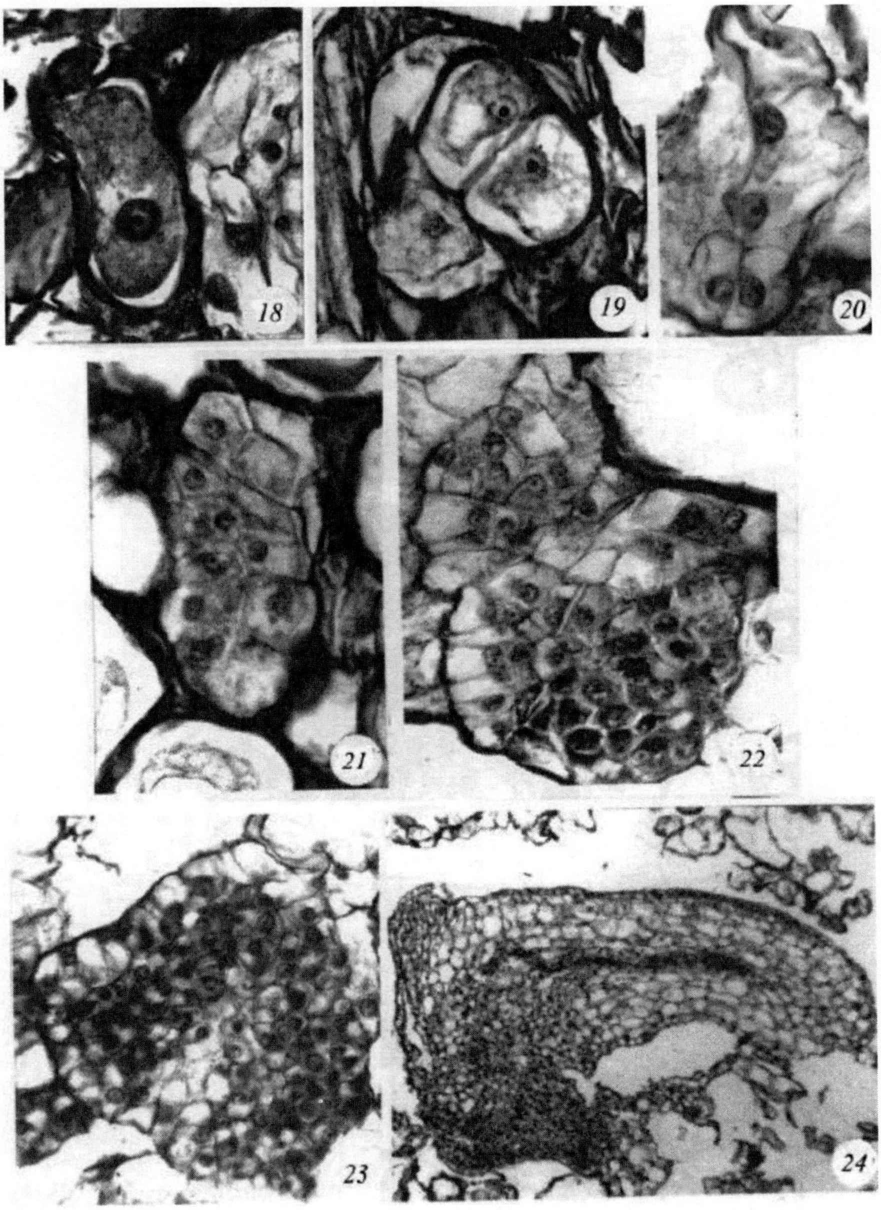

Figure 27. Development of sexual embryo in *Caltha palustris* (*Ranunculaceae*) in seed (*in vivo*) according to Onagrad-type (1–7); successive stages of development of embryoids *Aconitum heterophyllum* (*Ranunculaceae*) in culture *in vitro* according to Onagrad-type (8–15, 18–24), abnormalities can be found: early differentiation of xylem
in the area of central cylinder (16, 17, 23) (Ahuja, Titova, Batygina, orig. data).

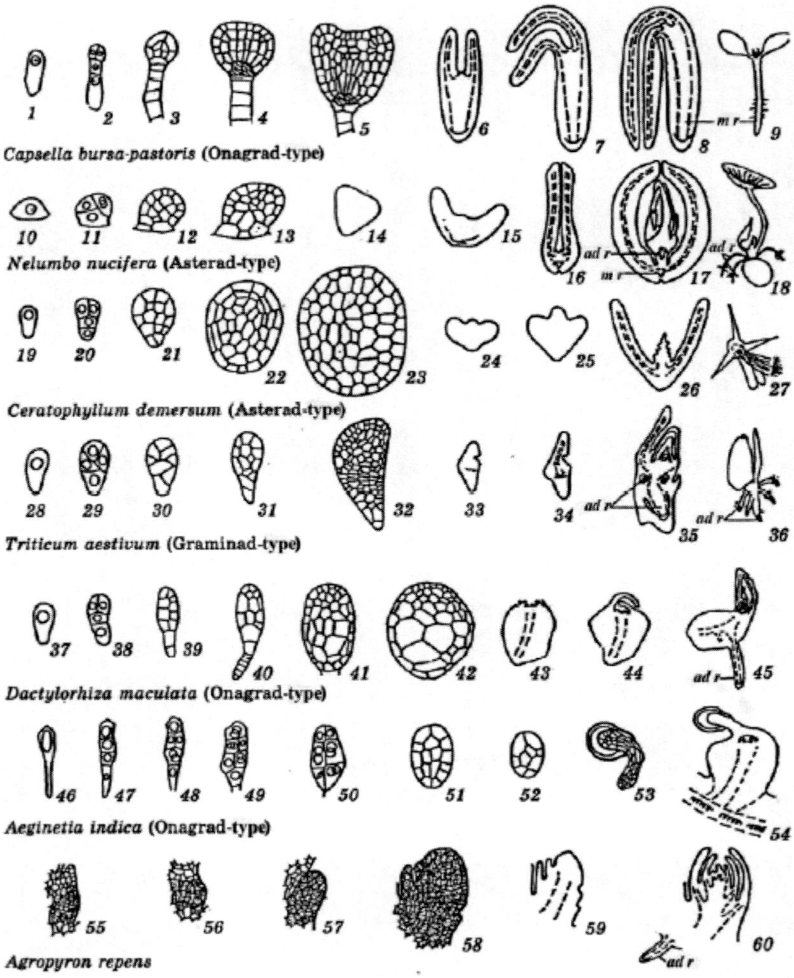

Figure 28. Parallelism of the first stages of morphogenesis of the sexual embryo and bud. (1–54) formation of the embryo and general view of the seedling: (1–9) *Capsella bursa-pastoris* (mesophyte); (10–18) *Nelumbo nucifera* (hydrophyte), the main root is underdeveloped, adventive roots can be found; (19–27) *Ceratophyllum demersum* (hydatophyte), the main root is undeveloped; (28–36) *Triticum aestivum* (xerophyte), advetive roots can be found; (37–45) *Dactylorhiza maculata* (mycorhizic plant), there are no shoot and root apices, they appear only after germination; (46–54) *Aeginetia indica* (parasite plant), the embryo is reduced, the shoot apex appears only after germination, proper roots are not formed; (55–60) development of a lateral bud in *Agropyron repens*; shading shows hypophysial centre; *ad r* – adventive root, *m r* – main root (from Batygina, Zakharova, 1997).

Thus, the complex of morphogenetic features examined above testifies not only the parallelism in the development of sexual and somatic embryos but also the parallelism in the development of embryo and bud. It permits speaking about possible evolutionary trends of transition from the bud to the classical sexual embryo (and/or back) throught "midway forms": brood buds, embryoids, nonclassical sexual embryos (Figure 28). Perhaps, these structures in the process of evolution have forming paralleled and realized in any extent together with sexual embryo in different taxa.

Functioning of Stem Cells as a General Mechanism of Switching Over Developmental Programs in Ontogenesis

The reproduction systems of seed plants could be considered in relation to the character of genetic information transferring to posterity and modes of its realization. Under the influence of environmental factors not only the genetic information changes but also the modes of its transferring the mechanisms of which in many respects continue to be vague.

The study of the phenomenon "stem cells" is to be one of the prospective trends in the research of this biological aspect. The Department of Embryology and Reproductive Biology of the Komarov Botanical Institute of RAS has for many years been working out the general biological problem of the origin and role of stem cells in the ontogenesis of plant and animal organisms.

The understanding of the term *stem cell* in the plant kingdom is believed to be debatable.

The origins of the problem of stem cells in plants is in the investigations of "dormant meristem" in a leaf (Sachs, 1887; Kerner von Marilaun, 1895; Yarbrough, 1932), ovule (Batygina, Freiberg, 1979; Batygina, 1991b), " quiescent centre" of root apex (Clowes, 1954), "méristème d'attente " in shoot apex (Buvat, 1955). For its solution the attracting of newest ideas and facts in the sphere of the study of animal and human stem cells is needed, discovered already in the beginning of XX century and obtaining a new resonance (Thomson, Marshal, 1998) in relation with their great theoretical and practical significance.

The analysis of very first ontogenesis stages, namely, the specifics of its embryonal period in the members of numerous taxa of flowering plants *permitted arriving at the conclusion (Batygina et al., 2004) that apical shoot and root meristems occur to be the product of functioning of zygote—the stem cell* ("proximate stem cell" after Barlow, 1997). *So, the stem cells represent not a part of apical shoot and root meristems, as many investigators considered* (Clark, 1997; Weigel, Jürgens, 2002; Groß-Hardt, Laux, 2003; Ivanov, 2003, 2007) but its source (Batygina et al., 2004; Batygina, Rudskiy, 2006; Batygina, Vinogradova, 2007; Batygina, 2007a-c)[3].

The study of reproductive biology, especially comparative embryology of members of flowering plants' different taxa allows considering that the stem cells form not only in shoot and root apices (méristème d'attente and quiescent centre). *The formation of stem cells derived from zygote is typical for all organs (flower, stem, leaf, root) and all stages of life cycle (sporophyte, gametophyte), moreover their functioning firstly depend on their localization and purpose.* It is the aspect of stem cells in which *initial cells of somatic embryos (embryoids) of different origin in natural conditions* and in culture *in vitro* should be examined. In relation with this the study of a special category of vegetative propagation—embryoidogeny—obtains a great significance (Batygina 1989a, b, 1993a, 29004, 2005, 2009; Batygina et al., 1979).

Plasticity in development and reproduction of plants is firstly related with multisided activity of cells of plant body possessing the property of stem cells. According to the analysis

---

[3] I assume that egg cell and zygote are likely correctly considered an in embryo sac and the formation of sexual or parthenogenetic embryo as well as nucellar, integumentary and cleavage embryoids a result of cell lines producing (in different directions) from the first stages of ovule development (epidermis formation, etc.). Owing

of literature (Pottten, Loeffler, 1990; Clark, 1997; Barlow, 1997; Weigel, Jürgens, 2002; Ivanov, 2003; Byrne et al., 2003) and original data the stem cells of plants are characterized by:

1) *the property of toti- or pluripotency*, i.e., the capacity for formation of not only different tissue and organ types but a new individual as well, by various morphogenesis pathways (*embryo-, embryoido-, gemmorhizogenesis*);
2) *the property of selfmaintenance, i.e., the creation of cell pool generally owing to symmetrical divisions and the system of intercellular interactions*;
3) *the capacity to proliferation and formation of cells-ancestors* of different tissue types ("niches")—owing to *asymmetrical divisions* under the definite signals;
4) *pulsatory and multistage character of the formation in tissue or organ and capacity to switch over the developmental program* that is provided by various molecular-genetic mechanisms (Batygina et al., 2004; Batygina, Rudskiy, 2006; Batygina, Vinogradova, 2007; Batygina, 2005, 2007a-c).

According to the series of stem cell features formulated above, we manifested the unity of morphogenetic and reproductive processes at all stages of plant life cycle.

In spite of structure diversity in shoots, roots and similar organs, all higher plants are characterized by the presence of quiescent centre or domain of undifferentiated cells in apical meristem that could consist even of a single cell as it does in sporiparous plants. *Dormant centres of shoot and root in some flowering plants appear in embryogenesis during epiphysis and hypophysis cell formation* (Figure 29A); *their differentiation usually occurs in fifth cell generation and is realized by means of asymmetrical divisions.* As a result of this, the united structural domain of undifferentiated cells, i.e. the initiation of all further organogenesis processes by definition could be created by cells of *independent lines.* The diversity of embryogenesis types known also demonstrates the independence of arising of shoot and root organs. For example, at Graminad-type of embryo development (distinguished by Batygina, 1969) in cereals, both the root and shoot tissues could arise in very different cell lines (Barlow, 1997; Batygina, personal report, XVII International Botanical Congress, 17–23 July, 2005, Austria, Vienna). However, the cells of the quiescent centre heterogeneous by origin are considered to be the stem ones in the traditional narrow sense of this term.

The notion of stem cell, both in botanical literature and in zoological and medical one, we propose to be the functional cell, because no universal genetic or epigenetic markers are known for identification of stem cells (Blau et al., 2001; Theise, Krause, 2002; Batygina, Rudskiy, 2006).

The complex WUS/CLV could be used as a marker for the shoot stem cells in the example of *Arabidopsis* genus (Veit, 2004), but it is not believed to be universal for the stem cells of a whole plant, in particular for the root where the products of the other genes fulfill the analogous function (Nakajima, Benfey, 2002). In one of the latest works on the model investigation of the role of CLV and WUS family proteins in regulation of shoot apical meristem structure the mechanism of maintenance of compartmental structure (space

to this, egg cell and zygote are believed to be the product of functioning of these cell lines, i.e., just the secondary stem cells, and the apical meristems in their turn are the product of zygote.

arrangement of zones of enzyme-marker synthesis) of shoot apical meristem was examined and the model of such mechanism was constructed (Nikolaev et al., 2007).

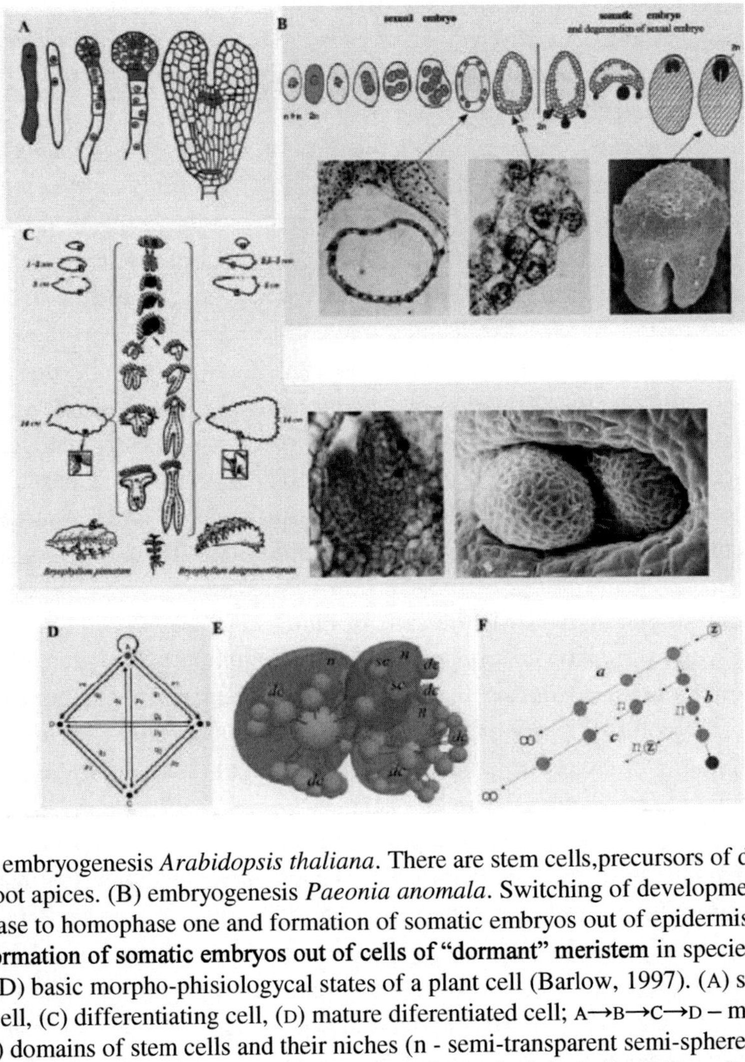

Figure 29. (A) embryogenesis *Arabidopsis thaliana*. There are stem cells,precursors of dormant centers of shoot and root apices. (B) embryogenesis *Paeonia anomala*. Switching of development programs from heterophase to homophase one and formation of somatic embryos out of epidermis of the sexual embryo. **(C) formation of somatic embryos out of cells of "dormant" meristem** in species of the genus *Bryophyllum*.(D) basic morpho-phisiologycal states of a plant cell (Barlow, 1997). (A) stem cell, (B) meristematic cell, (C) differentiating cell, (D) mature diferentiated cell; A→B→C→D – most probable transitions. (E) domains of stem cells and their niches (n - semi-transparent semi-spheres), which consist only of daughter cells (dc) cells, which differentiate in the initial of cell lines. In the surrounding (niche) of these cells sister cells (sc) can be found; these cells are at the same stage of differentiation, which principally distinguish them from the niche of the stem cells. (F) three components of morphogenetic processes in plants: (a) potentially infinite selfmaintenace of stem cells; (b) differentiation and loss of stemness up to the death of a cell; (c) restoration of stemness and transdifferentiation in viable cells with undamaged genome, with zygote (Z) being one of them. (from Batygina, Rudskiy, 2006).

Every plant cell is able to be in different morphophysiological states, the characteristics of one of them mostly correspond to the properties of stem cell, the others correspond to actively dividing meristematic cell. The transitions between these states are possess the stochastic character and in most cases histogenesis and organogenesis are connected with the loss of so called stemness but any other transitions are also probable (Barlow, 1997) (Figure 29D).

Besides the general space arrangement the stem cells are united by the processes proceeding *in domain*. Relatively symmetrical divisions are related to self-maintenance of the stem cell pool, the asymmetrical ones with transition of the daughter cell to the definite pathway of differentiation. *The derivatives of stem cells, the roots of tissue cell lines, create the niche* (Figure 29E). The surrounding of differentiated cell domains principally differs by the presence of sister cells (Figure 29E).

The zygote is not alone in its ability to be the initial cell of a plant. The diversity of morphogenetic processes comprehends *two components: potentially infinite meristem activity* that is conditioned by the functioning of dormant centre cells, being in differentiated state (Figure 29D, A) and *the loss of stemness*— senescence and apoptosis (Figure 29D, B).

Such processes can take place at different ontogenesis stages and in all structures, e.g., the formation of *somatic embryos* from epidermal cells of sexual embryo (new sporophyte) in *Paeonia* (Batygina, 2004) (Figure 29B). The plasticity of leaf tissues (*Bryophyllum*, *Gnetum*) provides the possibility of the formation of *somatic embryos* from a single or group of cells (*dormant meristem=stem cells*), in other words, the repeating of early ontogenesis stages in the tissues of adult plant is observed (Batygina et al., 1996; Bragina et al., 2005) (Figure 29C). Programmed destruction of definite part of leaf primordium cells (*Anubias*) leads to the formation of epidermis owing to the transdifferentiation of interior tissues (Rudskiy et al., 2005).

The plasticity of generative structure cells of *maternal plant* (*Poa*) permits the formation of somatic embryos from ovule covers (nucellus and integument) (Batygina, 1991b). Thereby the cells of somatic embryos during the development use the niche of the sexual one. The phenomenon of gametophytic apomixis widely spread among angiosperms consists in the initiation of somatic embryo formation from egg cell or synergid—the cells of female gametophyte without fertilization (Czapik, 1997).

In anthers of cereals at definite developmental conditions, *the morphologically competent (stem) cells* of microspore transfer from usual *gametophytic* pathway (formation of pollen grains with sperms) to *sporophytic* one—formation of haploid regenerants. In this case, the cells of microsporangium walls are likely to be considered as a niche. In *in vitro* culture, the anther cells realize their potential by different morphogenesis pathways: *primary embryoidogenesis* (*secondary embryoidogenesis* is also possible), *gemmorhizogenesis,* gemmogenesis, rhizogenesis, histogenesis (Figures 13, 24) (Batygina, 1987a, 1989a, 2005; Batygina et al., 1979).

The plasticity and polyvariety of these processes are conditioned by the diversity of the structure and origin of vegetative and generative structures in which the peculiar reserve of cells, possessing the property of stemness, is situated. Hence, *in plant morphogenetic processes, even the third component is present—the restoring of stemness* in many respects depending on *stemness degree* (Figure 29F, c).

Thus, the plasticity and tolerance in plant development and reproduction is related firstly with multisided activity of stem cells at all stages of ontogenesis.

Special attention should be paid to *the unique property of plant cell—totipotency,* characterizing all morphogenetic possibilities (i.e., all the potential), that is habitual to individual and realized by different morphogenesis pathways. The final result of this property manifesting, its modes and forms of realization can be different and are conditioned by *the totipotency degree of the cell* (cells) (Batygina, 1987a).

Not Every Cell Possesses Totipotency and Not Every Totipotent Cell Is Able to Become a Stem Cell

The degree of pluri-, omni- and totipotency of cells is taxon-specific and is determined by the totality of factors, in the first turn by the system (tissue, organ, organism) from which the cell was obtained. The sexual cells (sperms and egg cell) in the period of their differentiation and specialization and more mature cells as well as the zygote in the period of its establishment and development are not belonged to totipotent ones. The property of totipotency recreates every time from the moment of zygote proliferation and losses by most embryo cells in the course of their specialization. However in some cases the definite cells of embryo, seedling and plant at different ontogenesis stages stay the stem ones – so called **"méristème d'attente" or "dormant meristem"**, *that allows fulfilling the transition from one reproduction mode (heterophasic) into another (homophasic) and vice versa.*

In our opinion, *the phenomenon of multistage change of developmental program and creation of different subprograms in life cycle (probably repeated)* of most flowering plants is *universal*, and the *unique* property of plant cell—totipotency—lies in its base.

The direction of changes in developmental program, *the duration and the number of possible transitions are taxon specific* and in many respects *depend on external stress and internal* (hormonal, immunological) *factors.*

Thus, the changing in genetic program and polyvariety of transitions in plant *individual development* are *universal.* It is likely *conditioned by the alternation of of sexual and asexual modes of new individual formation in life cycle* that is conformed with G.P. Korotkova notions (1979) applying to animals.

The Phenomenon of Embryoidogeny and Its Role in Onto- and Phylogenesis

Comparative analysis of the results on the researches of the somatic embryo formation in natural conditions and in tissue culture permitted to formulate the notion of *embryoidogenesis as a process, in the base of which asexual mode of new sporophyte formation (i.e., without participation of gametes and fertilization) is laid and of embryoidogeny as a special type of vegetative propagation.* The propagation in this case is *realized by either embryoids or miniature seedlings, developing on the plant from embryoids, or by the seeds, in which embryoids form* (Batygina, 1987a, 2006a, 1009 etc.).

It is in conformity with the opinion of Winkler (1934), Battaglia (1963) and Grant (1981) that *adventive embryony (formation of adventive embryos in seeds) and viviparity (on leaves, stems, etc.) (homophasic, T.B.) are the forms of vegetative propagation.*

The phenomenon of embryoidogeny represented by different forms is of a *greater adaptive significance for flowering plants* than another type of vegetative propagation, i.e., *hemmorhizogeny.* During the evolution the origin of embryoids was likely conditioned by such their advantages *as a shorter and consequently more power favourable pathway of the formation of new individuals' numerous initials with uniparental heredity* (together with sexual embryos). It is clear that the formation of different *vegetative and generative diaspores*, specially seeds containing the sporophyte initials of "new" *daughter sexual generation (biparental* heredity) and "old" maternal generation (*uniparental* heredity), i.e., the formation of *agamosexual complexes* obviously expands the possibilities of *"struggle for existence", due to which these processes were picked up by natural selection and became the central item in preserving of biological diversity of plants.*

One of the significant advantages of embryoidogenous propagation type is that *the initial of a whole organism*, forming in the seed or on the vegetative organs, *occurs to be a unit of propagation.* A full-value *bipolar* embryo, forming at embryoidogeny in some plant species is able not only to root immediately, but also disperses like seeds by wind and water over small and large distances, i.e. the area of dispersal increases, which is very important for preserving resources and biological diversity.

The peculiarity of such an original embryoidogeneous type of reproduction (unlike the usual forms of vegetative propagation—particulation, sarmentation etc.) *owes to the same features that are typical for propagation by seeds: the plurality of new individual initials and their capacity to close and far dispersion.* Besides that, as it mentioned above, embryoidogeny *extends essentially the diapason of seed heterogeneity.* In consideration of polyembryony phenomenon and its different forms (monozygotic cleavage embryoidogeny, gametophytic apomixis, nucellar and integumentary embryony), sexual process, four categories of embryos and five types of genetic heterogeneity of seeds could be distinguished (Batygina, 1997, 1999a, b, 2000, 2006a, 2009; Batygina, Vinogradova, 2007).

It should be mentioned once again that there are somatic embryos, forming in seeds together with sexual ones, that determine the type of *heterogeneous seed.* Apart from *general seeds* (only with the sexual embryo), there are *seeds in which different modes of new individual formation—sexual and asexual—occur hand in hand.* Herewith, often no one mode of reproduction is clearly. For example, *the sexual process precedes the arising of "monozygotic twins" (or triplets, etc.).* In this case, one mode of new individual formation, sexual, is changed on another, asexual (change of morphogenetic developmental program).

Thus, in some cases, before "*agamospermous seeds*" in *"clear state"* were formed, i.e., containing only somatic embryos, the seeds in the process of their development were "*agamosexual*": they could contain nucellar, integumentary embryos and heterozygote, which subsequently could produce "monozygotic twins" by asexual mode—by the cloning (cleavage monozygotic embryoidogeny).

Embryodogenous types of reproduction in angiosperms could be considered as the mechanism of the polyembryony phenomenon (*sensu lato*) arising and the cloning of plants at different ontogenesis stages (Figure 30).

All these complex multistage processes taking place in the seed ensure the capacity to reproduce and propagate new daughter individuals (or new generation) with different genotypes, either uniparental or biparental inheritance (in the case of sexual embryo formation).

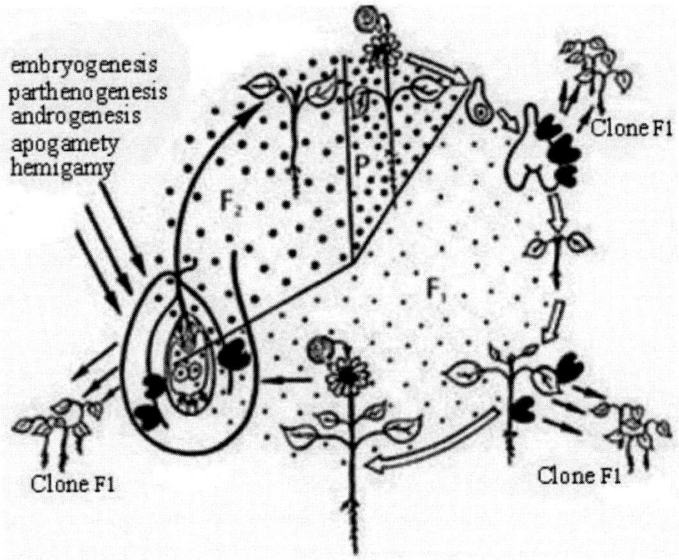

Figure 30. Embryoidogenic type of reproduction in angiosperm plants as a mechanism of appearing of the phenomenon of polyembryony and cloning of a plant organism at different stages of ontogenesis (from Batygina, 1993**a**).

Thus, the elaboration of a new conception of "embryoidogeny" permitted the presentation *of a nontraditional view of the system of flowering plant* reproduction. It is in verification of the traditional and distinguishing of new notions connected with renewal and propagation. So, there are *two modes of new individual formation—sexual and asexual; three pathways of morphogenesis in individual formation—embryogenesis, embryoidogenesis and gemmorhizogenesis; two forms of reproduction—heterophasic (embryogeny, gametophytic apomixis) and homophasic (reproductive and vegetative embryoidogeny, gemmorhizogeny—particulation, sarmentation and viviparity); two types of propagation—seed (gamospermy, agamospermy and gamo-agamospermy) and vegetative (which includes gemmorhizogenous and embryoidogenous modes of propagation).*

In conclusion, it should be mentioned that the introduction of a new highly specialized category (type) of vegetative propagation—*embryoidogeny—allowed uniting such phenomena that earlier seemed to be isolated: monozygotic cleavage, nucellar and integumentary, gametophytic, foliar, cauligenous and rhizogenous embryoidogeny.* The question on the position of these phenomena in the reproduction system and their evolutionary role for many decades has been *debatable.*

According to above-mentioned, one is hard to agree with the statement of some authors, who attribute the special role to definite forms of embryoidogeny in the reproduction system of flowering plants. It is not quite correct, as for *determination of the role* of every embryoidogeny form *it must be considered from the comparativepoint of view* under interactions with the other forms and also with *embryogeny, gemmorhizogeny and gametophytic apomixis.*

In every plant cell all genetic information on the morphogenesis pathways is laid down. During evolution the morphogenesis pathway and accordingly reproduction mode for every taxon have been fixed. Realization of the other morphogenetic potencies usually occurs in stress situation (tissue culture, hybridization, mutations, etc.) and is related with extraordinary properties of plant cell—its totipotency and stemness.

Analysis of a store of available data allowed to formulate the conclusion, that *namely the versatile activily of stem cells at all stages of plant organism ontogenesis causes the flexibility and tolerance of development of the organism and, correspondingly, of the plant reproduction systems.*

SUMMARY

- Stem cells in plants are formed in different organs (flower, stem, leaf, root) and at different stages of life cycle (sporophyte, gametophyte).
- The integrity of properties of stem cells—toti- and pluripotency, the ability for self-maintenance, pulsatory and multi-stage character of formation and especially the ability to switch over the developmental program—provides flexibility and tolerance of the reproduction system, which determines the functioning of a species-specific system of reserves of a plant at different stages of ontogenesis.
- The zygote—stem cell—is the "progenitor" of all stem cells of a plant.
- Stem cells are capable of forming a new individual by sexual and somatic embryos of various origin, and both various types of tissues and organs.
- Not every cell possesses the totipotency and not every totipotent cell is the stem cell, and not every stem cell is able to give new organism.
- The revealed role of stem cells gives new perspectives for
  - development of reproduction theory and fundamental problems of developmental biology: the phenomenon of polyembryony, genetic heterogeneity of seeds, apomixis and others;
  - solution to problems of plant evolution: origination of flowering plants, gametophyte evolution, morphological nature of the ovule, embryo organs and others;
  - further development of innovations into biotechnology, selection and medicine.

CONCLUSION

Perspective approaches in the Study of Stem Cells

- Use the system approach to the study of developmental biology and reproduction theory of plants and animals and, in particular, biology of stem cells: architectonics, critical cell mass, rate of cell divisions, peculiarities of DNA and RNA synthesis considering the phase of cell cycle, etc.
- Extending the range of model objects, contrasting by systematic position, centers of origination, peculiarities of ecology of germination and ethology in population and cenosis, considering optimal model systems: embryo-, embryoido- and gemmorhizogenesis, and considering taxon-specific localization of the system of stem cells in the organism

- Comparison of patterns of morphogenesis and differentiation of reproductive and vegetative structures—in particular, detailed study of polymorphism and parallelism of development of sexual and somatic embryos (*in situ, in vivo,* and *in vitro*)
- Revealing of morphogenetic correlations and critical periods in morphogenesis
- Revealing of biological meaning of doubling structures in development and reproduction at different levels of biohierarchy (failures and reserves), e.g., systems of reliability, homeostasis
- It is necessary to discuss and create a glossary for researchers, involved in studying of various aspects of stem cells.
  - A special attention should be paid to the terms «totipotency», «pluripotency», «multipotency», «omnipotency», «unipotency» defining degree of cell's morphogenic potential.
  - Is the ability to form an individual a necessary attribute of a totipotent stem cell?
  - In connection with heterogeneous system of stem cells and cell lines it is worth discussing the necessity of using the term "degree of stemness" and "change (loss and regain) of stemness" in connection with the different character of development of cells, their differentiation, specialization and dedifferentiation (Batygina, 1978, 1987, Batygina et al., 2004).

The perspectives mentioned above are the basis for the development of the theory of stem cell biology and will provide the opportunity to control separate stages of ontogenesis of plants and animals.

- Organizing the Center of Stem Cells and Reproduction in Russia in the Komarov Botanical Institute of RAS in Saint Petersburg. The Center of Reproduction has already been established on the basis of the Department of Embryology and Reproductive Biology, the leading Russian school of development of reproduction theory, whose members are scientists from different countries around the world.
- Organizing the Council on Stem Cells and Reproduction with representatives of leading laboratories of Russia and different countries of the world.

I suggest a new priority trend in developmental biology: investigation and combined use of plant and animal stem cells for the purposes of humanity.

Afterword

Analysis of developmental biology of plants and animals allows us to discuss some similarities in biology of stem cells.

Similarities

- structural–functional ensemble of integrated cell systems, including stem cells, provides a high degree of reliability of an individual:

- the phenomenon of polyembryony
- the phenomenon of viviparity
- the phenomenon of parthenogenesis, androgenesis
- the phenomenon of chimera formation
- mitosis, meiosis, cell cycle, cytomixis, mutations, polytene chromosomes, fertilization, etc.

– activity of stem cells that is realised through:

- multivariability of modes of formation of an individual—sexual, asexual and parthenogenesis, of pathways of morphogenesis: sexual embryogenesis, somatic embryogenesis (*embryoidogenesis) and gemmogenesis (*gemmorhizogenesis) and of morphological structural units of sexual and asexual reproduction

* - term used only for plants.

– in the evolution of each taxon there has developed its own system of "hidden" reproductive reserves and failures – stem cells, working at different levels of biohierarchy. It provides the flrxibility and tolerance in the development and reproduction of a plant and population, on the whole, at all stages of ontogenesis

– the ability of stem cells of plants and animals to proliferate which determines their inducing, substituting, organizing and regulating role with close interaction with cells of the niche

– ability of the system of stem cells to switch the morphogenetic program in the life cycle of plants and animals is one of the mechanisms of transition from sexual to asexual reproduction and vice versa (for example, monozygotic twins)

– presence of multiple ontogeneses in the life cycle of plants and animals, which key mechanism is the stem cell

– the revealed main properties of stem cells of plants that are similar to those in animals

Not every cell possesses the totipotency and not every totipotent cell is able to become the stem cell, and not every stem cell is able to give new organism..

REFERENCES

Afzelius K. Die Embryobildung bei *Nigritella nigra. Svensk. Bot. Tidskr*, 1928, vol. 22, № 1-2, 82-91.

Andronova E.V. Epiphysis. In: Batygina T.B., editor. *Embyology of Flowering Plants. Terminology and Concepts. Vol. 2. Seed.* St-Petersburg: Mir i Semya; 1997; 356-358.

Astaurov B.L. Genetics and the problems of individual development. *Ontogenesis.* 1972. vol.3. №6. c. 547-565

Barlow P.W. Stem cells and founder zones in plants, particularly their root. In: Poten C.S., editor. *Stem cells*. London; 1997; 29-57.

Barykina R.P. Sarmentation. In: Batygina T.B., editor *Embyology of Flowering Plants. Terminology and Concepts. Vol. 3. Reproductive systems.* St-Petersburg: Mir i Semya; 2000; 302-305.

Barykina R.P. Particulation. In: Batygina T.B., editor. *Embyology of Flowering Plants. Terminology and Concepts. Vol. 3. Reproductive systems.* St-Petersburg: Mir i Semya; 2000; 306-310.

Battaglia E. Apomixis. In: Maheshwari P., editor. *Recent Advances in the Embryology of Angiosperms*. Delhi: University of Delhi; 1963; 221-264.

Batygin N.F. Ontogenesis of higher plants. In: *Nikolay Fyodorovitch Batygin. Life and work.* St-Petersburg: NPO Professional, 2007, 23-140.

Batygin N.F. System approach in biology and agronomy. In: *Nikolay Fyodorovitch Batygin. Life and work.* St-Petersburg: NPO Professional, 2007, 151-196.

Batygina T.B. On the possibility of separation of a new type of embryogenesis in Angiospermae, *Revue cytologie et biologie vegetale*. 1969, vol. 32, №2, 335-341.

Batygina T.B. Embryology of wheat. L: Kolos. 1974 206 c.

Batygina T.B. Problems of morphogenesis in vivo and in vitro. Abstracts of Indian-Soviet Symposium "Embryology of Crop Plants", 23-26august 1977. Leningrad, 1977, 41-42.

Batygina T.B. On certain regularities of morphogenesis at plant regeneration. Abstracts of region scientific conference "Theoretical questions of plant regeneration", October, 1978, Makhachkala. 1978. C. 13-14.

Batygina T.B. Problems of morphogenesis *in situ*, *in vivo* and *in vitro*. Proceedings of the International Symposium "Plant tissue and cell culture application to crop improvement". 24-29 Sept. 1984. Olomouc, Chechoslovakia, Prague; 1984; 43-56.

Batygina T.B. The grain of cereals. Atlas. Leningrad: Nauka; 1987a.

Batygina T.B. New interpretation of asexual propagation and its classification in flower plants (in light of experimental data). Scientific reports of V All-Union School on theoretical plant morphology, Lvov; 1987b; 92-97.

Batygina T. B. New concept of asexual reproduction in flowering plants. Abstracts of XIV Int. Bot. Congress. Berlin; 1987c; P. 5.

Batygina T.B. A new approach to the system of reproduction in flowering plants. *Phytomorphology*, 1989a, vol. 39, № 1, 311-325.

Batygina T.B. New approach to the system of reproduction in flowering plants. *Apomixis Newslet*, 1989b, № 1, 52-55.

Batygina T.B. Embryoidogenic type – a new category of asexual reproduction in flowering plants. Proceedings of VII Intern. Congr. «Plant Tissue and Cell Culture», Amsterdam, 1990a, 244.

Batygina T.B. Embryoidogenic type of reproduction in flowering plants. *Apomixis Newsletter*, 1990b, № 2, 58-66.

Batygina T.B. New hypothesis about the initials and genesis of embryoids (somatic embryos) and a position of embryoidogeny in the reproduction system. *Apomixis Newsletter*, 1991a, № 3, 19-24.

Batygina T. B. Nucellar embryoidogeny in *Poa pratensis* (Poaceae). *Pol. Bot. Stud*, 1991b, vol. 2, 121-125.

Batygina T. B. Position of the phenomenon of embryoidogeny in the system of flowering plants reproduction. Proceedings of XI Int. Symp. «Embryology and seed reproduction» Leningrad, 1990; St-Petersburg; 1992a; 6-10.

Batygina T.B. Embryological bases of plasticity and adaptive possibility of reproductive systems in flowering plants/ In: Yurtzsev B.A., editor. *Biological diversity: approaches to research and preservation.* St-Petersburg, Russia: Publishing House of Zoological Institute RAS; 1992b; 201-212.

Batygina T.B. Embryoidogeny – new category of flowering plant propagation mode. In: Teryokhin E.S., editor. Problems of reproductive biology of seed plants. *Transactions of Komarov Botanical Institute RAS.* Vol.8; 1993a; 15-25.

Batygina T.B. Certain aspects of morphogenetic polarity in plant ontogenesis. Abstracts of III congress of Russian Society of Plant Physiologists. St-Petersburg, 1993b. C. 258.

Batygina T.B. Anther as a model of research of morphogenetic potencies and morphogenesis pathways. In: Batygina T.B., editor. *Embryology of Flowering Plants. Terminology and Concepts. Vol. 1. Generative organs of flower.* St-Petersburg: Mir i Semya; 1994a; 120-121.

Batygina T.B. Apomixis, agamoapermy and vivipary and their role in reproductive system of flowering plants. *Transactions of International symposium "Apomixis in plants: the state of problem and perspectives of studies"*; Saratov: Publishing House of Saratov State University; 1994b; 16-18.

Batygina T.B. Ovule and seed from the positions of biological system reliability. In: Batygina T.B., editor. *Embryology of Flowering Plants. Terminology and Concepts. Vol. 1. Generative organs of flower.* St-Petersburg: Mir i Semya; 1994c; 263-266.

Batygina T. B. Adaptive abilities of reproductive systems of flowering plants. Abstracts of Int. Conf. on Plants & Environmental Pollution, Lucknow, India ; 1996a ; 68.

Batygina T.B. Parallel development of somatic and sexual embryos. Abstracts of the 14 International Congress on Sexual Plant Reproduction, 18-23 February, 1996. Lorne, Australia; 1996b; 4.

Batygina T.B. Embryoidogeny. In: Batygina T.B., editor. *Embyology of Flowering Plants. Terminology and Concepts. Vol. 2. Seed.* St-Petersburg: Mir i Semya; 1997; 624-648.

Batygina T.B. Morphogenesis of somatic embryos developing in natural conditions. *Biologija*, 1998, № 3, 61-64.

Batygina T.B. Embryoidogenesis and morphogenesis of zygotic and somatic embryos. *Rus. J. Plant Physiol*, 1999a, vol. 46, № 6, 884-898.

Batygina T.B. Genetic heterogeneity of seeds. *Acta Biol. Cracov. Ser. Bot*, 1999b, vol. 41, № 1, 39-50.

Batygina T.B. Genetic heterogeneity of seeds. Plant physiology, 1999c, vol. 46, № 3, 438-453.

Batygina T.B. Embryoidogeny – new category of vegetative propagation. In: Batygina T.B., editor. *Embyology of Flowering Plants. Terminology and Concepts. Vol. 3. Reproductive systems.* St-Petersburg: Mir i Semya; 2000; 334-349.

Batygina T.B. Ovule and seed viewed from reliability of biological systems // Embryology of flowering plants. Terminology and concepts / Ed. Batygina T.B. Enfield (NH), USA-Plymouth, UK, 2002a. Vol. 1. P. 214-217.

Batygina T.B. Embryoidogeny // Embryology of flowering plants. Terminology and concepts / Ed. Batygina T.B. Enfield (NH), USA-Plymouth, UK, 2002b. Vol. 2.

Batygina T.B. Sexual and Asexual Processes in Reproductive Systems of Flowering Plants. *Acta Biol. Cracov. Ser. Bot*, 2005 Vol. 47, № 1, 51-60.

Batygina T.B. Theoretical bases of anther cultivation. In: Shamrov I.I., editor. *Embryological bases of wheat androcliny. Atlas.* Moscow: Nauka; 2005; 9-23.

Batygina T.B. Embryoidogeny. In: Batygina T.B., editor *Embyology of Flowering Plants. Terminology and Concepts. Vol. 2. Seed.* Enfield (NH), Plymouth, UK: Science Publishers; 2006a; 403-409.

Batygina T.B. Polymorphism of sexual and somatic embryos as manifestation of their developmental parallelism under natural conditions and in tissue culture. In: Batygina T.B., editor *Embyology of Flowering Plants. Terminology and Concepts. Vol. 2. Seed.* Enfield (NH), Plymouth, UK: Science Publishers; 2006b; 409-419.

Batygina T.B. Structure of reproductive organs and reproduction of plants. In: Serebryakova T.I., Voronin N.S., Elenevskiy A.G., Batygina T.B., Shorina N.I., Savinych N.P. *Botany with basics of phytocenology: Anatomy and morphology of plants. Textbook for colleges.* Moscow: IKC Akademkniga, 2006, 367-484.

Batygina T.B. Algorithms of morphogenesis. Polyembryony, gametophytic apomixis from position of stem cells. Abstracts of international symposium "Cellular, molecular and evolution aspects of morphohenesis. Moscow; 2007a. 15-18.

Batygina T.B. New algorithms of morphogenesis. VI congress of society of Russian plant physiologists. Materials of reports of international conference "Modern physiology of plants: from molecules to ecosystems", Syktyvkar Komi. Part I. 2007b. 10-12.

Batygina T.B. Polyvariety of development of gametophyte and double structures as a base of polyembryony and genetic heterogeneity of seeds. Materials of II International School for young scientists "Embryology, genetics and biotechnology". Ufa: Publishing House of Bashkir state university; 2007c. 20-22.

Batygina T.B. Embryoidogeny is a new type of vegetative propagation. In: Batygina T.B., editor *Embyology of Flowering Plants. Terminology and Concepts. Vol. 3. Reproductive systems.* Enfield (NH), Plymouth, UK: Science Publishers; 2009.

Batygina T.B.; Freiberg T.E. Polyembryony in *Poa pratensis* L. (Poaceae). *Botanical Journal,* 1979, vol. 64, №6, 793-804.

Batygina T.B.; Mametyeva T.B. On embryology of the genus *Poa* L. In: Yakovlev M.S., editor. Actual questions of angiosperm embryology. L: Nauka, 1979. C. 89-95.

Batygina T.B.; Rudskiy I.V. Role of stem cells in plant morphogenesis. *Reports of Russian Academy of Sciences,* 2006, Vol. 410, № 5, 1-3.

Batygina T.B., Shamrov I.I. Ovule and seed structure in Poa, Paeonia and orchids from the point of reliability. Abstr. XV Intern. Bot. Congr. Yokohama, Japan, 1993, p. 96.

Batygina T.B.; Vasilyeva V.E. Expediency of system approach to the problem of angiosperm embryo differentiation. Ontogenesis. 1983. Vol. 14, № 3. 304-311.

Batygina T.B.; Vasilyeva V.E. Sexual reproduction in flowering plants: periodization of egg cell and zygote development and possible types of caryogamy. In: B.Bhatia et al. (eds.), *Plant Form and Function*, Angkor Publishers, New Delhi, 1997, 170-198.

Batygina T.B.; Vasilyeva V.E. *In vivo* fertilization. In: Bhojwani S.S., Soh W.Y. (eds.), *Current Trends in the Embryology of Angiosperms*. Dordrecht, Boston, London: Kluwer Academic Publishers, 2001, 101-142.

Batygina T.B.; Valisyeva V.E. Plant reproduction. St-Petersburg: Publishing House of St-Petersburg State University; 2002.

Batygina T.B.; Vasilyeva V.E. Periodization of development of reproductive structures. Critical periods. *Acta Biol. Cracov. Ser. Bot*, 2003, Vol. 45, № 1, 27-36.

Batygina T.B., Titova G.E., Rudskiy I.V., Bragina E.A. Embryo morphogenesis – one of plant ontogenesis model. Book of Abstracts of the International Conference "Mathematical Models and Methods in Biology and Medicine", 29 May – 3 June, 2005. Bedlewo, Poland, 2005, p.17.

Batygina T.B.; Titova G.E.; Shamrov I.I., Bragina E.A., Vasilyeva V.E., Rudskiy I.V. Plant stem cell in term of embryology. *Acta Biol. Cracov. Ser. Bot*, 2005, Vol. 45, Supply 1, p. 51.

Batygina T.B.; Vinogradova G.Yu. Phenomenon of polyembryony. Genetic heterogeneity of seeds. *Ontogenesis*, 2007, Vol. 38, № 3, 166-191.

Batygina T.B.; Zakharova A.A. Polymorphism of sexual and somatic embryos as a evidence of their resemblance. *Bull. Pol. Acad. Sci. Biol. Sci*, 1997a, vol. 45, № 2-4, 235-255.

Batygina T.B.; Zakharova A.A. Parallels in the development of sexual and somatic embryos. In: Batygina T.B., editor. *Embyology of Flowering Plants. Terminology and Concepts. Vol. 2. Seed.* St-Petersburg: Mir i Semya; 1997b; 635-648.

Batygina T.B.; Bragina E.A.; Titova G.E. Morphogenesis of propagules in viviparous species *Bryophyllum daigremontianum* and *B. calycinum*. *Acta Soc. Bot. Polon.*, 1996, vol. 65, № 1-2, 127-133.

Batygina T.B.; Titova G.E.; Vasilyeva V.E. Plant reproduction: theoretical elaborations and innovation technologies. Innovations. 2007. №2 (100) 41-48.

Batygina T.B.; Vasilyeva V.E.; Mametyava T.B. Problems of morphogenesis *in vivo* and *in vitro* (embryoidogenesis in angiosperms). *Botanical Journal,* 1978, vol. 63, № 1, 87-111.

Batygina T.B.; Titova G.E.; Shamrov I.I.; Bragina E.A.; Vasilyeva V.E.; Rudskiy I.V. Problem of stem cells in plants (from position of embryology). Materials of X school on theoretical plant morphology "Constructional units in plant morphology". Kirov: Publishing House of Vyatka state university; 2004; 20-30.

Batygina T.B., Bragina E.A., Ereskovsky A.V., Ostrovsky A.N. Viviparity in plants and animals: invertebrates and lower chordares. PH of St.Petersburg universiti. 2006. 132.

Batygina T.B.; Kruglova N.N.; Gorbunova V.Yu.; Titova G.E.; Seldimirova O.A. From microspora to sort. Moscow: Nauka. 2009 in press.

Beklemishev V.N. On the general principles of life organization. *Bulletin of Moscow Society of Nature Experiments,* 1964, vol. 69, 22-38.

Beklemishev V.N. Methodology and systematics. Moscow: KMK Scientific press LTD; 1994.

Bhandari N.N.; Soman P.; Bhargava M. Histochemical studies on the female gametophyte of *Argemone mexicana* L. *Cytologia*, 1980, vol. 45, 1-2, 281-291.

Blau H.M.; Braselton T.R.; Weimann J.M. The evolving concept of a stem cell: entity or function? *Cell*, 2001, №105, 829-841.

Boris Lvovitch Astaurov: essays, memoirs, letters, materials. Moscow: Nauka, 2004, 427p.

Bosch T.G. Stem cells – from hydra to man.; Springer; 2008.

Bouman F. The ovule. In Johri B.M., editor. *Embryology of angiosperms*. Berlin ets.; 1984; 123-157.

Bragina E.A.; Titova G.E.; Batygina T.B. The models of the formation of new sporophyte in genus *Bryophyllum* (Crassulaceae). Abstracts of VII Conf. of Plant Embryologists of Czech Republic, Slovakia and Poland. Lublin; 1995; 8.

Bragina E.A.; Arnautov N.N.; Batygina T.B. The phenomenon of vivipary in gymnosperms and angiosperms. Abstracts of the XVII International Botanical Congress. 17-23 July 2005, Austria, Vienna; 2005; 287.

Brandt U., Fletcher J.C., Hobe M., Meyerowitz E.M., Simon R. Dependence of stem cell fate in Arabidopsis on a feedback loop regulated by *CLV3* activity // Science. 2000. Vol. 289. N 5479. P. 617-619.

Brien P. La croissance des spongillidae. Formation des choanocytes et des spicules. *Bull. Biol. France Belg*., 1956, 110, 211-252.

Brossard D. Le bourgennement èpiphylle chez le *Bryophylllum daigremontianum* Berger (Crassulaceae). Étude cytochimique, cytophotometrique et ultrastrucurale. *Ann. Soc. Nat. Bot*., 1973, 12 series, vol. 14, 93-214.

Brukhin V.V.; Batygina T.B. Embryo culture and somatic embryogenesis in culture of *Paeonia anomala*. *Phytomorphology*, 1994, vol. 44, № (3-4), 151-157.

Bugara A.M. Cytochemistry of meiosis. In: Batygina T.B., editor. *Embryology of Flowering Plants. Terminology and Concepts. Vol. 1. Generative organs of flower.* St-Petersburg: Mir i Semya; 1994; 76-81.

Button J.; Kochba J.; Borman C.H. Fine structure of an embryoid development from embryogenic ovular callus of "Shamouti" orange (*Citrus sinensis* Osb.). *J. Exp. Bot*., 1974, vol. 25, №85, 446-457.

Buvat R. Le méristème apical de la tige. *Ann. Biol. Ser*., 1955, vol. 31, 595-656.

Byrne M.E.; Kidner C.A.; Martienssen R.A. Plant stem cells: divergent pathways and common themes in shoots and roots. *Current Opinion in Genetics and Development*, 2003, vol. 13, Issue 5, № 10, 551-557.

Carmichael J.S., Friedman W.E. Double fertilization in *Gnetum gnemon*: The relationship between the cell cycle and sexual reproduction, *Plant cell*, 1995, 7, 1975-1988.

Carmichael J.S., Friedman W.E. Double fertilization in *Gnetum gnemon* (*Gnetaceae*): Its bearing on the evolution of sexual reproduction within the *Gnetales* and the anthophyte clade, *Am. J. Bot*., 1996, 83, 767-780.

Clark S.E. Organ formation at the vegetative shoot meristem. *Plant Cell*, 1997, vol. 9, № 7, 1067-1076.

Clowes F.A.L. The promeristems and the minimal constructional center in grass root apices. *New Phytol*., 1954, vol. 53, 108-116.

Cook R. A note on embryo rejuventation. *Jour. Hered*., 1938, 29, 419-422.

Créte P. La polyembryonie chez le *Lobelia syphiliyica* L. *Bull. Soc. Bot. France.*, 1938, vol. 85, № 7/8, 580-538.

Czapik R. Theoretical aspects of apogamety in angiosperms. *Bull. Pol. Acad. Sci., Biol. Sci.*, 1997, vol. 45, №2-4, 57-64.

Dogra P.D. The Embryology, Breeding Systems and Sterility in Cupressaceae – A Monograph; 1984; V. VI; 1-126.

Dondua A.K. Developmental Biology. In two volumes. St-Petersburg: Publishing House of St-Petersburg State University; 2005.

Dostal R. Growth correlations in *Bryophyllum* leaves and exogenous growth regulations. *Biol. Plantarum*, 1970, vol. 12, 2, 125-133.

Erdelska O. Polyembryony in maize – histological analysis. *Polish J. Bot.*, 1996, vol. 65, №1-2, 123-124.

Ernst A. Bastardierung als Ursache der Apogamie im Pflanzenreich. Eine Hypothese zur experimentellen Vererbungus-und Abstammungslehre. Jena: G. Fischer; 1918.

Fagerlind F. Die Samenbildung und die Zytologie bei agamospermischen und sexuellen Artenvon *Elatostema* und einigen nahestetehenden Gattungen nebst Beleuchtung einiger damit zusammenhängender Probleme. *K. Svenska Vet.-Akad.*, 1944, Handl. III, Bd. 21, Heft 4, 1-130.

Fagerlind F. Sporogenesis, Embryosackentwiclung und pseudogame Samenbildung bei *Rudbeckia laciniata* // Acta Horti. Berg., 1946. 14)3). 39-90.

Fischer Ch.; Speth V.; Fleig- Eberenz S.; Neuhaus G. Induction of zygotic polyembryos in wheat: influence of auxin polar transport. *The Plant Cell*, 1997, vol. 9, 1767-1780.

Frost H.B. Nucellar embryony and juvenile characters in clonal varieties of *Citrus. Jour. Hered.*, 1938, vol. 29, 423-432.

Gamborg O.L.; Miller R.A.; Ojima K. Nutrient requirements of suspension cultures of soybean root cells. *Exp Cell Res*, 1968, vol. 50, 151-158.

Gamburg K.Z. Phytohormones and cells. Moscow: Nauka; 1970; 104p.

Garcês H.M.P.; Champagne C.E.M.; Townsley B.T.; Park S.; Malhó R.; Pedroso M.C.; Harada J.J.; Sinha N.R. Evolution of asexual reproduction in leaves of the genus *Kalanchoë. PNAS.*, 2007, vol. 104, №39, 15578-15583.

Gerassimova E.N. Fertilization in Crepis capillaris. Cellule, 1933, 42(1), 103-148.

Gerassimova-Navashina E.N. The mitotic hypothesis of double fertilization, *Rept. Acad. Sci. USSR*, 1947, 57, 395-492.

Gerassimova-Navashina E.N. On the gametophyte and main features of development and functioning of reproducing elements in the angiosperms. Problems of Botany, 1958, 3, 125-167.

Gerassimova-Navashina H.N. A contribution to the cytology of fertilization in flowering plants, *Nucleus*, 1960, 3, 111-120.

Gerassimova-Navashina H.N. On the periods of mitotic cycle of gametes in angiosperms, *Bull. Moscow State Univ.*, Ser. Biol., 1971a, 3, 93-94.

Gerassimova-Navashina E.N. Process of double fertilization in angiosperms and mitotic cycle of the cell, *Phytomorphology*, 1982, 32, 222-233.

Gilbert S.F.; Opitz J.M.; Raff R.A. Resynthesizing evolutionary and developmental biology. *Developmental Biology,* 1996, vol. 173, 2, 357-372.

Gilbert S.F. Development biology. 8-th ed. Sunderland, MA: Sinauer Associates, Inc., 2006, 751p.

Gordeeva O.F., Mitalipova M.M. Pluripotent stem cells: maintenance of genetic and epigenetic stability and prospects of cell technologies. *Russian Journal of Developmental Biology,* 2008, vol. 39, 6, 325-336.

Gordeeva O.F.; Mitalipova M.M. Pluripotent stem cells: maintenance of genetic and epigenetic stability and prospects of cell technologies. *Ontogenesis,* 2008, vol. 39, 6, 325-336.

Grant V. Population structure in relation to macroevolution. *Biol. Zentralblatt*, 1977, Bd. 96, 129-139.

Grant V. Plant Speciation. N.-Y.: Columbia Univ. Press; 1981.

Groβ-Hardt R.; Laux T. Stem cell regulation in the shoot meristem. *J. Cell Sci.*, 2003, vol. 116, № 9, 1659-1666.

Guttenberg H. Grundzüge der Histogenese der höherer Pflanzen. I. Die Angiospermen. Berlin-Nikolassee; 1960.

Guignard J.L., Mestre J.C. L'embryon des Gramineés. Phytomorphology, 1971, 20, 2, 190-197.

Haccius B. Weitere Untersuchungen über Somatogenese aus den Suspensorenzellen von *Eranthis hiemalis* Embryonen. *Planta*, 1965, vol. 64, № 3, 219-224.

Haccius B. Question of unicellular origin of non-zygotic embryos in callus cultures. *Phytomorphology*, 1978, vol. 28, 74-81.

Hall J.G. An embryological study of *Limnocharis emarginata. Bot. Gaz.*, 1902, № 33, 214-219.

Hanstein J. Die Entwicklung des Keimes der Monokotylen und Dikotylen // Bot. Abh. 1870. Bd 1. P. 1-122.

Higashiyama T., Kuroiwa H., Kawano S., Kuroiwa T. Kinetics of double fertilization in Torenia fournieri based on direct observations of the naked embryo sac, Planta, 1997, 203, 101-110.

Hodgson R.W.; Cameron S.H. Effects of reproduction by nucellar embryony on clonal characteristics in *Citrus. Jour. Hered.*, 1938, vol. 29, 417-419.

Hofmeister W. Vergleichende Untersuchungen der Keimung, Entfaltung und Fruchtbildung der höherer Kryptogamen und der Samenbildung der Coniferen. Leipzig: Verlag von Friedrich Hofmeister; 1851.

Howe M.D. A morphological study of the leaf notches of *Bryophyllum calycinum. Amer. J. Bot.*, 1931, vol. 18, № 5, 387-390.

Isaeva V.V.; Akhmadieva A.V.; Aleksandrova Ya.N., Shukalyuk A.I. Morphofunctional organization of reserve stem cells providing for asexual ans sexual reproduction in Invertebrates. *Russian Journal of Developmental Biology,* 2009, vol. 40, 2, 83-96.

Ilyina G.M. Comparative embryological study of poppies family (Papaveraceae Juss.) in connection with its position in angiosperm system. PHd theses. Moscow; 1968.

Ivanov V.B. Problem of stem cells in plants. *Ontogenesis,* 2003, vol. 34, № 4, 253-261.

Ivanov V.B. Stem cells in the root and the problem of stem cells in plants. *Ontogenesis,* 2007, vol. 38, № 6, 406-419.

Ivanova-Kazas O.M. Asexual propagation. Leningrad: Publishing House of Leningrad State University; 1977, 240p.

Jeffrey E.C. Polyembryony in *Erythronium americanum. Ann. Bot.*, 1895, № 9, 537-541.

Johansen D.A. Plant embryology. Embryogeny of the Spermatophyta. Waltham;1950.

Johnson M. A. The origin of the pseudo-bulbils *Kalanchoë daigremontiana. Bull. Torrey Bot. Club.*, 1934, vol. 61, 355-366.

Kanaev I.I. Twins and genetics. Leningrad: Nauka; 1968.

Kerner von Marilaun A. Editor E.W. Oliver. The natural history of plants; their forms, growth, reproduction and distribution. New York: H. Holt and Co.; 1895-1896; vol. 2.

Kerstetter R.A., Hake S. Shoot meristem formation in vegetative development // Plant Cell. 1997. Vol. 9. P. 1001-1010.

Khan R. Contributions to the morphology of *Ephedra foliate* Bioss. II. Fertilization and Embryogeny // *Proc. Nat. Acad. Sci. Ind.* 1943. Vol. 13, № 4, P. 357-375.

Kolesova G.E.; Batygina T.B. Nelumbonaceae family. In: *Comparative anatomy of seeds.* Leningrad: Nauka; 1988; Vol. 2; 157-162.

Koltsov N.K. Selected works. Moscow: Nauka; 2006, 295p.

Konar R.N.; Nataraja K. Experimental studies in *Ranunculus sceleratus* L. Development of embryos from the stem epidermis. *Phytomorphology*, 1965, Vol. 15, №2, 132-137.

Konar R. N.; Thomas E.; Street H. E. Origin and structure of embryoids arising from epidermal cells of the stem of *Ranunculus sceleratus* L. *J. Cell Sci.*, 1972, vol. 11, 77-93.

Kordyum E.L. Peculiarities of early ontogenesis of ovule with different type of archesporium in some angiosperms. // *Cytology and genetics*. 1968. Vol. 2. № 5. P. 415-428.

Korobova S.N. The movement of sperms in a pollen tube and an embryo sac of angiosperms, in M.S. Yakvlev (ed.) *Actual Problems of Embryology of Angiosperms*, Nauka, Leningrad, 1979, pp. 5-19.

Korotkova G.P. Origin and evolution of ontogenesis. Leningrad: Publishing House of Leningrad State University; 1979.

Korotkova G.P., Tokin B.P. Embryology and genetics. Disputable questions. Leningrad: Publishing House of Leningrad State University; 1977, 63p.

Kruglova N.N., Batygina T.B., Gorbunova V.Yu., Titova G.E., Seldimirova O.A. Embryological bases of wheat androcliny:Atlas / Ed. by Shamrov. Moscow: Nauka, 2005, 99p.

Land W.J. Spermatogenesis and oogenesis in *Ephedra trifurcate* // *Bot. Gaz.* 1904. Vol. 38, № 1, P. 1-18.

Levina R.P. Reproductive biology of seed plants. Review of the problem. Moscow: Nauka; 1981.

Maheshwari P. An Introduction to the Embryology of Angiosperms. N.-Y.: McGraw-Hill Book Company; 1950.

McVeigh J. Regeneration in *Crassula multicava. Amer. J. Bot.*, 1938, vol. 25, 7-11.

Medvedev S.S. Physiology of plants. St-Petersburg: Publishing House of St-Petersburg State University; 2004, 335p.

Meyen S.V. Basics of paleobotany. Reference manual. Moscow: Nedra; 1987, 403p.

Modilevksy Ya.S. On embryos of *Allium odorum* L. *Reports of Kiev Botanical Garden*, 1931, №(12-13). 27-48.

Mogensen H.L. Double fertilization in barley and cytological explanation for haploid embryo formation, embryoless caryopses, and ovule abortion, Carlsberg Res. Commun., 1982, 47, 313-354.

Nakajima, K., Benfey, P.N. Signalling in and out: control of cell division and proliferation in the shoot and root. *The Plant Cell*, 2002, vol. 14 (Suppl.), 265-S276.

Naumova T.N. Apomixis in angiosperms. Nucellar and integumental embryony. Boca Raton: CRC Press Inc., 1993. 144 p.

Navashin, S.G. New observations of fertilization in *Fritillaria tenella* and *Lilium martagon*, *The Diary of X Symp. of Russian naturalists and physicians*, Kiev, 1898, 6, 16-21.

Navashin S.G. On fertilization in Asteraceae and Orchidaceae, *Bull. Imp. l'Acad. Sci. St. Petersburg*, 1900, 13, 335-340.

Navashin, S.G. Results of revision of the fertilization processes in *Lilium martagon* and *Fritillaria tenella*, Selected works, Moscow-Leningrad, Publishing House of AS USSR, 1951, v.1, 188-192.

Naylor E.E. The morphology of regeneration in *Bryophyllum calycinum*. *Amer. J. Bot.*, 1932, vol. 19, № 1, 32-40.

Nikolaev S.V.; Penenko A.V.; Lavrekha V.V.; Melsness E.D.; Kolchanov L.A. Model investigation of CLV1, CLV2, CLV3 and WUS proteins in regulation of shoot apical meristem structure. *Ontogenesis,* 2007, vol. 38, № 6, 457-462.

Norstog K. Early development of the barley embryo: Fine structure. *Am. J. Bot.*, 1972, 59, 123-132.

Periasamy K. A new approach to the classification of angiosperm embryos. *Proc. Indian Acad. Sci.*, 1977, vol. 9, 1001-1010.

Pisyaukova V.V. Elements of morphological plant evolution. Textbook. Leningrad Publishing House of Leningrad state pedagogical institute; 1980.

Pijl L. van der. Princeples of dispersal in higher plants. Berlin etc.: Springer-Verlag; 1969.

Poddubnaja-Arnoldi V.A. Polyembryony in the Orchids. *Bull. Main Bot. Garden USSR Acad. Sci*, 1960, vol. 36, 56-61.

Poddubnaja-Arnoldi V.A. Cytoembryology of angiosperms. Moscow: Nauka; 1976.

Polevoy V.V. Physiology of plants. Moscow: Vyschaya Shkola, 1989, 464p.

Potten C.S.; Loeffler M. Stem cells: attributes, cycles, spirals, pitfalls and uncertainties. Lessons for and from the crypt. *Development*, 1990, vol. 110, 1001-1020.

Raghavan V. Origin of the quiescent center in the root of *Capsella bursa-pastoris* (L.) Medik. *Planta*, 1990, vol. 181, 62-70.

Ray S., Park S.S., Ray A. Pollen tube guidance by the female gametophyte, *Development*, 1997, 124, 2489-2498.

Reinert J.; Bajaj Y.P.S. Plant cell, tissue, and organ culture. New York: Springer-Verlag, 1977.

Resende F. Variabilidade fenoe genotipicamente determinada na fisiologia do desenvolvimento dos pseudo-bolbilhos (*Bryophyllum*). *Bol. Soc Port. Ci. Nat.*, 1954, vol. 20, №1, 78.

Rodin A. Reflections about an art. Memoirs of contemporaries. Moscow: Respublika, 2000, 367p.

Rudskiy I.V.; Batygina T.B.; Titova G.E. Determination mechanisms of mature plant construction in embryogenesis and germination in three species of Araceae Juss. family. Abstracts of the XVII International Botanical Congress, 17-23 July 2005, Austria, Vienna; 2005; 305.

Sachs J. On the physiology of plants. 1887.

Savina G.I. Fertilization in *Cypripedium calceolus* L. *Botanical Journal,* 1964, vol. 49, №9; 1317-1322.

Saxton W.T. Contributions to the life-history of *Actinostrobus pyramidalis. // Ann. Bot.* 1913. Vol. 27, P. 321-345.

Seydoux G., Braun R.E. Pathways to totipotency: lessons from germ cells. Cellule, 2006,127 (5), 891-904.

Schmitz F. Die Blüten-Entwicklung der Piperaceen // Hanstein Bot. Abh. 1872. Bd 2. S. 1-74.

Seldimirova O.A.; Titova G.E. Peculiarities of secondary embryogenesis in the culture *in vitro* of spring soft wheat anthers. Materials of II International School for young scientists "Embryology, genetics and biotechnology". Ufa: Publishing House of Bashkir state university; 2007. 105-106.

Serebryakova T.I. Bud as the stage of the stem development // Abstract of the VII RBS congress. L. 1983. 232-233.

Serebryakova T.I.; Voronin N.S.; Elenevsky A.G.; Batygina T.B.; Shorina N.I.; Savinykh N.P. Botany with the bases of phytocoenology: anatomy and morphology of plants. Moscow IKC "Academkniga". 2006.

Severtzov, A.N. Morphological regularities of evolution, Moscow-Leningrad, Publishing House of AS USSR, 1939, 601p.

Severtzov A.S. Directionality of evolution. Moscow: Publishing House of Moscow State University; 1990, 270p.

Shamrov I.I. Ovule primordium. In: Batygina T.B., editor. *Embryology of Flowering Plants. Terminology and Concepts. Vol. 1. Generative organs of flower.* St-Petersburg: Mir i Semya; 1994; 132-134.

Shamrov I.I., Dyachuk T.I., Batygina T.B., Dyachuk P.A. Embryoidogenous type of asexual propagation and classification of anomalies in culture of anther on the example of wheat. Materials of International Conference "Biology of Cultivated Cells and Biotechnology", Novosibirsk, 1988, 210-211.

Shmalgausen I.I. Regulation of morphogenesis in individual development. Moscow, 1964, 136p.

Shmalgausen I.I. Factors of evolution. Theory of stabilizing selection. Moscow: Nauka; 1968, 451p.

Singh H., Oberoi J.P. A contribution to the life history of *Biota orientalis* Endl. // *Phytomorphology*. 1962. V. 12, №4, 373-393.

Sinnott E.W. Plant morphogenesis. New York, Toronto, London: McGraw-Hill Book Company, Inc., 1960, 603p.

Smirnova O.V. Peculiarities of vegetative propagation of herbs in oak forests in connection with questions of self-maintenance of population. In: Age content of flowering plant populations in relation to their ontogenesis. Moscow: Publishing House of Moscow state pedagogical institute; 1974; 168-195.

Smith A. A glossary for stem-cell biology. Edinbourgh UK: Nature, 2006, 441. p. 1060.

Snegirevskaya N.S. Materials towards morphology and systematics of genus *Nelumbo* Adans. *Transactions of Komarov Botanical Institute of USSR AS,* 1964, vol. 1, №13, 104-172.

Souèges R. L'hypophyse et l'epiphyse: les problems d'histogenese qui leur sont lies // *Bull. Soc. Bot. France*. 1934. Vol. 81. P. 737-748, 769-778.

Souèges R. Exposes d'embryologie et de morphologie végétales. VII. Les lois du dèveloppement // *Act. Sci. Industr.* 1937. Vol. 521. P. 1-94.

Souèges R. Exposes d'embryologie et de morphologie végétales. X. Embryogénie et classification. Deuxieme fascicule: Essai d'un systeme embryogénique (Partie générale). *Act. Sci. Industr.*, 1939, 85.

Swamy B.G.L. Gametogenesis and embryogeny of *Eulophea epidendrae* Fischer. *Proc. Natl. Inst. Sci., India.*, 1943, № 9, 59-65.

Swamy B.G.L. Agamospermy in *Spiranthes cernua. Lloydia*, 1948, vol.11, №3, 149-162.

Takhtajan A.L. Bases of evolutionary morphology of angiosperms. Moscow-Leningrad: Nauka; 1964.

Takhtajan A.L. Principles of organization and transformation of complex systems: system approach. St-Petersburg; 1998.

Tchaylachyan M.Ch. Regulation of flowering in higher plants. Moscow, 1988, 560p.

Teryokhin E.S. Parasitic flowering plants. Evolution of ontogenesis and mode of life. Leningrad: Nauka; 1977.

Teryokhin E.S. Problems of evolution seed plant ontogenesis. In: Petrov Yu.E., editor. *Transactions of Komarov Botanical Institute of USSR AS.* vol.2. 1991.

Teryokhin E.S., editor. Problems of reproductive biology of seed plants. *Transactions of Komarov Botanical Institute RAS.* Vol.8; 1993a; 15-25

Teryokhin E.S. Seed and seed propagation. St-Petersburg: Mir i Semya; 1996, 376p.

Theise N.D.; Krause D.S. Toward a new paradigm of cell plasticity. *Leukemia*, 2002, № 16, 542-548.

Thomson J.A., Marshal V.S. Primate embryonic stem cells. *Cur. Top Develop. Boil.*, 1998, vol. 38, 133-165.

Timofeev-Resovsky N.V. From the history of problem of interrelation of micro- and macroevolution. Materials of symposium "Micro- and macroevolution", Tartu, September 2-5, 1980. Tartu, 1980, 7-12.

Titova G.E. Hypophysis. In: Batygina T.B., editor. *Embyology of Flowering Plants. Terminology and Concepts. Vol. 2. Seed.* St-Petersburg: Mir i Semya; 1997; 352-356.

Tokin B.P. Regeneration and somatic embryogenesis. Leningrad: Leningrad state University Press; 1959.

Tokin B.P. General embryology. Moscow: Vyschaya Shkola, 1987, 480p.

Tyrnov V.S. Haoloidy in plants: scientific and applied significance. Moscow: Nauka, 1998, 53p.

Upadhyay N., Batygina T.B. Embryogenesis and gemmorhizogenesis in *Rauwolfia*. Abstracts of XI International symposium "Embryology and Seed Reproduction", 3-7 July, 1990, Leningrad; 1990; 178.

Upadhyay N., Makoveychuk A. Yu., Nikolaeva L.A.; Batygina T.B. Organogenesis and somatic embryogenesis in leaf callus culture of *Rauwolfia caffra* Sond. // Journal of Plant Physiology, 1992. Vol. 140, № 2. P. 218-222.

Van den Berg C., Willemsen V., Hage W. Cell fate in *Arabidopsis* root meristem determined by directional signaling // Nature. 1995. Vol. 378. P. 62-65.

Van den Berg C., Weibeek P., Scheres B. Cell fate and cell differentiation status in the *Arabidopsis* root // Planta. 1998. Vol. 205. P. 483-491.

Veit B. Determination of cell fate in apical meristems. *Curr. Opin. Plant Biol.,* 2004, vol. 7, 57-64.

Vinogradova G. Polyembryony in Allium ramosum L. and Allium Schoenoprasum L. (Alliaceae) // PhD. Thesis. Komarov Botanical Institute RAS. S.Petersburg 2009. 22p.

Waddington C.H. New patterns in genetics and development. New York, London: Columbia University Press, 1962, 259p.

Warden J. Cytological observation of first phases of induced leaf-plantlet development in *Bryophyllum crenatum*. Their relation to gene activation. *Rev. Biol.*, 1968, vol. 6, 357.

Weigel D.; Jürgens G. Stem cells that make stems. *Nature*, 2002, vol. 415, 751-754.

Winkler H. Fortpflanzung der Gewächse. VII. Apomixis. Handwörterbuch der Naturwissenschaft. Jena, 1934, vol. 4, 451-461.

Winslow T. Stem cells. Scientific progress and future research directions. Bethesda: Nat. Inst. Health, 2001

Yakovlev M.S. On unity of embryogenesis of angiosperms and gimnosperms. *Transactions BIN AS USSR, ser. VII, 1951*, 2, 231-242.

Yakovlev M.S. Paeoniaceae family. In: Yakovlev M.S., editor. *Comparative embryology of flowering plants. Phytolaccaceae-Thymelaeaceae*. Leningrad: Nauka; 1983. 70-77.

Yakovlev M.S.; Ioffe M.D. On some peculiar features in the embryogeny of *Paeonia* L. Phytomorphology. 1957. V. 7 N 1. 74-82.

Yakovlev M.S.; Ioffe M.D. Further research of new type of angiosperm embryogenesis. *Botanical Journal,* 1961, vol. 46, №10, 1402-1421.

Yarbrough J.A. Anatomical and developmental studies of the foliar embryos of *Bryophyllum calycinum. Amer. J. Bot.*, 1932, vol. 19, № 6, 443-453.

Yarbrough J.A. History of leaf development in *Bryophyllum calycinum* // Amer. J. Bot. 1934. 21(8). 467-482

In: Daughter Cells
Editors: A. Hitomi et al., pp. 129-161
ISBN: 978-1-60876-790-8

Chapter III

MORPHOGENESIS OF *DROSOPHILA MELANOGASTER* MACROCHAETES: CELL FATE DETERMINATION FOR BRISTLE ORGAN

***D. P. Furman*[1,2] *and T. A. Bukharina*[1]**
[1]Institute of Cytology and Genetics, Siberian Branch, Russian Academy of Sciences, Novosibirsk, Russia;
[2]Novosibirsk State University, Novosibirsk, Russia.

ABSTRACT

Formation of specialized spatial structures comprising various cell types is most important in the ontogenesis of multicellular organisms. An example is the *D. melanogaster* bristle organs. Bristles (micro- and macrochaetes) are external sensory organs, elements of the peripheral nervous system, playing the role of mechanoreceptors. Their comparatively simple organization comprising only four specialized cells and a common origin of these cells make macrochaetes a convenient model for studying cell differentiation.

The four cells forming bristle organ result from two successive divisions of a single cell, sensory organ precursor (SOP) cell. The number of macrochaetes on drosophila body corresponds to the number of SOP cells.

The morphogenesis of macrochaetes comprises three stages, the first two determining a neural fate of the cells. The third stage is cell specialization into components of the bristle organ—neuron, thecogen, tormogen, and trichogen.

Development of each bristle commences from segregation of proneural clusters, of 20–30 cells, from the massif of undifferentiated cells of the wing imaginal disc. At this stage, each cluster cell can potentially become a SOP cell. At the second stage, the only SOP cell and its position are determined within each cluster. Finally, two asymmetric divisions of the SOP cell with subsequent differentiation of the daughter cells gives the bristle organ.

Several dozens genes are involved in the control of macrochaete morphogenesis. The main component of this system is the proneural genes of *achaete-scute* complex (*AS-C*). An increased content of proneural proteins fundamentally distinguished the cells that will follow the neural developmental pathway from the disc epidermal cells.

A local *AS-C* expression, initiated at specified disc sites by specific transcription factors, determines the number and topology of proneural clusters. The expression of *AS-C* genes, continuing in the cells of the cluster, increases the difference in proneural protein content, first, between the cluster cells and then, between the cluster cells and the single SOP cell, where it reaches the maximum level. This process is provided by both the intracellular regulation of *AS-C* gene activity and intercellular events mediated via the EGFR and Notch signaling pathways.

The third stage in macrochaete morphogenesis comprises two successive asymmetric SOP cell divisions, determining the final specialization. The selector genes, in particular, *numb*, *neuralized*, *tramtrack*, and *musashi*, play the key role in cell type specification.

This review systematizes the data on molecular genetic system controlling drosophila bristle morphogenesis and proposes an integral scheme of its functioning.

Introduction

The main mechanism for creation of cell diversity in multicellular organisms is an asymmetric cell division, which produces the daughter cells differing from both the parental cell and one another in their ability to differentiate in a certain direction. The asymmetric division can be autonomous or directed by external signals between the daughter cells or from neighboring cells [1]. Drosophila macrochaetes (bristles) are widely used as a model system for studying the mechanisms involved in the genetic control of asymmetric division. This review systematizes the experimental data on the main genes and their products that control the main stages in macrochaete morphogenesis—formation of the proneural clusters in the ectoderm of wing imaginal disc; determination of the sensory organ precursor cell; and specialization of the cells forming the bristle organ. The roles of the genes of *achaete–scute* complex, EGFR and Notch signaling pathways, and selector genes in implementation of these processes are considered. An integral scheme describing the function of the molecular genetic system controlling the morphogenesis of *D. melanogaster* macrochaetes is proposed.

General Information about the Structure and Morphogenesis of Drosophila Macrochaetes

The bristles (their total number reaches 6000) are a component of the drosophila nervous system, where they play the role of mechanoreceptors.

In their size and localization pattern on the fly head and body, the bristles are divided into macro- and macrochaetes. These two types of mechanoreceptors are somewhat morphologically and functionally different [2,3] yet display the common patterns of development and genetic control.

Microchaetes (the bristles of small size) are numerous, have no stringent localization but are rather located in more or less regular rows. Unlike microchaetes, the number and localization of macrochaetes (large bristles) is stringently fixed, and the bristle pattern which they form is a species-specific for drosophila, which allows this feature to be used as a classification criterion [4,5]. The bristle pattern characteristic of each species has been formed during the evolution via the loss of part of macrochaetes from the common ancestral set [2,3,6]. In particular, 11 pairs of macrochaetes are located on the head, thorax, and scutum. Their positions are so constant that each bristle got its individual name depending on the position.

The genetic control of bristle pattern formation is realized via the system of prepattern, which predetermines the bristle topography, and the system responding to it [7].

The bristle organ (hereinafter referred to as bristle, macrochaete, or mechanoreceptor) comprises the bristle itself, shaft; socket, which encompasses its base; bipolar neuron; and sheath. All these components are the results of specialization of the four cells originating from the only SOP (*S*ensory *O*rgan *P*recursor) cell (Figure 1). The spatial localization of bristles on the body of imago is identical to the localization of SOP cells in the ectodermal layer of the wing imaginal disc, so that the preciseness of bristle pattern depends on the correct positioning of the SOP cells [8].

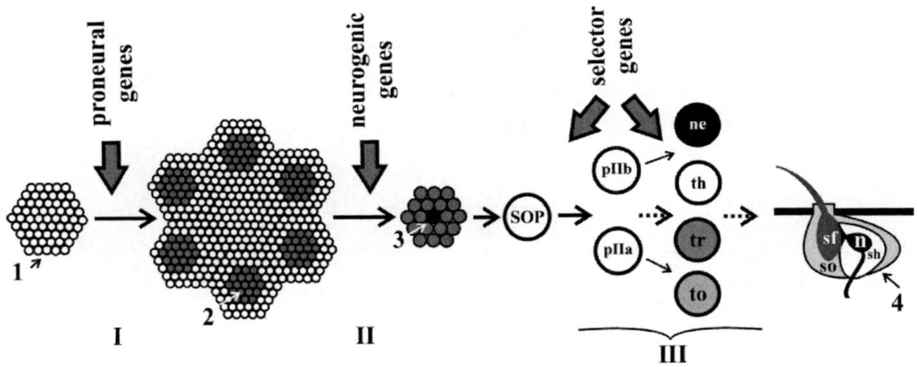

Figure 1. Scheme of development of *D. melanogaster* mechanoreceptors. I, II, and III are the stages of mechanoreceptor development (see text for details); (1) ectodermal cells of the wing imaginal disc; (2) proneural clusters (gray) in the wing imaginal disc; (3) proneural cluster with SOP cell (in the center, shown black); SOP cell; daughter cells of SOP cell (ne, neuron; th, thecogen; tr, trichogen; and to, tormogen); and (4) bristle organ (n, bipolar neuron; sh, sheath; sf, shaft; and so, socket).

Two asymmetric divisions of each SOP cell produce four daughter cells: trichogen, tormogen, neuron, and thecogen. Two daughter cells, trichogen and tormogen, later form the shaft and the socket; these components are well visible on the fly body. The other two cells give the bipolar neuron and sheath, which are located in the body under the bristle. The period of bristle development from the appearance of SOP cells to the completion of its formation takes about 55 h and occurs at the late larval and early pupal stages [8-10].

The formation of macrochaetes comprises three stages, two of which are connected with the decisive moment of their morphogenesis, determination of the SOP cell.

The first stage (Fig. 1, I) consists in segregation of the so-called proneural clusters, groups of 20-30 cells, from the massif of ectodermal cells, forming the wing imaginal discs. The isolation of proneural clusters commences in the late third instar and early pupa, approximately 14 h after puparium formation [9]. At the second stage (Figure 1, II), the SOP cell is determined, and its position within the proneural cluster is specified. The neighboring cells of the cluster become ectodermal. The final stage (Figure 1, III) is division of the SOP cell and specialization of the bristle organ cells.

Each stage has its own genetic control. Three gene groups are involved in the bristle morphogenesis, namely, proneural, which determine the segregation and location of proneural clusters; neurogenic, determining and positioning the SOP cell within the cluster; and selector, which specify the differentiation type for each daughter cell (Figure 1).

The critical factor predetermining the neural cell fate is the threshold level of proneural proteins, the products of *achaete–scute* (*AS-C*) gene complex. The control of this level is provided, on the one hand, by an intracellular regulation of the *AS-C* activity and, on the other, by intercellular events mediated via the EGFR and Notch signaling pathways. This process involves dozens of genes.

This is a brief scheme of the succession of events leading to formation of the bristle pattern.

Some details of these processes are considered below.

THE FIRST STAGE IN BRISTLE ORGAN FORMATION: THE ROLE OF PRONEURAL GENES AND EGFR SIGNALING PATHWAY

The first stage in development of the bristle organ is formation of the proneural cluster, a group of cells with a potential neural fate. The proneural genes of *AS-C* complex play the key role at this stage. Two of these genes, *achaete* (*ac*) and *scute* (*sc*), are key genes in the hierarchy of the genes providing for the bristle morphogenesis. Their expression determines the corresponding competence of the cluster cells [11-17].

Inactivation of the *ac* and *sc* genes causes disappearance of some or all macrochaetes in imagoes. An ectopic expression of these genes due to switching of the ontogenetic mechanism from the epidermal to neural fate results in development of additional bristles at ectopic positions.

AS-C genes encode the bHLH family transcription factors, containing helix–loop–helix amino acid sequences (bHLH) and basic domains, through which they bind to the specific sites CANNTG in the regulatory regions of the genes they control, the so-called E boxes [18]. Both the proneural genes themselves and a number of other genes, in particular *Delta, scabrous, E*(*spl*)*-C, charlatan, senseless,* and *phyllopod* (*phyl*) [19-23], are the targets for these transcription factors.

The gene *cut* plays a special role in the macrochaete morphogenesis. Its expression is also activated by the proneural proteins of *AS-C* complex. The transcription factor Cut is the particular agent that determines the type of developing sensory organ. The presence of this protein in the SOP cell predetermines its development into mechanoreceptor, a bristle. In the other case, the SOP cell gives another element of the peripheral nervous system, a chordotonal organ [24-27].

AS-C occupies approximately 90 kbp in the genome and contains nine transcription units separated by untranscribed regions. The presence of the transcripts T5 (*ac*), T6 (*sc*), T3 (*lethal of scute, l'sc*), and T8 (T1a; *asense, ase*) plays an important role in the macrochaete morphogenesis. Each transcript has its own time and spatial distribution profiles. The expression patterns of the *AS-C* genes are determined by enhancers, located at a distance of up to 100 kbp from this complex [28-32].

One type of enhancers initiates the expression of *ac* and *sc* genes in the cells of individual proneural cluster, and enhancers of the second type trigger this process in each SOP cell [22].

Activities of the enhancers of both types depend on a local combination of the transcription factors, or prepattern factors in the context of Stern's hypothesis [7,11,33]. These factors are the products of both the proneural genes themselves and other genes, in particular, *u-shaped, pannier,* and *iroquois* complex genes (*araucan, caupolican,* and *mirror*) as well as some proteins of the EGFR signaling pathway [34-38].

For example, the *AS-C* complex in the mid-notum is activated by the protein Pannier, whereas in the lateral part, by the proteins encoded by the *iroquois* complex genes. In its turn, the expression of *pannier* and *iroquois* genes is regulated by the products of the genes *decapentaplegic* and *wingless,* respectively, belonging to the EGFR signaling pathway cascade [36,39-41].

Thus, the preciseness in bristle positioning is achieved through a coordinated joint spatially limited expression of *AS-C* genes determined, on the one hand, by the prepattern—a set of the corresponding transcription factors—and, on the other, the system responding to the prepattern—the *AS-C* genes with their set of enhancers.

The cells of proneural cluster differ from the surrounding ectodermal cells in the content of AC–SC proteins: it is essentially higher in the proneural cluster as compared with the neighboring ectodermal cells and reaches the maximal values in the SOP cell. In addition, the SOP cells also accumulate the protein ASE.

Several dozens of genes interconnected by mutual regulation and self-regulation involving the signaling pathways are responsible for the establishment and maintenance of the differences in the content of proneural proteins.

Expression Regulation of AS-C Genes

As AS-C proteins are transcription factors, they are able to regulate transcription, including transcription of the genes that code for them. As transcription factors, they function either as homodimers or within heterodimers with certain other proteins belonging to the HLH class. An efficient interaction with the regulatory regions of target genes requires two basic domains; depending on the presence of a basic domain in the second subunit forming heterodimer, such complexes act as either active positive or active negative regulators of *AS-C* gene expression (Figure 2).

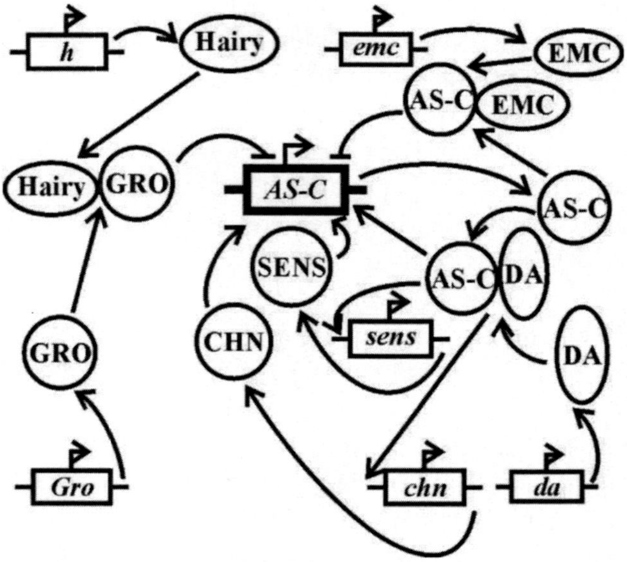

Figure 2. Scheme of intracellular regulation of *AS-C* genes. The gene names are italicized and protein names are given in Roman type: AS-C, achaete-scute complex; sens, senseless; h, hairy; gro, groucho; emc, extramacrochaete; da, daughterless; and chn, charlatan. Arrows denote activation and the lines with stud ends, repression of the corresponding processes.

The heterodimers of AC and SC with the protein DA, the product of gene *daughterless* (*da*), also a bHLH protein, are positive regulators of the *AS-C* gene transcription. The transcription is activated through the binding of such heterodimers to the three E boxes in *AS-C* regulatory region immediately before the TATA box upstream of the transcribed region [12,13,42,43].

The heterodimers of the proteins AS-C and EMC, the product of gene *extramacrochaete,* are passive negative regulators of the *AS-C* expression. EMC is an HLH protein, deprived of the DNA-binding basic domain; therefore, the complexes formed by proneural proteins and EMC are incapable of binding to DNA. Competing with DA for binding to AS-C proteins, EMC decreases the concentration of functional heterodimers, thereby decreasing the transcription level of *AS-C* genes [43-46].

Activity of the proneural genes is regulated not only by the heterodimers containing AC and SC, but also by other factors.

Charlatan (CHN) is a direct transcription activator for the proneural genes. This transcription factor contains zinc finger domains and binds to the cluster-specific enhancers in *AS-C* regulatory region. An abnormal expression of gene *chn* leads to either loss of the macrochaetes (in the case of protein CHN deficiency) or development of extra macrochaetes (when surplus CHN is produced). In turn, the *chn* transcription in the cells of proneural clusters is activated by AS-C proteins [22].

The proteins produced by the neurogenic genes of the *Enhancer of split* (*E*(*spl*)-*C*) complex and *hairy* are direct negative activity regulators of the proneural genes.

E(*spl*)-*C* contains at least 11 transcription units [47]. It is assumed that transcription of the genes belonging to this complex is activated with the involvement of proneural proteins [48]. Seven *E*(*spl*)-*C* transcripts encode the proteins of bHLH type carrying the WRPW tetrapeptides at their C end [49]. On the one hand, this structure makes them able to bind to

DNA and, on the other, to form protein–protein complexes [47]. The target genes are repressed by a direct binding to their regulatory regions of both E(SPL)-C homo- and heterodimers and the E(SPL)-C proteins within the heterodimeric complexes with [SC–DA] [50].

In this process, the product of the neurogenic gene *groucho* (*gro*) is involved as a corepressor. GRO interacts with E(SPL)-C proteins with involvement of seven repeats in the highly conserved C-terminal domain WD4 (Trp–Arg–Pro–Trp) of GRO protein and the WRPW region of E(SPL)-C proteins [37,51,52].

Hairy (H), a bHLH protein with the WRPW C-terminal region, is a direct repressor of *AS-C* gene transcription activity. This transcription factor binds to the so-called C boxes (CACNNG) in the regulatory regions of the target genes [53]. In addition, Hairy needs a cofactor, GRO, for its functional activity [51,54,55]. It is assumed that the [Hairy–GRO] complex is involved in chromatin remodeling or interacts with the complex transcribing the target gene [56-59].

The transcription factor Senseless (SENS) plays a dual role in the activity regulation of proneural genes; this factor contains four zinc finger domains, through which it is capable of binding to both DNA and proneural proteins, direct activators of *sens* gene expression. SENS is an activator or a repressor of the proneural gene transcription depending on its content in the cell. At a low concentration, it acts as a repressor of the proneural gene activity, directly binding to DNA at the corresponding sites of *AS-C* enhancer regions, the so-called S boxes; at a high concentration, when the S boxes are saturated with this protein, free SENS molecules form complexes with proneural proteins and DA, binding within these complexes to the E boxes of *AS-C* enhancer regions and acting as a coactivator of the proneural gene expression. It is assumed that the functional duality of SENS is connected with the conformational state of zinc finger domains and their different affinity for DNA and proneural proteins (Figure 3). Thus, SENS acts as a switch of the proneural gene activity and, consequently, as a switch for the proneural cluster cell fate, thereby enhancing the SOP cell determination [60-64].

It has been recently demonstrated that *senseless* and *daughterless* can assume the functional role of proneural genes in mutants of the *achaete-scute* complex [62].

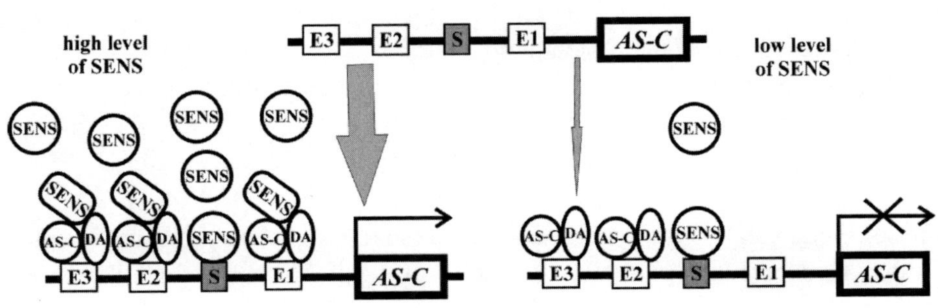

Figure 3. A model of the action of Senseless transcription factor: E1–E3 and S denote E and S boxes, respectively; the remaining designations are as in Figure 2.

EGFR (Epidermal Growth Factor Receptor) Signaling Pathway and its Role in the Regulation of Drosophila Macrochaete Development

Along with a direct intracellular regulation of the proneural gene activity, the EGFR signaling pathway plays an important role in the macrochaete morphogenesis; this pathway brings about the so-called lateral cooperation. The genes constituting this signaling cascade are switched on at all the three stages of the sensor organ development, namely, formation of proneural clusters, SOP cell determination within this clusters, and cell specialization [37,65].

Depending on the performed functions, the proteins of the EGFR signaling pathway can be divided into the following groups:

- Drosophila epidermal growth factor receptor, EGFR or DER;
- Its ligands, Spitz (SPI) and Argos (AOS);
- The proteins involved in the ligand processing, Star (S) and Rhomboid (RHO); and
- The proteins transducing signal from the cell surface into the nucleus (Ras/Raf/MAP kinase cascade and Pointed).

The transmembrane receptor DER, the only representative of the EGFR receptors in drosophila, belongs to the tyrosine kinase receptor family, the proteins with intrinsic kinase activity. The extracellular protein fragment comprises four domains; two of the domains, which provide the binding to ligand, are cysteine-rich [66]. The receptor has two isoforms; however, the precise functions of these isoforms are yet vague [67].

The ligands for this receptor are SPI and AOS. An intracellular signal transduction is activated or blocked depending on the bound ligand [68,69].

Initially, SPI is synthesized as an inactive precursor and accumulated in the endoplasmic reticulum [67,70,71].

Then the precursor is transferred by the protein Star to the Golgi apparatus, where it maturates. In the Golgi apparatus, the [precursor–Star] complex is cleaved by the RHO protease; then the mature ligand is transferred to the cell membrane [67,71,72].

The signal transduction commences from the SPI binding to the DER extracellular domain. Then the intracellular domain of receptor in the recipient cell is phosphorylated and the Ras/Raf/MAP kinase cascade is activated. The intracellular signal transduction from the cell membrane to the nucleus initiates the transcription of gene *pointed* and the subsequent synthesis of two isoforms of the protein Pointed, PNT-P1 and PNT-P2 [73,74]. The details of the intracellular signal transduction are still vague. Both isoforms of Pointed can play the role of transcription factors for proneural genes of the *AS-C* complex, because they are capable of binding to the same sites within the enhancers of *sc* and *ase* genes [75-78]. The corresponding succession of events is shown in Figure 4.

The secreted ligand Argos is a repressor of the EGFR signaling pathway. The gene *argos* is activated simultaneously with the activation of proneural genes, and its expression is observed exclusively in the proneural cluster cells. Secretion of the ligand and its binding to the receptor block the EGFR signal transduction into the cells neighboring the cells actively expressing AS-C proteins [37,79] (Figure 4).

Thus, the local differential expression of *AS-C* genes and the EGFR signaling pathway determine a precise location of the proneural cluster and provide accumulation of proneural proteins in the cells of the cluster.

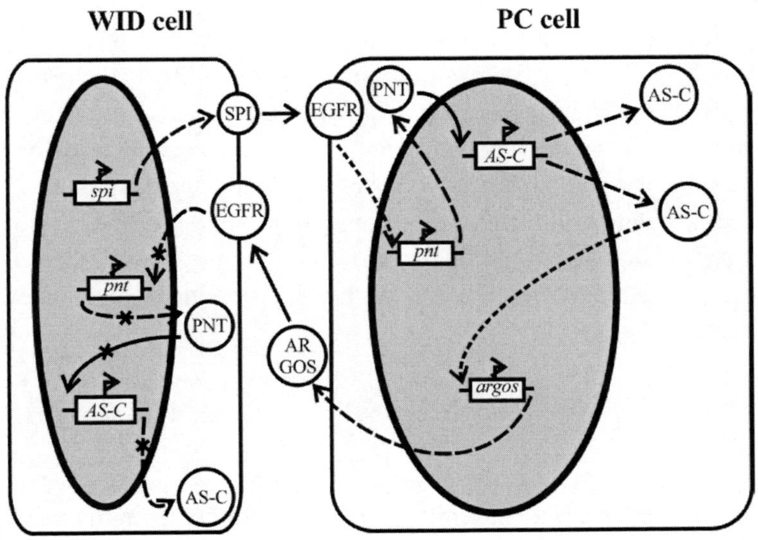

Figure 4. Scheme of involvement of the EGFR signaling pathway in the activation of *AS-C* genes. Ovals denote the cell nuclei; gene names are italicized; protein names are given in Roman type; PC, proneural cluster; WID, wing imaginal disc; solid lines, direct interactions; and dashed lines, indirect.

THE SECOND STAGE IN BRISTLE ORGAN FORMATION: THE ROLE OF NOTCH SIGNALING PATHWAY

The second stage in the macrochaete formation comprises SOP determination and precise positioning within the proneural cluster and is genetically controlled by a group of neurogenic genes. Abnormalities in their function make the majority of the cluster cells or even all cells neural [80]. The obligatory condition for SOP cell determination is that the concentration of AS-C proteins reaches the maximal value as compared with the neighboring cells. The cells with insufficient concentration of proneural proteins remain ectodermal [81-82].

The lateral inhibition, which is mediated by the Notch signaling pathway leading to determination of only one cell as neural, is the master factor at this stage [81,83,84]. In the remaining cells, the activity of proneural genes is repressed by a direct interaction of the E(spl) protein from the Notch cascade with *AS-C* enhancer regions [77].

Several dozens of proteins, products of neurogenic genes, are involved in the Notch signaling cascade; the main proteins can be divided into the following groups:

- The gene encoding Notch receptor, *Notch* (*N*);
- The genes encoding Notch ligands *Delta* (*Dl*) and *Serrate* (*Ser*);
- The genes whose products provide the receptor processing and intercellular signal transduction, namely, *Presenilin* (*Ps*), *kuzbanian* (*kuz*), *polychaetoid* or *tamou* (*pyd* or *tam*), *big brain* (*bib*), and *sanpodo* (*spdo*);
- The genes whose products are involved in the receptor and ligand internalization, namely, *neuralized* (*neur*), *Suppressor of deltex* (*Su*(*dx*)), *shibire*, and *numb;* and

- The genes whose products provide the intercellular signal transduction, namely, genes of the *Enhancer of split* (*E*(*spl*)) and *Bearded* (*Brd-C*) complexes, *mastermind* (*mam*), *Hairless* (*H*), *Suppressor of Hairless* (*Su*(*H*)), and *deltex* (*dx*).

The first three gene groups encode mainly transmembrane proteins and the proteins located on the cell surface and the other two, cytoplasmic and nuclear proteins.

The main components of the Notch signaling pathway are the Notch receptor, its ligand Delta, and the intracellular target, the final point of its action—genes of the *Enhancer of split* (*E*(*spl*)-*C* complex. The products of these genes are the particular repressors of the proneural gene transcription.

Notch Receptor

Notch receptor is a typical transmembrane protein with extra- and intracellular domains. The large extracellular domain contains 36 tandem conservative EGF-like repeats, involved in binding ligands, and three repeats of the cysteine-rich sequence N/LIN 12 [85]. The intracellular domain contains six tandem ankyrin repeats, the region containing 30 glutamine residues (the so-called opa repeat), and the PEST sequence, rich with proline, glutamine, serine, and threonine. It is assumed that the opa and PEST sequences are important for the regulation of protein stability [86].

Initially, Notch is synthesized as a protein with a molecular weight of about 300 kDa. It is then processed by proteases in the Golgi apparatus to give two subunits—*N*otch *E*xtra*C*ellular *D*omain (NECD) and *N*otch *T*rans*M*embrane (NTM) domain. NTM comprises a short extracellular sequence and *N*otch *I*ntra*C*ellular *D*omain (NICD). Then the fragments NECD and NTM form a complex, and the mature receptor composed of two subunits appears on the cell surface [87-90].

Ligands of Notch Receptor

Delta is a transmembrane protein with a large extracellular domain containing nine EGF repeats and the conserved DSL (*D*elta-*S*errate-*L*AG-2) repeat [49,91,92].

The ligand SER, functionally related to the protein DL, has the extracellular domain containing 14 EGF-like repeats, transmembrane domain, and a small intracellular part [93]. SER and DL are alternative ligands for the Notch receptor, because they interact with the same Notch extracellular fragment; however, the possibility of their mutual substitution is essentially limited [85,94]. The glycosyltransferase Fringe, which inhibits the Notch ability to bind SER and enhances its binding to DL, selects the particular ligand [95-98].

The interactions in the pair N–Dl are the key interactions in the intercellular signal transduction within the proneural cluster, providing a correct progress of the process [83]. It is known that the embryos homozygous at the mutations in *Dl* locus die as a result of nervous tissue hyperplasia. *Dl* expression is activated by the proneural proteins AC-SC. The accumulation of DL ligand in the future SOP cell and its binding to N receptors, localized on the membranes of neighboring cells, trigger the mechanism of lateral inhibition [83,99].

Intercellular Signal Transduction

This process is controlled by a large group of genes encoding the proteins mainly localized to the surface of the cell membrane or within the membrane. The mechanisms of the action of *Presenilin* (*Ps*), *kuzbanian* (*kuz*), *polychaetoid* (*pyd*) or *tamou* (*tam*), *big brain* (*bib*), and *sanpodo* (*spdo*) genes are most well studied.

The genes *Ps, kuz,* and *pyd* encode the corresponding proteases, whose functions in the Notch signaling pathway are in the cleaving of the mature N receptor into its extra- and intracellular domains after ligand binding. It is known that the protein KUZ is involved in the cleavage and separation of the complex between Notch extracellular domain and Delta from the cell membrane. After this, PS protease cleaves the Notch intracellular domain from the membrane, and it is transported into the nucleus [56,97,100-104].

The gene *big brain* (*bib*) codes for a transmembrane domain belonging to the tunnel proteins and homologous to aquaporins, involved in the channel formation in the cytoplasmic membrane. It has been demonstrated that mutants at gene *bib* contain a doubled number of sensor neurons, i.e., similar to the other neurogenic genes, its role is in determination of the fate for the proneural cluster cells. It has been also shown that BIB protein is necessary for receiving the lateral inhibition signal or responding to it rather than for generating such signal. The BIB action is synergistic with the DL and N and, presumably, enhances their binding or is involved in the next stage of signal transduction; however, the precise mechanism is yet unknown [105,106].

The gene *spdo* encodes a transmembrane protein, an activator of the Notch signaling pathway. The mutants at this gene develop two neurons instead of the neuron and the glial cell [107]. The protein contains four transmembrane domains at its C end. According to one of the hypotheses, the function of SPDO is in formation of [N–SPDO] complex, the conformation of which allows the protease PS to correctly cleave the N receptor [108]. According to another opinion, the function of SPDO protein is in the regulation of N endocytosis [109].

Some of the neurogenic genes encode the proteins that are not only involved in the Notch signaling pathway, but also directly influence the expression of regulator genes for proneural gene activity. In particular, *pyd* inhibits the proneural gene expression being a direct activator of *extramacrochaete* gene transcription. Mutations in the *pyd* gene (its synonym is *tamou*) leads to the development of numerous additional bristles on the fly head and body [56].

Internalization of Receptor and Ligands

The ubiquitin ligases Neuralized (NEUR) and Suppressor of deltex (SU(DX)) are involved in this process as well as the proteins Dynamin (DYN), encoded by the gene *shibire,* and Numb, which function as activators or inhibitors of this signaling pathway.

NEUR and DYN maintain an inductive state of the signal-sending cell and are positive regulators of the Notch cascade. It has been demonstrated that NEUR acts as an ubiquitin ligase [110,111]. Attaching ubiquitin molecules to the complex of the ligand DL with the

Notch extracellular domain, NEUR initiates endocytosis of the complex by the inducer cell [111-113].

The internalization of the [DL–N extracellular domain] complex into the inducer cell depends also on the protein DYN, displaying a GTPase activity and cleaving the vesicle with the transported proteins from the membrane. Thus, the membrane of inducer cell acquires a vacant place for new ligand molecules. This allows for formation of new ligand–receptor complexes, thereby prolonging the cell inductive state [112,113]. It has been demonstrated that the internalization of this complex influences the intracellular signal transduction in the recipient cell as well; however, the mechanism of this effect is still vague [112,114].

The ubiquitin ligase SU(DX) and protein Numb are negative regulators for the Notch cascade, acting via the internalization and transformation of the N receptor in the recipient cell.

SU(DX) assists the full-sized receptor in entering the recipient cell. In the cell, the complex [SU(DX)–N] triggers the receptor degradation mechanisms. Thus, the receptor outflow from the cell membrane is provided [112,113].

The protein Numb interrupts the Notch signal transduction and blocks the overall signaling pathway. It has been demonstrated that this effect is connected with the inactivation of N receptor caused by its direct interaction with Numb [115]. Recent data demonstrate that Numb induces endocytosis of the full-sized N receptor by the recipient cell. The adapter protein α-adaptin and, presumably, SPDO, forming the complex with receptor, are involved in this process [108,113,116-118].

Intracellular signal transduction

Intracellular signal transduction is provided by the genes of *Enhancer of split* (*E(spl)-C*) and *Bearded* (*Brd-C*) complexes, *mastermind* (*mam*), *Suppressor of Hairless* (*Su*(*H*)), *Hairless* (*H*), and *deltex* (*dx*).

The genes of *E*(*spl*) complex are the intracellular target and final component of the Notch cascade. In the neurogenesis, *E*(*spl*)*–C* acts as an antagonist of proneural genes as activators of neural cell development. The E(spl)-C proteins play a repressor role, inhibiting transcription of proneural genes. It has been demonstrated that the role of E(spl)-C proteins in neurogenesis is in repression of not only the proneural genes themselves, but also their target genes, in particular, *deadpan, neuralized,* and *scabrous*. Moreover, a necessary condition for the activity of E(spl)-C proteins is the presence of cofactor, GRO [48,52].

The *Bearded* complex (*Brd-C*) unites six genes encoding small proteins with α-helix at their N end [119,120]. It is assumed that the BRD family proteins are involved in the regulation of the Notch cascade via the influence on DL endocytosis [121]. The SU(H) and proneural proteins activate *Brd-C* expression [122,123].

The neurogenic gene *mastermind* (*mam*) codes for a nuclear protein composed of alternating acid and basic domains, which suggests its DNA-binding activity [124]. Within the complex with SU(H), which MAM enters only in the presence of the intracellular Notch domain, it acts as a strong transcription coactivator for the target genes of the Notch signaling pathway, in particular, *E*(*spl*)-*C* genes [87,125,126].

The antagonistic genes *Suppressor of Hairless* (*Su*(*H*)) and *Hairless* (*H*) play an exclusively important role in the Notch signaling pathway [127]. The mutants at *H* gene display an abnormal determination of SOP cells, which leads to the absence of bristles at the corresponding positions. *Su*(*H*) acts as a dominant suppressor of the *H* phenotype. An increased expression of these genes phenotypically appears in a similar manner as development of additional bristles.

The SU(H) occupies a key position in the Notch signaling pathway, since it is involved in the signal transduction from the cell membrane to nucleus and is a direct activator of *E*(*spl*)–*C* transcription.

The signal transduction is performed by the [SU(H)–NICD] complex, formed when SU(H) binds to the ankyrin repeats of the receptor [128]. After this complex is transferred to the nucleus to bind the MAM protein, the SU(H) within the complex specifically binds to the consensus sequence 5'-GTGRGAR-3' in the regulatory regions of *E*(*spl*)–*C* genes to initiate their transcription [129]. It is assumed that the binding specificity is determined by the integrase domain of SU(H) [130].

The basic protein Hairless acts as an antagonist of SU(H) as a transcription activator for *E*(*spl*)–C genes. The functionally active repressor of its target genes is the complex formed of H, SU(H), and the corepressor dCtBP (*d*rosophila *C*-*t*erminal *b*inding *p*rotein) and GRO [126]. Hairless interacts with the corepressor dCtBP via the C-terminal sequence PLNLS and with GRO via the sequence YSIHSLLG (the so-called eh1 motif) [131-133]. Both corepressors provide for recruiting histone deacetylase into the complex, thereby decreasing the level of chromatin acetylation in the corresponding regions and, as a result, decreasing their transcription activity. However, it is still unclear whether the interactions of both corepressors with the complex are alternative or the corepressors provide for quantitatively distinct repression levels of the target genes [134].

An important function of the GRO protein is in switching the activities of differently directed signaling pathways. It has been demonstrated that the GRO phosphorylation by MAPK (EGFR signaling pathway) weakens the GRO-dependent repression of *E*(*spl*)-*C* (Notch signaling pathway) [135,136]. GRO is a corepressor for many regulatory molecules; therefore, a change in its activity can influence a wide range of genes whose expression it regulates. In the macrochaete morphogenesis, GRO plays a dual role: it inhibits the activity of *E*(*spl*)-C genes in the complex with [SU(H)–H] in the cells determined as SOP cells, whereas in the complex with the E(spl)-C proteins, GRO represses the *AS-C* genes in the encompassing cells of proneural cluster. In the former case, the content of proneural proteins in the cells of proneural cluster increases and in the latter, decreases [125,133].

Deltex, the product of *deltex* gene, is a basic protein; it contains three domains separated by the regions of glutamine-rich sequence. At the C-end, it has a zinc finger domain, which provides for the interaction with other proteins, in particular, binding with the N intracellular domain in the region of ankyrin repeats [137-139]. It is assumed that the interaction between N and DX enhances accelerated transfer of the complex [SU(H)–NICD] into the nucleus of recipient cell [139]. It has been demonstrated that DX can stabilize the receptor, preventing its degradation in lysosomes and thereby enhancing the preservation of the pool of functionally adequate receptor molecules [113,140]. Thus, DX acts as a positive regulator in the Notch signaling pathway. In addition, Deltex can also play the role of repressor for the

Notch signaling pathway, because it prevents the formation of the complex between Su(H) and Notch intracellular domain, which is necessary to induce expression of the *E(spl)* complex genes [139].

Summarizing these data, we propose the following scheme of the functioning of the Notch signaling pathway in the drosophila macrochaete morphogenesis (Figure 5).

Since all the cells of ectodermal proneural cluster express the proneural proteins AS-C, N receptor, and DL ligand, each of them has a potential to follow either a neural or an epidermal fate and each can be either signal transducer or signal recipient. Random fluctuations in the content of these proteins in cells are increased via feedback loops; consequently, one cell acquires a suprathreshold concentration of the proneural proteins, which activate the synthesis of DL. This cell will become a SOP cell. In the remaining cells, the lateral inhibition, mediated by the Notch signaling pathway, is launched.

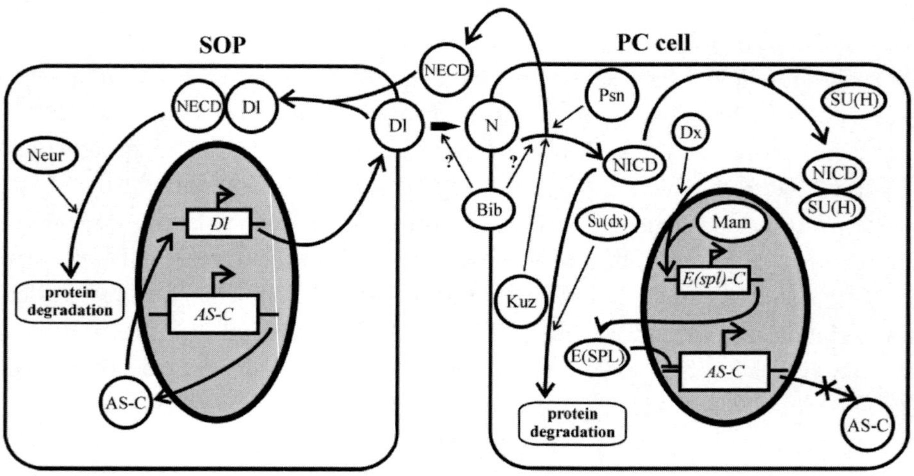

Figure 5. Scheme of the functioning of the Notch signaling pathway in the regulation of transcription activity of *AS–C* genes. See text for abbreviations of gene and protein names; NECD, *N*otch *E*xtra*C*ellular *D*omain; and NICD, *N*otch *I*ntra*C*ellular *D*omain.

The Notch signaling pathway is activated with the binding of the extracellular domains of N receptor, localized on the surface of a signal-receiving cell, and ligand DL, localized on the membrane of a signal-inducing cell. The receptor–ligand interaction takes place in the intercellular space between two neighboring cells of proneural cluster [95,98].

In the recipient cell, the proteases KUZ and PS cleave the N intracellular domain [103,104,112,141]. Then the intracellular domain within the complex with SU(H) is transported to the nucleus, where MAM joins the complex. The signal transduction via the Notch signaling pathway ends with the *E(spl)–C* transcription activation after a site-specific binding of SU(H) to the corresponding regulatory regions. The E(SPL)-C proteins inhibit the transcription of their target genes, first and foremost, the proneural genes, making the recipient cell unable to differentiate according to a neural type [87,142].

The complex between the extracellular domain of receptor and the ligand is transported into the inducer cell, a future SOP cell, where it undergoes a complete degradation. Dynamin and NEUR are involved in internalization of the complex [112,114,143]. A unilateral transduction of Notch signal prohibits appearance of the E(SPL)-C proteins in inducer cell, and the cell continues to synthesize proneural proteins to the level providing for its determination as a SOP cell.

Thus, the functioning of the Notch signaling pathway determines the single cell of proneural cluster that will follow a neural fate, whereas the remaining cells will become epidermal.

The lateral inhibition is efficient for the cells immediately neighboring the presumptive SOP cell. As for the fate of the more distant cells in the proneural cluster, the gene *scabrous* (*sca*) plays an exclusively important role in its determination; the expression of this gene is activated by the heterodimers of AS-C and DA proteins [144,145].

SCA is a secreted protein, with the sequence similar to fibrinogen β- and γ-chains at its carboxyl end [144,146]. It has been found that SCA is necessary for determining and maintaining the characteristics of ectodermal cells. SCA is able to bind N; however, it is not an active ligand for N. In the absence of SCA, the neural fate of the cells not directly contacting the future SOP cell is not blocked. On the other hand, SCA is not necessary for the lateral inhibition of the cells tightly contacting with the SOP cell; the presence of DL is quite enough for this. Since a gradient of SCA distribution is observed within the proneural cluster, it is assumed that this is the factor that determines the size of the region where the inhibitory signal is active. The precise mechanism of the SCA involvement in lateral inhibition is yet vague. Presumably, it is necessary for stabilization of the [N–DL] complex [145].

The Third Stage in Bristle Organ Formation: Asymmetric Cell Division and the Role of Selector Genes

The lateral inhibition ends with determination of the only cell in the proneural cluster as a SOP cell. This is followed by two successive divisions producing four specialized cells—trichogen, tormogen, neuron, and thecogen, which will develop into the components of the bristle organ—the bristle itself, shaft; socket, which encompasses its base; bipolar neuron; and sheath (Figure 1). The main mechanism determining the differences of the daughter cells from one another and the SOP cell is an asymmetric division. The determinative conditions for an asymmetric division of the SOP cell are the time when it enters mitosis, specific positioning of the Numb and Neuralized proteins, and a correct orientation of the spindle relative to these protein determinants.

The selector genes *phyllopod* (*phyl*), *tramtrack* (*ttk*), *musashi* (*msi*), and *prospero* (*pros*) are involved in the asymmetric division. At this stage, two neurogenic genes, *numb* and *neuralized* (*neur*), also play the role of selector genes.

The time moment when the SOP cell enters mitosis is essential for the further fate of daughter cells. The later the division starts, the lower is the probability for daughter cells to follow a neural fate. The duration of premitotic phase depends on the accumulation rate of the adaptor protein Phyllopod in the SOP cell. The expression of *phyllopod* (*phyl*) gene is controlled by the proneural AS-C proteins via the interaction with E boxes in its regulatory region [147].

The genes *numb* and *neuralized* (*neur*) code for membrane proteins. The role of Numb and NEUR in specialization of the daughter cells is determined by their localization in the SOP cell—they are concentrated at only one of its poles. Thus, the very first mitotic division is asymmetric: the distribution of Numb and NEUR in the daughter cells is different due to a predominant segregation of these proteins to one of the cells [113,148,149]. Consequently, the daughter cells also differ in the contents of other proteins involved in the macrochaete morphogenesis, including regulatory proteins, which in turn entails modulations in the activities of the corresponding target genes.

The daughter cell that received Numb and NEUR (pIIb) will follow a neural fate, since Numb blocks the transduction of a Notch signal into this cell, while NEUR enhances transducing the signal into the second cell free of both Numb and NEUR. In the next division, this cell will give rise to the neuron and thecogen (Figure 6).

The absence of Numb and NEUR in the other daughter cell (pIIa) predetermines its ability to receive a Notch signal, which will close a neural fate for it; correspondingly, this cell will give trichogen and tormogen after the next division [113,115] (Figure 6).

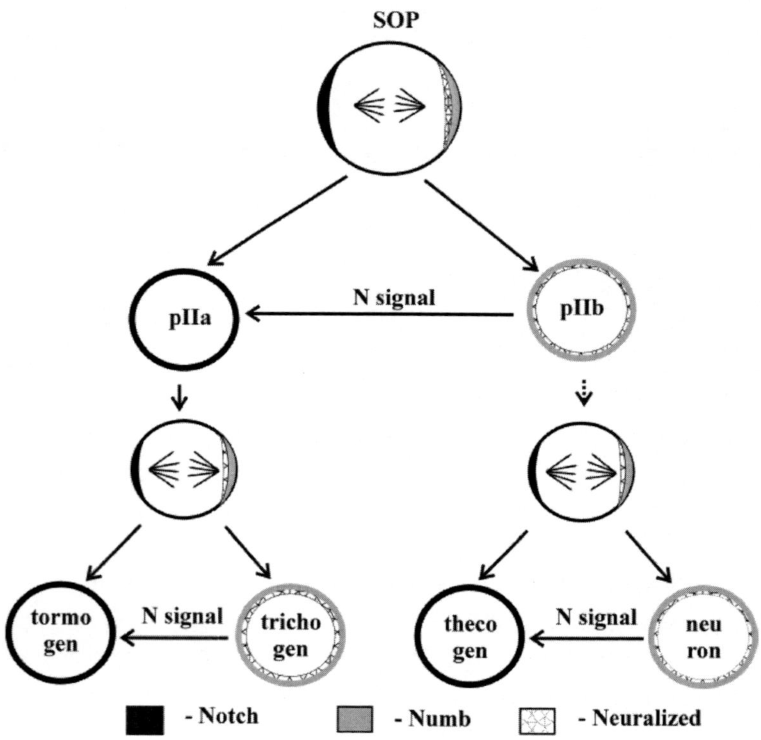

Figure 6. Scheme of the distribution of Numb, Neuralized, and Notch proteins and the directions of N-signal transduction in the asymmetric divisions of the SOP cell and its daughter cells; pIIa and pIIb are the daughter cells of the SOP cell.

The positioning of Numb and NEUR protein determinants in the SOP cell is determined by the specific protein complexes formed at its poles [150]. This positioning is most comprehensively studied by the example of Numb.

One of the mechanisms providing for the asymmetric distribution of Numb in the SOP cell is connected with the protein Partner of Inscuteable (Pins), proteins of G complex, and the transmembrane protein Frizzled (FZ) [150,151-153].

The G proteins have a heterotrimeric structure. They comprise a large α-subunit and two smaller β- and γ-subunits; the bond between α-subunit and βγ dimer is weaker than that between the subunits of the dimer. The complex can be either in a free state or bound to the transmembrane protein Frizzled via the βγ heterodimer.

Under certain conditions, α-subunit can detach from the heterotrimer, namely, when Frizzled binds its ligand Wingless or the α-subunit within the complex interacts with the protein Pins [152]. The result is the βγ heterodimer **(Gβγ)** and a free α-subunit **(Gαi)**, which then can bind to Pins.

In its turn, the dimer **Gαi/Pins** is able to form the complex with a number of proteins, including Numb and a nonphosphorylated form of the protein Lethal(2) giant larvae (LGL), which is stably bound to the cell membrane. Due to this binding, Numb within the complex is kept near the cell membrane [151].

Concurrently, the complex of proteins is assembled at the opposite pole; this complex contains the atypical protein kinase C (aPKC). This enzyme is involved in phosphorylation of LGL, which changes the LGL conformation and, as a consequence, leads to the loss of its binding to the cell membrane, interfering with the Numb localization in this region [150].

The necessary condition of an asymmetric division is a stringently determined position of the spindle relative to the unequal polar distribution of the protein determinants, first and foremost, Numb. The complex of proteins Pins, **Gαi,** and Mud (*Mu*shroom body *d*efect), determining the initial point of spindle formation, play the key role in this process. Due to a polar position of the **Gαi/Pins** dimer, the complex (and, consequently, the spindle) is properly oriented relative to Numb [154].

A correct course of all the described processes provides for an asymmetric pattern of the SOP cell division, which determines further fate of the daughter cells.

The further fate of the daughter cells also depends on the particular selector genes expressing in them.

The gene *ttk* encodes a nuclear protein. Mutations in this gene lead to development of additional neurons at the expense of main bristle organ components. The protein TTK first appears in one of the two daughter cells produced by the SOP cell division, pIIa (Figures 6 and 7). This particular cell after the next division produces trichogen and tormogen. Thus, *ttk* is never expressed in the sensory organ cells following a neural fate. In addition, it has been demonstrated that there is a time lag between the end of pIIa cell division and the commencement of *ttk* expression [155]. The TTK appearance in the cell is determined by activation of the Notch signaling pathway; however, the precise mechanism of this correlation is vague [156,157].

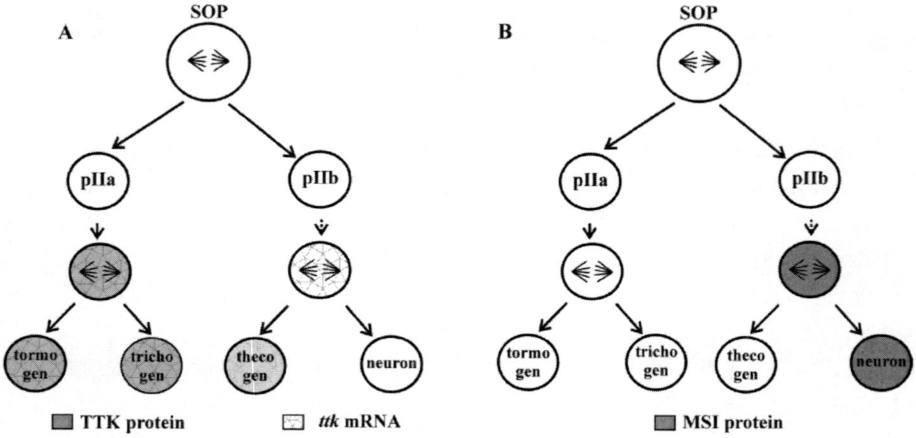

Figure 7. Scheme of the distribution of TTK and MSI proteins and *ttk* mRNA in the asymmetric divisions of the SOP cell and its daughter cells. Distribution of (A) mRNA and TTK protein and (B) MSI protein in the cells of developing bristle organ.

Although TTK is undetectable in the second daughter cell, the contents of *ttk* mRNA in both cells are approximately equal. It has been demonstrated that the differences in the TTK content are connected with the action of another selector gene, *musashi* (*msi*) [156,157].

The gene *msi* codes for a nuclear protein expressed at all the stages of mechanoreceptor development and able to prevent the translation of *ttk* mRNA, specifically binding to its 3'-UTR. Mutations in this gene lead to development of additional glial cell instead of the neuron [156,158,159]. The protein MSI is detectable in both daughter cells after the first division of SOP cell; however, it interferes with the translation of *ttk* transcript only in one of these cells, pIIb, the cell which will give the neural components of the bristle organ, neuron and thecogen. This is the cell that received Numb in the asymmetric division and where the Notch signal is blocked. In the second daughter cell, pIIa, where the Notch signal is not blocked, MSI is degraded and *ttk* mRNA is translated; this cell will give trichogen and tormogen [156].

The gene *pros* encodes a transcription factor containing a homeodomain and the conserved PROS domain at its carboxyl end [160]. Both these domains are currently considered as a single homeo-Pros domain, which is necessary for binding to specific DNA sites in the regulatory regions of target genes [161,162]. Prospero regulates activities of the genes that determine a neural fate of the SOP cell derivatives and cell cycle genes. Among its targets are *achaete, scute, asense, miranda, aPKC, Cyclin E,* and several other genes [163].

The protein PROS is detectable in both the cell nucleus and cytoplasm, and its distribution between the nucleus and cytoplasm is a dynamic process. It has been found that the homeo-Pros domain is responsible for release of this protein from the nucleus; however, the detailed mechanism of this process requires further studies [164]. The rate of the transport of this protein from the nucleus to the cytoplasm correlates with an increase in its phosphorylation level [161].

After division of the SOP cell, this protein is detectable in the nucleus of only one daughter cell, pIIb, which will give the neural components of the bristle organ. In the mitotic division of pIIb cell, PROS is transferred to the membrane at one of the poles, where it colocalizes with Numb. Then PROS appears in both daughter cells of pIIb, neuron and thecogen. The dynamics of PROS content in these cells is opposite: it gradually disappears in the neuron and accumulates in the thecogen. PROS is never detected in the SOP cell; its daughter cell after the first division, pIIa, which later gives tormogen and trichogen; and tormogen and trichogen themselves [165,166] (Figure 8).

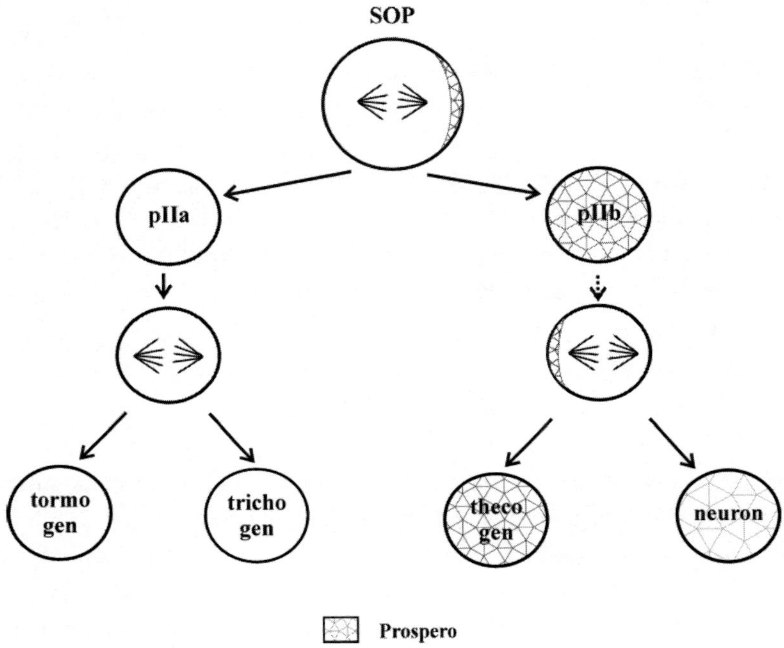

Figure 8. Localization of the protein Prospero in the SOP cell and its derivatives.

The products of the genes *miranda* (*mira*) and *staufen* (*stau*) are involved in the PROS distribution in pIIb and its daughter cells.

The protein Staufen contains the so-called double-stranded RNA binding domain (dsRBD), which can provide for the binding to double-stranded RNA, including the *prospero* mRNA [167]. The protein Miranda joins the formed complex [168]. MIRA interacts with the complex [Staufen/pros mRNA] via its the so-called cargo-binding domain, which is a double helix [169].

MIRA is able to bind not only the complex [Staufen/*pros* mRNA], but also the Prospero protein itself [170] and Numb protein [171].

Since Numb is localized to only one pole of the dividing cell, the fact that Miranda within the complexes [Miranda/Staufen/*pros* mRNA] and [Miranda/Prospero] binds to this protein determines the asymmetric distribution of this protein and *prospero* mRNA in the dividing pIIb cell and an inequality of the daughter cells in their contents [171].

It has been demonstrated that the loss in functional activity of the gene *miranda* changes the pattern of PROS localization in the cell and that *mira* expression coincides with the time of asymmetric division [172]. Thus, Miranda is the particular protein responsible for the Prospero distribution pattern in daughter cells.

As has been demonstrated, the cells that via asymmetric divisions are predetermined by their developmental pathways to different functions in the bristle organ also differ in the expression pattern of genes *bereft (brf)* and *Bar*.

The expression of *brf* is characteristic of pIIa, trichogen, and tormogen, which further give the cuticular structures of bristle organ, shaft, and socket; and the *brf* expression is more pronounced in trichogen. The activity of *bereft* depends on the proteins Cut and TTK. Since the protein BRF has not been detected, it is considered that *brf* encodes an RNA able to bind to mRNA of other genes, preventing their translation; however, its targets are yet unknown [173].

Unlike *brf,* expression of the gene *Bar* is observed in the neural derivatives of the SOP cell, which will later develop into the neuron and its envelope, glia [173,174].

Thus, the asymmetric cell divisions and activities of selector genes determine the further specialization of the SOP daughter cells as different components of the sensory organs.

Conclusions

Development of a bristle pattern on drosophila body is a result of successive limitation of the ontogenetic potential of ectodermal cells in the imaginal disc.

The macrochaete development is controlled via the system of dynamic intracellular and intercellular processes. Functioning of this system is provided by a wide gene network, the components of which are interconnected by the mechanisms of cross-regulation and self-regulation, providing for a fine tuning of their activities. A correct functioning of this system guarantees the development of a full-fledged bristle pattern—a fixed number of macrochaetes at stringently determined positions.

The control of three stages in macrochaete development can be described by the following scheme (Figure 9).

The genes of *achaete-scute* complex are key players in the regulation of macrochaete development performing two functions. First, they initiate the overall bristle development, providing for its first stage—separation of the proneural cell clusters in the ectoderm of imaginal disc. The topography of proneural clusters is specified by a set of certain transcription factors. A mosaic of these factors creates the so-called Stern's prepattern of bristle pattern [7]. The transcription factors activate the expression of *achaete–scute* genes in an addressed manner. The cell competence within the cluster is determined by the threshold level of proneural protein contents in individual cells, which is created and maintained by the control of *achaete–scute* gene expression via the EGFR signaling pathway, self-regulation of their activities by the AS-C/DA heterodimers, and trans-regulation by the interaction with positive (SENSE, CHA, and PNT) and negative (Hairy and EMC) regulators of the transcription activity.

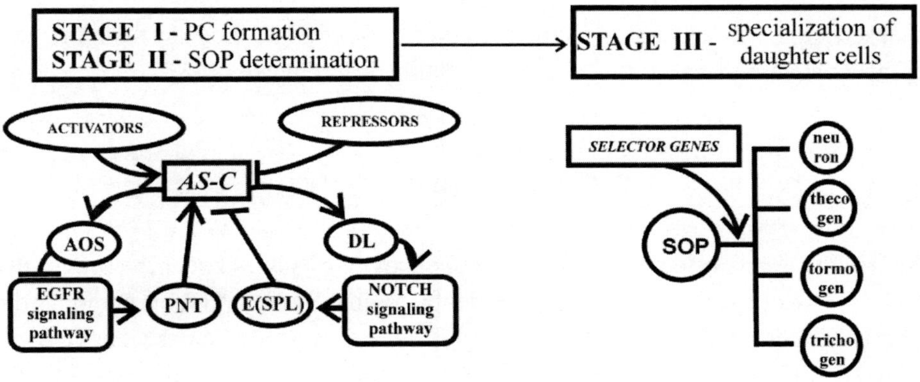

Figure 9. Scheme of the system controlling the macrochaete development in *D. melanogaster:* PC, proneural cluster; SOP, sensory organ precursor cell; *AS-C,* genes of *achaete–scute* complex; PNT, Pointed; E(SPL), proteins of *Enhancer of split* complex; AOS, Argos; and DL, Delta. Arrows indicate the activation events and lines with stub ends, repressor events.

Second, *AS-C* genes are involved in the determination of the SOP cell within the proneural cluster. To determine a cell as a SOP cell, it is necessary that the content of AS-C proneural proteins in this cell exceeds a certain threshold level. Being transcription factors, the proneural proteins activate expression of the gene *Delta,* which triggers a cascade of the Notch signaling pathway genes. The final event in this pathway is the inhibition by the E(SPL)-C repressor proteins of *AS-C* genes and/or their target genes in all the cells of proneural cluster except for the SOP cell, where the content of proneural proteins reaches the required suprathreshold values.

Thus, the circuit of mutual activation–inhibition interactions in the network of proneural and neurogenic genes determining the conditions for local expression in the ectodermal cells of imaginal discs and determining the SOP cells closes at the *AS-C* complex.

An asymmetric division of the SOP cell and specialization of the daughter cells are controlled by selector genes.

The drosophila macrochaetes as a model system for studying the mechanisms of cell specialization are used for over 50 years. During this time, the understanding of molecular genetic organization of the *AS-C* region and its role in the mechanoreceptor morphogenesis has been acquired; the expression pattern of proneural genes has been clarified; the concept of prepattern signals, regarding as transcription factors that interact with *AS-C* enhancers, has been partially specified; the signaling pathways and genes providing for their implementation has been identified; a certain progress in understanding of the mechanisms underlying lateral cooperation, lateral inhibition, and asymmetric division has been reached; the list of genes involved in macrochaete morphogenesis has been considerably expanded; and the functions of many known participants of this process have been detailed.

Nonetheless, neither the complete set of genes involved in the establishment of bristle pattern nor the mechanisms of these programs have been fully clarified. The precise temporal parameters and quantitative characteristics are still unknown for the majority of these processes.

A large volume of experimental data has been so far accumulated; their comprehensive analysis is possible only using the state-of-the-art bioinformatics methods providing for correct description, formalization, and modeling of the overall macrochaete and bristle pattern development. The first steps in this direction are already attempted [175].

Acknowledgments

The work was supported by the Russian Academy of Sciences under the programs no. 15 "The Origin of Biosphere and Evolution of Geobiological Systems" (subprogram 2, direction 5, project "Computer and Experimental Study of Evolution of Molecular Genetic Systems of Organisms") and no. 22 "Molecular and Cell Biology" (project 8 "Systems Biology: Computer and Experimental Approaches"); the program for support of leading scientific schools (grant no. NSh-2447.2008.4 "Bioinformatics and Systems Computer Biology"); Siberian Branch of the Russian Academy of Sciences (interdisciplinary integration project no. 119 "Postgenomic Bioinformatics: Computer Analysis and Modeling of Molecular Genetic Systems"); Federal Agency for Science and Innovations (state contract no. 02.512.11.2332 of June 01, 2009); and Russian Foundation for Basic Research (grant no. 09-04-12209-ofi_m).

References

[1] Hawkins, N; Garriga, G. Asymmetric cell division: from A to Z. *Genes Dev.*, 1998, 12, 3625-3638.

[2] Usui, K; Pistillo, D; Simpson, P. Mutual exclusion of sensory bristles and tendons on the notum of dipteran flies. *Curr. Biol.,* 2004, 14, 1047-1055.

[3] Simpson, P; Marcellini, S. The origin and evolution of stereotyped patterns of macrochaetes on the nota of cyclorraphous *Diptera. Heredity*, 2006, 97, 148-156.

[4] Plunkett, CR. The interaction of genetic and environmental factors in development. *J. Exp. Zool.,* 1926, 46, 181-244.

[5] Sturtevant, AH. Studies on the bristle pattern of Drosophila. *Dev. Biol.,* 1970, 21, 48-61.

[6] Garcia-Bellido, A. From the gene to the pattern: chaeta differentiation. In: Lloyd CW, Rees DA editors. *Cellular controls in differentiation* NewYork: Acad Press; 1981; 281-304.

[7] Stern, C. Two or three bristles. *Am. Sci.*, 1954, 42, 213-247.

[8] Campuzano, S; Modolell, J. Patterning of the *Drosophila* nervous system: the *achaete-scute* gene complex. *Trends Genet.*, 1992, 8, 202-206.

[9] Hartenstein, V; Posakony, JW. Development of adult sensilla on the wing and notum of *Drosophila melanogaster. Development,* 1989, 107, 389-405.

[10] Huang, F; Dambly- Chaudiere, C; Ghysen, A. The emergence of sense organs in the wing disc of Drosophila. *Development*, 1991, 111, 1087-1095.

[11] Ghysen, A; Dambly-Chaudiere, C. From DNA to form: the achaete-scute complex. *Genes Dev.,* 1988, 2, 495-501.

[12] Cabrera, CV; Alonso, MC. Transcriptional activation by heterodimers of the *achaete-scute* and *daughterless* gene products of Drosophila. *EMBO J.,* 1991, 10, 2965-2973.

[13] Villares, R; Cabrera, CV. The *achaete-scute* gene complex of *D. melanogaster*: conserved domains in a subset of genes required for neurogenesis and their homology to *myc*. *Cell*, 1987, 50, 415-424.

[14] Gonzalez, F; Romani, S; Cubas, P; Modolell, J; Campuzano, S. Molecular analysis of the *asense* gene, a member of the *achaete-scute* complex of *Drosophila melanogaster*, and its novel role in optic lobe development. *EMBO J.*, 1989, 8, 3553-3562.

[15] Bertrand, N; Castro, DS; Guillemot, F. Proneural genes and the specification of neural cell types. *Nat. Rev. Neurosci.*, 2002, 3, 517-530.

[16] Kiefer, JC. Proneural factors and neurogenesis. *Dev Dyn.,* 2005, 234, 808-813.

[17] Reeves, N; Posakony, JW. Genetic programs activated by proneural proteins in the developing Drosophila PNS. *Dev. Cell*, 2005, 8, 413-425.

[18] Powell, LM; Zur Lage, PI; Prentice, DR; Senthinathan, B; Jarman, AP. The proneural proteins Atonal and Scute regulate neural target genes through different E-box binding sites. *Mol. Cell Biol.*, 2004, 24, 9517-9526.

[19] Mlodzik, M; Baker, NE; Rubin, GM. Isolation and expression of *scabrous*, a gene regulating neurogenesis in Drosophila. *Genes Dev.*, 1990, 4, 1848-1861.

[20] Hu, X; Lee, EC; Baker, NE. Molecular analysis of *scabrous* mutant alleles from *Drosophila melanogaster* indicates a secreted protein with two functional domains. *Genetics*, 1995, 141, 607-617.

[21] Jafar-Nejad, H; Acar, M; Nolo, R; Lacin, H; Pan, H; Parkhurst, SM; Bellen, HJ. Senseless acts as a binary switch during sensory organ precursor selection. *Genes Dev.*, 2003, 17, 2966-2978.

[22] Escudero, LM; Caminero, E; Schulze, KL; Bellen, HJ; Modolell, J. Charlatan, a Zn-finger transcription factor, establishes a novel level of regulation of the proneural *achaete/scute* genes of Drosophila. *Development*, 2005, 132, 1211-1222.

[23] Chang, PJ; Hsiao, YL; Tien, AC; Li, YC; Pi, H. Negative-feedback regulation of proneural proteins controls the timing of neural precursor division. *Development*, 2008, 135, 3021-3030.

[24] Blochlinger, K; Bodmer, R; Jan, LY; Jan, YN. Patterns of expression of cut, a protein required for external sensory organ development in wild-type and *cut* mutant Drosophila embryos. *Genes Dev.,* 1990, 4, 1322-1331.

[25] Blochlinger, K; Jan, LY; Jan, YN. Postembryonic patterns of expression of *cut*, a locus regulating sensory organ identity in Drosophila. *Development*, 1993, 117, 441-450.

[26] Vervoort, M; Zink, D; Pujol, N; Victoir, K; Dumont, N; Ghysen, A; Dambly-Chaudière, C. Genetic determinants of sense organ identity in *Drosophila*: regulatory interactions between *cut* and *poxn*. *Development*, 1995, 121, 3111-3120.

[27] Brewster, R; Hardiman, K; Deo, M; Khan, S; Bodmer, R. The selector gene *cut* represses a neural cell fate that is specified independently of the *Achaete-Scute-*Complex and *atonal*. *Mech. Dev.*, 2001, 105, 57-68.

[28] Romani, S; Campuzano, S; Macagno, ER; Modolell, J. Expression of *achaete* and *scute* genes in Drosophila imaginal discs and their function in sensory organ development. *Genes Dev.*, 1989, 3, 997-1007.

[29] Cubas, P; de Celis, JF; Campuzano, S; Modolell, J. Proneural clusters of *achaete-scute* expression and the generation of sensory organs in the Drosophila imaginal wing disc. *Genes Dev.*, 1991, 5, 996-1008.

[30] Skeath, JB; Carroll, SB. Regulation of *achaete-scute* gene expression and sensory organ pattern formation in the Drosophila wing. *Genes Dev.*, 1991, 5, 984-995.

[31] Gomez-Skarmeta, JL; Rodriguez, I; Martinez, C; Culi, J; Ferres-Marco, D; Beamonte, D; Modolell, J. Cis-regulation of *achaete* and *scute*: shared enhancer-like elements drive their coexpression in proneural clusters of the imaginal discs. *Genes Dev.*, 1995, 9, 2598-2608.

[32] Modolell, J; Campuzano, S. The *achaete-scute* complex as an integrating device. *Int. J. Dev. Biol.*, 1998, 42, 275-282.

[33] Gomez-Skarmeta, JL; Campuzano, S; Modolell, J. Half a century of neural prepatterning: the story of a few bristles and many genes. *Nat. Rev. Neurosci.*, 2003, 4, 587-598.

[34] Leyns, L; Gomez-Skarmeta, JL; Dambly-Chaudiere, C. *iroquois*: a prepattern gene that controls the formation of bristles on the thorax of Drosophila. *Mech. Dev.,* 1996, 59, 63-72.

[35] Haenlin, M; Cubadda, Y; Blondeau, F; Heitzler, P; Lutz, Y; Simpson, P; Ramain, P. Transcriptional activity of pannier is regulated negatively by heterodimerization of the GATA DNA-binding domain with a cofactor encoded by the *u-shaped* gene of Drosophila. *Genes Dev.*, 1997, 11, 3096-3108.

[36] Garcia-Garcia, MJ; Ramain, P; Simpson, P; Modolell, J. Different contributions of *pannier* and *wingless* to the patterning of the dorsal mesothorax of Drosophila. *Development*, 1999, 126, 3523-3532.

[37] Culi, J; Martin-Blanco, E, Modolell, J. The EGF receptor and N signalling pathways act antagonistically in Drosophila mesothorax bristle patterning. *Development*, 2001, 128, 299-308.

[38] Fujise, M; Izumi, S; Selleck, SB; Nakato, H. Regulation of *dally*, an integral membrane proteoglycan, and its function during adult sensory organ formation of Drosophila. *Dev. Biol.*, 2001, 235, 433-448.

[39] Tomoyasu, Y; Nakamura, M; Ueno, N. Role of dpp signalling in prepattern formation of the dorsocentral mechanosensory organ in *Drosophila melanogaster. Development*, 1998, 125, 4215-4224.

[40] Phillips, RG; Warner, NL; Whittle, JR. Wingless signaling leads to an asymmetric response to decapentaplegic-dependent signaling during sense organ patterning on the notum of *Drosophila melanogaster. Dev. Biol.*, 1999, 207, 150-162.

[41] Calleja, M; Renaud, O; Usui, K; Pistillo, D; Morata, G; Simpson, P. How to pattern an epithelium: lessons from *achaete-scute* regulation on the notum of Drosophila. *Gene,* 2002, 292, 1-12.

[42] Van Doren, M; Ellis, HM; Posakony, JW. The Drosophila Extramacrochaetae protein antagonizes sequence-specific DNA binding by daughterless/achaete-scute protein complexes. *Development*, 1991, 113, 245-255.

[43] Van Doren, M; Powell, PA; Pasternak, D; Singson, A; Posakony, JW. Spatial regulation of proneural gene activity: auto- and cross-activation of *achaete* is antagonized by *extramacrochaetae*. *Genes Dev.*, 1992, 6, 2592-2605.

[44] Vaessin, H; Brand, M; Jan, LY; Jan, YN. *daughterless* is essential for neuronal precursor differentiation but not for initiation of neuronal precursor formation in Drosophila embryo. *Development*, 1994, 120, 935-945.

[45] Cabrera CV; Alonso, MC; Huikeshoven, H. Regulation of scute function by extramacrochaete in vitro and in vivo. *Development*, 1994, 120, 3595-3603.

[46] Smith, JE; Cronmiller, C. The Drosophila *daughterless* gene autoregulates and is controlled by both positive and negative *cis* regulation. *Development*, 2001, 128, 4705-4714.

[47] Giebel, B; Campos-Ortega, JA. Functional dissection of the Drosophila Enhancer of split protein, a suppressor of neurogenesis. *Proc. Natl. Acad. Sci. USA*, 1997, 94, 6250-6254.

[48] Giagtzoglou, N; Koumbanakis, KA; Fullard, J; Zarifi, I; Delidakis, C. Role of the Sc C terminus in transcriptional activation and E(spl) repressor recruitment. *J. Biol. Chem.*, 2005, 280, 1299-1305.

[49] Artavanis-Tsakonas, S; Matsuno, K; Fortini, ME. Notch signaling. *Science*, 1995, 268, 225-232.

[50] Heitzler, P; Bourouis, M; Ruel, L; Carteret, C; Simpson, P. Genes of the *Enhancer of split* and *achaete-scute* complexes are required for a regulatory loop between *Notch* and *Delta* during lateral signalling in Drosophila. *Development*, 1996, 122, 161-171.

[51] Paroush, Z; Finley, RL; Kidd, T; Wainwright, SM; Ingham, PW; Brent, R; Ish-Horowicz, D. Groucho is required for Drosophila neurogenesis, segmentation, and sex determination and interacts directly with hairy-related bHLH proteins. *Cell*, 1994, 79, 805-815.

[52] Giagtzoglou, N; Alifragis, P; Koumbanakis, KA; Delidakis, C. Two modes of recruitment of E(spl) repressors onto target genes. *Development*, 2003, 103, 259-270.

[53] Van Doren, M; Bailey, AM; Esnayra, J; Ede, K; Posakony, JW. Negative regulation of proneural gene activity: hairy is a direct transcriptional repressor of *achaete*. *Genes Dev.*, 1994, 8, 2729-2749.

[54] Fisher, AL; Caudy, M. The function of hairy-related bHLH repressor proteins in cell fate decisions. *BioEssays,* 1998, 20, 298-306.

[55] Bianchi-Frias, D; Orian, A; Delrow, JJ; Vazquez, J; Rosales-Nieves, AE; Parkhurst, SM. Hairy transcriptional repression targets and cofactor recruitment in Drosophila. *PLoS Biol.*, 2004, 2, 0975-0990.

[56] Chen, CM; Freedman, JA; Bettler, DR; Manning, SD; Giep, SN; Steiner, J; Ellis, HM. *polychaetoid* is required to restrict segregation of sensory organ precursors from proneural clusters in Drosophila. *Mech. Dev.*, 1996, 57, 215-227.

[57] Courey, AJ; Jia, S. Transcriptional repression: the long and the short of it. *Genes Dev.*, 2001, 15, 2786-2796.

[58] Zhang, H; Emmons, SW. *Caenorhabditis elegans unc-37/groucho* interacts genetically with components of the transcriptional mediator complex. *Genetics,* 2002, 160, 799-803.

[59] Nuthall, HN; Joachim, K; Stifani, S. Phosphorylation of serine 239 of Groucho/TLE1 by protein kinase CK2 is important for inhibition of neuronal differentiation. *Mol. Cell. Biol.*, 2004, 24, 8395-8407.

[60] Nolo, R; Abbott, LA; Bellen, HJ. Senseless, a Zn finger transcription factor, is necessary and sufficient for sensory organ development in Drosophila. *Cell*, 2000, 102, 349-362.

[61] Jafar-Nejad, H; Acar, M; Nolo, R; Lacin, H; Pan, H; Parkhurst, SM; Bellen, HJ. Senseless acts as a binary switch during sensory organ precursor selection. *Genes Dev.*, 2003, 17, 2966-2978.

[62] Jafar-Nejad, H; Tien, AC; Acar, M; Bellen, HJ. Senseless and Daughterless confer neuronal identity to epithelial cells in the Drosophila wing margin. *Development*, 2006, 133, 1683-1692.

[63] Acar, M; Jafar-Nejad, H; Giagtzoglou, N; Yallampalli, S; David, G; He, Y; Delidakis, C; Bellen, HJ. Senseless physically interacts with proneural proteins and functions as a transcriptional co-activator. *Development*, 2006, 133, 1979-1989.

[64] Pi, H; Chien, CT. Getting the edge: neural precursor selection. *J. Biomed. Sci.*, 2007, 14, 467-473.

[65] Freeman, M. Complexity of EGF receptor signalling revealed in Drosophila. *Curr. Opin. Genet. Dev.*, 1998, 8, 407-411.

[66] Livneh, E; Glazer, L; Segal, D; Schlessinger, J; Shilo, BZ. The Drosophila *EGF receptor* gene homolog: conservation of both hormone binding and kinase domains. *Cell*, 1985, 49, 599-607.

[67] Shilo, BZ. Signaling by the Drosophila epidermal growth factor receptor pathway during development. *Exp. Cell Res.*, 2003, 284, 140-149.

[68] del Alamo, D; Terriente, J; Diaz-Benjumea, FJ. Spitz/EGFr signalling via the Ras/MAPK pathway mediates the induction of bract cells in Drosophila legs. *Development*, 2002, 129, 1975-1982.

[69] Klein, DE; Nappi, VM; Reeves, GT; Shvartsman, SY; Lemmon, MA. Argos inhibits epidermal growth factor receptor signalling by ligand sequestration. *Nature*, 2004, 430, 1040-1044.

[70] Lee, JR; Urban, S; Garvey, CF; Freeman, M. Regulated intracellular ligand transport and proteolysis control EGF signal activation in Drosophila. *Cell*, 2001, 107, 161-171.

[71] Tsruya, R; Schlesinger, A; Reich, A; Gabay, L; Sapir, A; Shilo, BZ. Intracellular trafficking by Star regulates cleavage of the Drosophila EGF receptor ligand Spitz. *Genes Dev.*, 2002, 16, 222-234.

[72] Urban, S. Rhomboid proteins: conserved membrane proteases with divergent biological functions. *Genes Dev.*, 2006, 20, 3054-3068.

[73] Gabay, L; Seger, R; Shilo, BZ. *In situ* activation pattern of Drosophila EGF receptor pathway during developmenmt. *Science*, 1997, 277, 1103-1106.

[74] Kumar, JP; Hsiung, F; Powers, M.A; Moses, K. Nuclear translocation of activated MAP kinase is developmentally regulated in the developing Drosophila eye. *Development*, 2003, 130, 3703-3714.

[75] Kolodkin, AL; Pickup, AT; Lin, DM; Goodman, CS; Banerjee, U. Characterization of *Star* and its interactions with *sevenless* and *EGF receptor* during photoreceptor cell development in Drosophila. *Development*, 1994, 120, 1731-1745.

[76] Albagli, O; Klaes, A; Ferreira, E; Leprince, D; Klambt, C. Function of ets genes is conserved between vertebrates and Drosophila. *Mech.Dev.*, 1996, 59, 29-40.

[77] Culi, J; Modolell, J. Proneural gene self-stimulation in neural precursors: an essential mechanism for sense organ development that is regulated by *Notch* signaling. *Genes Dev.*, 1998, 12, 2036-2047.

[78] zur Lage, PI; Powell, LM; Prentice, DR; McLaughlin, P; Jarman, AP. EGF receptor signaling triggers recruitment of Drosophila sense organ precursors by stimulating proneural gene autoregulation. *Dev. Cell.*, 2004, 7, 687-696.

[79] Golembo, M; Schweitzer, R; Freeman, M; Shilo, BZ. *argos* transcription is induced by the Drosophila EGF receptor pathway to form an inhibitory feedback loop. *Development*, 1996, 122, 223-230.

[80] Wright, TR. The genetics of embryogenesis in Drosophila. *Adv. Genet.*, 1970, 15, 261-395.

[81] Artavanis-Tsakonas, S; Matsuno, K; Fortini, ME. Notch signaling. *Science*, 1995, 268, 225-232.

[82] Kimble, J; Simpson, P. The LIN-12/Notch signaling pathway and its regulation. *Annu. Rev. Cell Dev. Biol.*, 1997, 13, 333-361.

[83] Heitzler, P; Simpson, P. The choice of cell fate in the epidermis of Drosophila. *Cell*, 1991, 64, 1083-1092.

[84] Ghysen, A; Thomas, R. The formation of sense organs in Drosophila: a logical approach. *Bioessays*, 2003, 25, 802-807.

[85] Fortini, ME; ArtavanisTsakonas, S. *Notch*: neurogenesis is only part of the picture. *Cell*, 1993, 75, 1245-1247.

[86] Wharton, KA; Johansen, KM; Xu, T; Artavanis-Tsakonas, S. Nucleotide sequence from the neurogenic locus Notch implies a gene product that shares homology with proteins containing EGF-like repeats. *Cell*, 1985, 43, 567-581.

[87] Mumm, JS; Kopan, R. Notch signaling: from the outside in. *Dev. Biol.*, 2000, 228, 151-165.

[88] Rand, MD; Grimm, LM; Artavanis-Tsakonas, S; Patriub, V; Blacklow, SC; Sklar, J; Aster, JC. Calcium depletion dissociates and activates heterodimeric Notch receptors. *Mol. Cell Biol.*, 2000, 20, 1825-1835.

[89] Kopan, R. Notch: a membrane-bound transcription factor. *J. Cell Sci.*, 2002, 115, 1095-1097.

[90] Shimizu, K; Chiba, S; Saito, T; Takahashi, T; Kumano, K; Hamada, Y; Hirai, H. Integrity of intracellular domain of Notch ligand is indispensable for cleavage required for release of the Notch2 intracellular domain. *EMBO J.*, 2002, 21, 294-302.

[91] Haenlin, M; Kramatschek, B; Campos-Ortega, JA. The pattern of transcription of the neurogenic gene *Delta* of *Drosophila melanogaster*. *Development*, 1990, 110, 905-914.

[92] Sun, X; Artavanis-Tsakonas, S. Secreted forms of DELTA and SERRATE define antagonists of Notch signaling in Drosophila. *Development*, 1997, 124, 3439-3448.

[93] Fleming, RJ; Scottgale, TN; Diederich, RJ; Artavanis-Tsakonas, S. The gene Serrate encodes a putative EGF-like transmembrane protein essential for proper ectodermal development in *Drosophila melanogaster*. *Genes Dev.*, 1990, 4, 2188-2201.

[94] Gu, Y; Hukriede, NA; Fleming, RJ. *Serrate* expression can functionally replace Delta activity during neuroblast segregation in the Drosophila embryo. *Development*, 1995, 121, 855-865.

[95] Panin, VM; Papayannopoulos, V; Wilson, R; Irvine, KD. Fringe modulates Notch-ligand interactions. *Nature*, 1997, 387, 908-912.

[96] Ju, BG; Jeong, S; Bae, E; Hyun, S; Carroll, SB; Yim, J; Kim, J. Fringe forms a complex with Notch. *Nature*, 2000, 405, 191-195.

[97] Lai, EC. Notch signaling: control of cell communication and cell fate. *Development*, 2004, 131, 965-973.

[98] Schweisguth, F. Regulation of Notch signaling activity. *Curr. Biol.*, 2004, 14, R129-138.

[99] Kunisch, M; Haenlin, M; Campos-Ortega, JA. Lateral inhibition mediated by the Drosophila neurogenic gene *Delta* is enhanced by proneural proteins. *Proc. Natl. Acad. Sci. USA*, 1994, 91, 10139–10143.

[100] Struhl, G; Greenwald, I. Presenilin is required for activity and nuclear access of Notch in Drosophila. *Nature*, 1999 398, 522-525.

[101] Guo, Y; Livne-Bar, I; Zhou, L; Boulianne, GL. *Drosophila presenilin* is required for neuronal differentiation and affects notch subcellular localization and signaling. *J. Neurosci.*, 1999, 19, 8435-8442.

[102] Struhl, G; Adachi, A. Requirements for presenilin-dependent cleavage of notch and other transmembrane proteins. *Mol. Cell*, 2000 6, 625-636.

[103] Struhl, G; Greenwald, I. Presenilin-mediated transmembrane cleavage is required for Notch signal transduction in Drosophila. *Proc. Natl. Acad. Sci. USA*, 2001, 98, 229-234.

[104] Lieber, T; Kidd, S; Young, MW. *kuzbanian*-mediated cleavage of Drosophila Notch. *Genes Dev.*, 2002, 16, 209-221.

[105] Rao, Y; Bodmer, R; Jan, LY; Jan, YN. The *big brain* gene of Drosophila functions to control the number of neuronal precursors in the peripheral nervous system. *Development*, 1992, 116, 31-40.

[106] Doherty, D; Jan, LY; Jan, YN. The Drosophila neurogenic gene *big brain*, which encodes a membrane-associated protein, acts cell autonomously and can act synergistically with *Notch* and *Delta*. *Development*, 1997, 124, 3881-3893.

[107] Dye, CA; Lee, JK; Atkinson, RC; Brewster, R; Han, PL; Bellen, HJ. The Drosophila *sanpodo* gene controls sibling cell fate and encodes a tropomodulin homolog, an actin/tropomyosin-associated protein. *Development*, 1998, 125, 1845-1856.

[108] O'Connor-Giles, KM; Skeath, JB. Numb inhibits membrane localization of Sanpodo, a four-pass transmembrane protein, to promote asymmetric divisions in Drosophila. *Dev. Cell*, 2003, 5, 231-243.

[109] Hutterer, A; Knoblich, JA. Numb and alpha-Adaptin regulate Sanpodo endocytosis to specify cell fate in *Drosophila* external sensory organs. *EMBO Rep.*, 2005, 6, 836-842.

[110] Yeh, E; Zhou, L; Rudzik, N; Boulianne, GL. Neuralized functions cell autonomously to regulate Drosophila sense organ development. *EMBO J.*, 2000, 19, 4827-4837.

[111] Lai, EC; Deblandre, GA; Kintner, C; Rubin, GM. Drosophila Neuralized is a ubiquitin ligase that promotes the internalization and degradation of Delta. *Dev. Cell*, 2001, 1, 783-794.

[112] Seto, ES; Bellen, HJ; Lloyd, TE. When cell biology meets development: endocytic regulation of signaling pathways. *Genes Dev.*, 2002, 16, 1314-1336.

[113] Le Borgne, R; Bardin, A; Schweisguth, F. The roles of receptor and ligand endocytosis in regulating Notch signaling. *Development*, 2005, 132, 1751-1762.

[114] Parks, AL; Klueg, KM; Stout, JR; Muskavitch, MA. Ligand endocytosis drives receptor dissociation and activation in the Notch pathway. *Development*, 2000, 127, 1373-1385.

[115] Frise, E; Knoblich, JA; Younger-Shepherd, S; Jan, LY; Jan, YN. The Drosophila Numb protein inhibits signaling of the Notch receptor during cell-cell interaction in sensory organ lineage. *Proc. Natl. Acad. Sci. USA*, 1996, 93, 11925-11932.

[116] Santolini, E; Puri, C; Salcini, AE; Gagliani, MC., Pelicci, PG; Tacchetti, C; Di Fiore, PP. Numb is an endocytic protein. *J. Cell Biol.*, 2000, 151, 1345-1352.

[117] Berdnik, D; Torok, T; Gonzalez-Gaitan, M; Knoblich, JA. The endocytic protein alpha-Adaptin is required for Numb-mediated asymmetric cell division in Drosophila. *Dev. Cell*, 2002, 3, 221-231.

[118] Jafar-Nejad, H; Norga, K; Bellen, H. Numb: "Adapting" Notch for endocytosis. *Dev. Cell*, 2002, 3, 155-156.

[119] Lai, EC; Posakony, JW. The Bearded box, a novel 3' UTR sequence motif, mediates negative post-transcriptional regulation of *Bearded* and *Enhancer of split* Complex gene expression. *Development*, 1997, 124, 4847-4856.

[120] Leviten, MW; Lai, EC; Posakony, JW. The Drosophila gene *Bearded* encodes a novel small protein and shares 3' UTR sequence motifs with multiple *Enhancer of split* complex genes. *Development*, 1997, 124, 4039-4051.

[121] Bardin, AJ; Schweisguth, F. Bearded family members inhibit Neuralized-mediated endocytosis and signaling activity of Delta in Drosophila. *Dev. Cell*, 2006, 10, 245-255.

[122] Wurmbach, E; Wech, I; Preiss, A. The *Enhancer of split* complex of *Drosophila melanogaster* harbors three classes of Notch responsive genes. *Mech. Dev.*, 1999, 80, 171-180.

[123] Lai, EC; Bodner, R; Kavaler, J; Freschi, G; Posakony, JW. Antagonism of Notch signaling activity by members of a novel protein family encoded by the *Bearded* and *Enhancer of split* gene complexes. *Development*, 2000, 127, 291-306.

[124] Petcherski, AG; Kimble, J. Mastermind is a putative activator for Notch. *Curr. Biol.*, 2000, 10, R471-473.

[125] Castro, B; Barolo, S; Bailey, AM; Posakony, JW. Lateral inhibition in proneural clusters: cis-regulatory logic and default repression by Suppressor of Hairless. *Development*, 2005, 132, 3333-3344.

[126] Maier, D. Hairless: the ignored antagonist of the Notch signalling pathway. *Hereditas*, 2006, 143, 212-221.

[127] Lyman, DF; Yedvobnick, B. Drosophila Notch receptor activity suppresses *Hairless* function during adult external sensory organ development. *Genetics*, 1995, 141, 1491-1505.

[128] Schweisguth, F. *Suppressor of Hairless* is required for signal reception during lateral inhibition in the Drosophila pupal notum. *Development*, 1995, 121, 1875-1884.

[129] Bailey, AM; Posakony, JW. Suppressor of Hairless directly activates transcription of *Enhancer of split* complex genes in response to Notch receptor activity. *Genes Dev.*, 1995, 9, 2609-2622.

[130] Schweisguth, F; Nero, P; Posakony, JW. The sequence similarity of the Drosophila Suppressor of Hairless protein to the integrase domain has no functional significance *in vivo*. *Dev. Biol.*, 1994, 166, 812-814.

[131] Morel, V; Lecourtois, M; Massiani, O; Maier, D; Preiss, A; Schweisguth, F. Transcriptional repression by *Suppressor of Hairless* involves the binding of a Hairless-dCtBP complex in Drosophila. *Curr. Biol.*, 2001, 11, 789-792.

[132] Barolo, S; Walker, RG; Polyanovsky, AD; Freschi, G; Keil, T; Posakony, JW. A Notch-independent activity of Suppressor of Hairless is required for normal mechanoreceptor physiology. *Cell*, 2000, 103, 957-969.

[133] Barolo, S; Stone, T; Bang, AG; Posakony, JW. Default repression and Notch signaling: Hairless acts as an adaptor to recruit the corepressors Groucho and dCtBP to Suppressor of Hairless. *Genes Dev.*, 2002, 16, 1964-1976.

[134] Lai, EC. Keeping a good pathway down: transcriptional repression of Notch pathway target genes by CSL proteins. *EMBO Rep.*, 2002, 3, 840-845.

[135] Hasson, P; Egoz, N; Winkler, C; Volohonsky, G; Jia, S; Dinur, T; Volk, T; Courey, AJ; Paroush, Z. EGFR signaling attenuates Groucho-dependent repression to antagonize Notch transcriptional output. *Nat. Genet.*, 2005, 37, 101-105.

[136] Hasson, P; Paroush, Z. Crosstalk between the EGFR and other signalling pathways at the level of the global transcriptional corepressor Groucho/TLE. *Br. J. Cancer.*, 2006, 94, 771-775.

[137] Busseau, I; Diederich, RJ; Xu, T; Artavanis-Tsakonas, S. A member of the Notch group of interacting loci, *deltex* encodes a cytoplasmic basic protein. *Genetics*, 1994, 136, 585-596.

[138] Diederich, RJ; Matsuno, K; Hing, H; Artavanis-Tsakonas, S. Cytosolic interaction between deltex and Notch ankyrin repeats implicates deltex in the Notch signaling pathway. *Development*, 1994, 120, 473-481.

[139] Matsuno, K; Diederich, RJ; Go, MJ; Blaumueller, CM; Artavanis-Tsakonas, S. Deltex acts as a positive regulator of Notch signaling through interactions with the Notch ankyrin repeats. *Development*, 1995, 121, 2633-2644.

[140] Hori, K; Fostier, M; Ito, M; Fuwa, TJ; Go, MJ; Okano, H; Baron, M; Matsuno, K. Drosophila Deltex mediates suppressor of Hairless-independent and late-endosomal activation of Notch signaling. *Development*, 2004, 131, 5527-5537.

[141] Ye, Y; Lukinova, N; Fortini, ME. Neurogenic phenotypes and altered Notch processing in Drosophila *presenilin* mutants. *Nature*, 1999, 398, 525-529.

[142] Portin, P. General outlines of the molecular genetics of the Notch signalling pathway in *Drosophila melanogaster*: a review. Hereditas, 2002, 136, 89-96.

[143] Pavlopoulos, E; Pitsouli, C; Klueg, KM; Muskavitch, MA; Moschonas, NK; Delidakis, C. *neuralized* encodes a peripheral membrane protein involved in Delta signaling and endocytosis. *Dev. Cell*, 2001, 1, 807-816.

[144] Mlodzik, M; Baker, NE; Rubin, GM. Isolation and expression of *scabrous*, a gene regulating neurogenesis in Drosophila. *Genes Dev.*, 1990, 4, 1848-1861.

[145] Renaud, O; Simpson, P. *scabrous* modifies epithelial cell adhesion and extends the range of lateral signalling during development of the spaced bristle pattern in Drosophila. *Dev. Biol.*, 2001, 240, 361-376.

[146] Hu, X; Lee, EC; Baker, NE. Molecular analysis of *scabrous* mutant alleles from *Drosophila melanogaster* indicates a secreted protein with two functional domains. *Genetics*, 1995, 141, 607-617.

[147] Chang, PJ; Hsiao, YL; Tien, AC; Li, YC; Pi, H. Negative-feedback regulation of proneural proteins controls the timing of neural precursor division. *Development*, 2008, 135, 3021-3030.

[148] Knoblich, JA; Jan, LY; Jan, YN. Asymmetric segregation of Numb and Prospero during cell division. *Nature*, 1995, 377, 624-627.

[149] Knoblich, JA; Jan, LY; Jan, YN. The N terminus of the Drosophila Numb protein directs membrane association and actin-dependent asymmetric localization. *Proc. Natl. Acad. Sci. USA*, 1997, 94, 13005-13010.

[150] Betschinger, J; Mechtler, K; Knoblich, JA. The Par complex directs asymmetric cell division by phosphorylating the cytoskeletal protein Lgl. *Nature*, 2003, 422, 326-330.

[151] Knust, E. G protein signaling and asymmetric cell division. *Cell*, 2001, 107, 125-128.

[152] Schaefer, M; Petronczki, M; Dorner, D; Forte, M; Knoblich, JA. Heterotrimeric G proteins direct two modes of asymmetric cell division in the Drosophila nervous system. *Cell*, 2001, 107, 183-194.

[153] Katanaev, VL; Tomlinson, A. Dual roles for the trimeric G protein Go in asymmetric cell division in Drosophila. *Proc. Natl. Acad. Sci. USA*, 2006, 103, 6524-6529.

[154] Bowman, SK; Neumüller, RA; Novatchkova, M; Du, Q; Knoblich, JA. The Drosophila NuMA Homolog Mud regulates spindle orientation in asymmetric cell division. *Dev. Cell*, 2006, 10, 731-742.

[155] Ramaekers, G; Usui, K; Usui-Ishihara, A; Ramaekers, A; Ledent, V; Ghysen, A; Dambly-Chaudière, C. Lineage and fate in Drosophila: role of the gene *tramtrack* in sense organ development. *Dev. Gen. Evol.*, 1997, 207, 97-106.

[156] Okabe, M; Imai, T; Kurusu, M; Hiromi, Y; Okano, H. Translational repression determines a neuronal potential in Drosophila asymmetric cell division. *Nature*, 2001, 411, 94-98.

[157] Badenhorst, P; Finch, JT; Travers, AA. Tramtrack co-operates to prevent inappropriate neural development in Drosophila. *Mech. Dev.*, 2002, 117, 87-101.

[158] Sakakibara, S; Okano, H. Expression of neural RNA-binding proteins in the postnatal CNS: implications of their roles in neuronal and glial cell development. *J. Neurosci.*, 1997, 17, 8300-8312.

[159] Okano, H; Imai, T; Okabe, M. Musashi: a translational regulator of cell fate. *J. Cell Sci.*, 2002, 115, 1355-1359.

[160] Hirata, J; Nakagoshi, H; Nabeshima, Y; Matsuzaki, F. Asymmetric segregation of the homeodomain protein Prospero during Drosophila development. *Nature*, 1995, 377, 627-630.

[161] Ryter, JM; Doe, CQ; Matthews, BW. Structure of the DNA binding region of prospero reveals a novel homeo-prospero domain. *Structure*, 2002, 10, 1541-1549.

[162] Yousef, MS; Matthews, BW. Structural basis of Prospero-DNA interaction: implications for transcription regulation in developing cells. *Structure*, 2005, 13, 601-607.

[163] Choksi, SP; Southall, TD; Bossing, T; Edoff, K; de Wit, E; Fischer, BE; van Steensel, B; Micklem, G; Brand, AH. Prospero acts as a binary switch between self-renewal and differentiation in Drosophila neural stem cells. *Dev. Cell*, 2006, 11, 775-789.

[164] Bi, X; Kajava, AV; Jones, T; Demidenko, ZN; Mortin, MA. The carboxy terminus of Prospero regulates its subcellular localization. *Mol. Cell Biol.*, 2003, 23, 1014-1024.

[165] Manning, L; Doe, CQ. Prospero distinguishes sibling cell fate without asymmetric localization in the Drosophila adult external sense organ lineage. *Development*, 1999, 126, 2063-2071.

[166] Roegiers, F; Younger-Shepherd, S; Jan, LY; Jan, YN. Bazooka is required for localization of determinants and controlling proliferation in the sensory organ precursor cell lineage in Drosophila. *Proc. Natl. Acad. Sci. USA*, 2001, 98, 14469-14474.

[167] Ramos, A; Grünert, S; Adams, J; Micklem, DR; Proctor, MR; Freund, S; Bycroft, M; St Johnston, D; Varani, G. RNA recognition by a Staufen double-stranded RNA-binding domain. *EMBO J.*, 2000, 19, 997-1009.

[168] Bardin, AJ; Le Borgne, R; Schweisguth, F. Asymmetric localization and function of cell-fate determinants: a fly's view. *Curr. Opin. Neurobiol.*, 2004, 14, 6-14.

[169] Yousef, MS; Kamikubo, H; Kataoka M. Kato R., Wakatsuki S. Miranda cargo-binding domain forms an elongated coiled-coil homodimer in solution: implications for asymmetric cell division in Drosophila. *Protein Sci.*, 2008, 17, 908-917.

[170] Ikeshima-Kataoka, H; Skeath, JB; Nabeshima, Y; Doe, CQ; Matsuzaki, F. Miranda directs Prospero to a daughter cell during Drosophila asymmetric divisions. *Nature*, 1997, 390, 625-629.

[171] Shen, CP; Knoblich, JA; Chan, YM; Jiang, MM; Jan, LY; Jan, YN. Miranda as a multidomain adapter linking apically localized Inscuteable and basally localized Staufen and Prospero during asymmetric cell division in Drosophila. *Genes Dev.*, 1998, 12, 1837-1846.

[172] Shen, CP; Jan, LY; Jan, YN. Miranda is required for the asymmetric localization of Prospero during mitosis in Drosophila. *Cell*, 1997, 90, 449-458.

[173] Hardiman, KE; Brewster, R; Khan, SM; Deo, M; Bodmer, R. The *bereft* gene, a potential target of the neural selector gene *cut*, contributes to bristle morphogenesis. *Genetics*, 2002, 161, 231-247.

[174] Sato, M; Kojima, T; Michiue, T; Saigo, K. *Bar* homeobox genes are latitudinal prepattern genes in the developing Drosophila notum whose expression is regulated by the concerted functions of *decapentaplegic* and *wingless*. *Development*, 1999, 126, 1457-1466.

[175] Furman, DP; Bukharina, TA. The gene network determining development of *Drosophila melanogaster* mechanoreceptors. *Comput. Biol. Chem.*, 2009, 33, 321-234.

In: Daughter Cells
Editors: A. Hitomi et al., pp. 163-196
ISBN: 978-1-60876-790-8

Chapter IV

GENERATION OF DAUGHTER CELLS IN THEIR NICHE

Guadalupe Aparicio Gallego[1], Moisés Blanco Calvo[1], Angélica Figueroa[1], Manuel Valladares Ayerbes[2] and Luis M. Antón Aparicio[2,3,*]

[1]Biomedical Research Institute of A Coruña, Coruña, Spain
[2]Medical Oncology Service, A Coruña University Hospital, Coruña, Spain
[3]University of A Coruña, Coruña, Spain

ABSTRACT

Stem cells (SCs) are thought to reside in many tissues within a niche formed by a group of surrounding cells and their extracellular matrices, which provide an optimal microenvironment for the SCs function. In general, the niche consists of a highly organized microenvironment in which various factors, such as signals coming from secreted cytokines, extracellular matrix (ECM), and intercellular adhesions are thought to work cooperatively to maintain the undifferentiated stem cell phenotype. Among SCs, adult SCs are often localized into specific niches where they utilize many, but not necessarily all, the external and intrinsic factors used by the embryonal counterparts to select a specific fate. Within the niche, SCs can maintain their ability for self-renewal as well as their potential so that, consequently, their detachment from the niche compartment induces their differentiation and loss of self-renewal capacity. Thus, when a stem cell begins to divide, it is thought that one daughter cell remains in a niche to replace the original stem cell, whereas another daughter cell is expelled out of the niche and starts its process of differentiation. In this process, a cell retains self-renewal properties and differentiation inhibitory factors, so that it stays as SC, whereas another daughter cell is destined to proliferate during a certain number of divisions for final differentiation along

* Correspondence concerning this chapter should be addressed to: Luis M. Antón Aparicio, MD, PhD, Head of Medical Oncology Service, A Coruña University Hospital, As Xubias de Arriba, 84, 15006 – A Coruña, Spain. E-mail: Luis.M.Anton.Aparicio@sergas.es.

a particular lineage. This latter daughter cell will receive too few *stemness* factors to maintain its identity as a stem cell, and/or will inherit differentiation factors that can overcome its stem cell phenotype. To maintain tissue homeostasis and correct functioning of an organism, the number of daughter cells that retain stem cell identity must be strictly controlled such that differentiated cells can be generated in response to any injury. Likewise, the rate of division of SCs into a niche must be tightly controlled, since an overproduction of daughter cells destined to be differentiated may be harmful because this may result in cancer generation. In the present chapter, we will analyze distinct types of niches for distinct tissues as well as the mechanisms that influence the generation of daughter cells from each niche.

Keywords: adult stem cells, niche, daughter cells, self-renewal, differentiation, plasticity, pathways

INTRODUCTION

SCs modulate the formation, the maintenance and the reparation of tissues based on a complex interaction of cell-autonomous and cell-non autonomous regulatory mechanisms. SC populations have fixed architectural positions so that these cells are embedded in tissues and therefore cannot be readily transplanted, a key element in defining SC function. The ability to self-renewal is a characteristic property of SCs, affording a theoretical limitless ability to produce daughter cells and their descendents. In contrast to their more vigorously proliferative daughter cells, SCs reside in relative quiescence, dividing infrequently. SC populations are established in niches, the specific anatomical locations constituting the microenvironment that regulates the SC function. The location of SCs of many adult organs is being now characterized. However, research into the regulatory functions of cell types in the niche has provided several lines of experimental evidence that serve as guides to their knowledge: (1) the number of SCs in a niche is tightly regulated and highly constrained from two to four; (2) SCs are undifferentiated, have self-renewal capabilities, and can produce offspring with explosive proliferative capability; (3) physical interaction among heterologous types of cells is important for the maintenance of the SC state; (4) products of the niche provide the molecular basis for intercell interactions so that it is necessary a balance of inhibitory and stimulatory signals; and (5) the niche can impose SC-like characteristics on some cells (even if they are not SCs) so that an empty niche can remain viable if other cell types occupy the niche [1-3].

SCs can either enter a self-renewal cycle, undergoing symmetrical division to generate one SC, or give rise to one differentiated cell through asymmetrical division, depending on signals arising from the local microenvironment or other tissues within the organism [4]. Most SCs in a tissue are either in the resting phase of cell cycle or are slowly cycling, but can give rise to a pool of transit amplifying cells in response to external stimuli [5-7]. Presumably, the evolutionary advantage of this amplification step is to maintain only a very small number of SCs with self-renewal capacity.

If an SC hierarchy is present in the organs, it should be possible to recover cells with SC properties from tissues. These properties generally include longevity, self-renewal, short-term cloning ability, a lineage-negative and side population phenotype, and the ability to perform asymmetric cell division.

Longevity may tend to the selection of a certain progenitor in a culture-dependent manner [8,9]. Normal human breast epithelium (i.e., from reduction mammoplasties) retain the capacity to replicate for up to 50 generations in culture [10] before senescence.

A hallmark of SCs is the ability to divide from a single cell to yield a large clone [11]. Short-term clonal culture is more likely to unravel multiple progenitors by its ability to support different clones without negative selection. By clonal analysis of breast epithelium, it has been depicted the presence of two clones which appears to be either lineage-restricted luminal cells or lineage-restricted myoepithelial cells [12]. These clones express lineage markers either co-ordinately or in a mixed pattern to give a final type that was negative for the lineage markers used [13]. By use of fluorescence activated cell sorting, it is possible to separate the two lineage-restricted cells [14].

Another approach to claiming the presence of SCs in tissues is the identification of a so-called side population. Side populations have been observed in a variety of human tissues including breast [15-17], where its reported size ranges from 0.01% to about 1% of cells [15,16].

When stem cells cannot be identified or isolated in a particular organ, their existence may be inferred from kinetics studies of 5'-bromo-2'-deoxyuridine (BrdU) incorporation. Adult SCs are often relatively slow-cycling cells able to respond to specific environmental signals and generate new SCs or select a particular differentiation. Because of SCs slowly divide, the presence of labelled-retaining cells (LRC) can identify the anatomical location of a stem cell niche.

On the other hand, after asymmetrical division of SCs are generated transit-amplifying cells or progenitors [18]. In tissues, the possible SCs have been characterized by their ability to remain quiescent after a number of cell divisions so that they represent a reservoir for further cell division upon renewal stimulation. A hallmark of adult tissue SCs is their capacity to divide asymmetrically, providing committed progenitors while maintain their SCs properties. The self-renewal capacity of these cells is shown by dissociation into single cells, which then give to a second generation of cells with similar characteristics [15]. This is a strong indicator of the presence of cells with intrinsic self-renewing capacity [19].

It has been predicted that there is a continuous crosstalk between SCs and their machinery into the niche. The stromal population in the niche microenvironment mediates a potential protective network to provide a healthy uninterrupted SC generation. The stromal cells constitute a group of cells that act as the supportive 'mattress' on which the mature precursors, or in other words the stem and the progenitor cells, rest directly. The shaping of the niche structure is under a continuous dynamics, most possibly due to the need of localizing constituent factors within the niche entity [20]. The niche is critical in maintaining the SC characteristics. Thus, the undifferentiated character and the intrinsic self-renewal capacity of resident SCs are provided by the niche's microenvironment which is both proliferation- and differentiation-inhibitory. This microenvironment is established by signals from other cells (stroma) which normally surround the niche. These signals arising from microenvironment of niche seem to be important for the slow-cycling properties of LRCs and their maintenance in reserve.

Features and functions common to all SC niches have emerged as follows [21]:

- The SC niche is composed of a group of cells in a special location that functions to maintain SCs.
- The niche is a physical anchoring site for SCs, and adhesion molecules are involved in the interaction between SCs and the niche and between SCs and the ECM.
- The niche generates extrinsic factors to control the number, the proliferation, and the fate determination of SCs.
- The niche controls normal asymmetrical division of SCs. This has been shown in invertebrates; whether it is conserved in mammalians is still an open question.

NICHE GENERAL CONCEPT

The concept of SC niche was first proposed in 1978 by Schofield [22], as a specific microenvironment in which adult SCs reside in their tissue of origin. Within the niche, SCs are able to maintain their ability for self-renewal, as well as their multipotentially. Consequently, detachment from the niche compartment induces SC differentiation and loss of self-renewal.

The ability of SCs to reside within niches is an evolutionarily conserved phenomenon. Adult SCs are often localized to specific niches where they utilize many, but not necessarily all, of the external and intrinsic cues used by the embryonal counterparts in selecting a specific fate.

The regulation of the SC niche can dictate directly characteristics of an organ. It is common that the regulation of the SC niche have a major influence on the function and morphology of an organ. The flexible regulation of the SC niche could have been a relatively easy way to acquire radically different SCs types during evolution.

In general, the niche consists of a highly organized microenvironment in which various factors, such as secreted cytokines, interactions with ECM, and the intercellular adhesion are thought to work cooperatively to maintain the undifferentiated SC phenotype [23].

The identification of a niche within any tissue involves knowledge of the location of the SCs. In accordance with the literature, to prove that a niche is present, SCs must be removed and subsequently replaced, while the niche persists, which is indicative of a role for stromal cells [24].

Conceptually, a niche is a recess in a supporting medium that provides protection and nourishment to a SC, avoiding its exposition to molecules that may cause differentiation or mutation. Thus, where the niche is well defined, SCs are virtually enveloped by differentiated cells, specialized to protect and interact with the SCs [24-26].

There are now strong evidence that SCs of many tissues are located in physically delineated and physiologically specialized structures, termed niches.

Inside the niche, SCs are often quiescent. Outside the niche, SCs must have intrinsic factors to overcome differentiation or succumb to their fate.

Among SC niches, there are considerable variations in their design. Some SCs of adult mammals don't seem to have a specified niche within their respective tissue (i.e., skeletal muscle). In other cases, however, a SC compartment is established within a developing tissue (i.e., skin, hair follicles, epidermis, mammary gland, lung, brain). Cells within this niche are then activated in response to specific environment cues.

The niche operates to regulate symmetric and asymmetric mitosis of SCs. This regulation is carried out by the action of various signalling pathways such as Wnt, Hh, Notch, Bmp, and probably others (Figure 1). Niche-forming cells are stimulated by growth factors to produce ligands which act on SC receptors to initiate mitosis or specify differentiation. Stromal cells, and the microenvironment they create, including the space between them, are features of niche that allow it to maintain the SCs, preventing their differentiation and directing the growth and renewal of tissues through its daughters [27].

Structural Architecture of the Niche

The nature of the niche in terms of its composition and its microenvironment is still not understood. The microenvironment of the SCs and the SC niche define the properties of SCs as much as SCs itself. The niche can be identified as the environment that sustains the population of SCs and that is instructive in the differentiation and proliferation of the progeny (Figure 2).

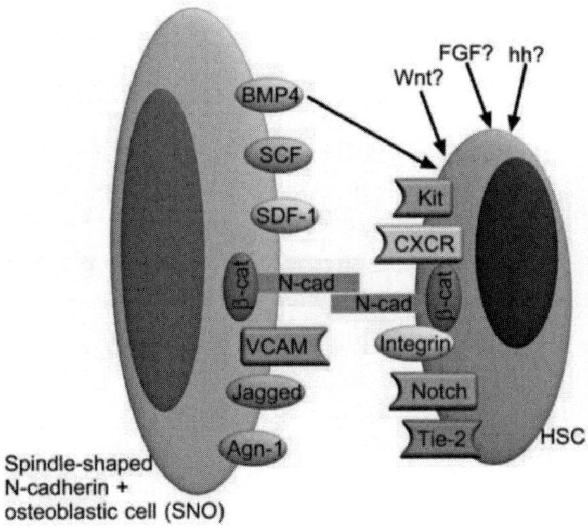

Figure 1. Example of signaling involved in stem cell niche regulation. Here, a hematopoietic stem cell (HSC) receives signals from stromal cell, a specialized osteoblast, in its niche.

Emerging evidence indicates that a specialized microenvironment, the SC niche, is one of the factors regulating normal SC maintenance and their self-renewal. The niche regulates SC maintenance and the crucial choice between self-renewal and the initiation of differentiation [24]. Thus, SCs appear to require paracrine signals from the niche in which they reside to maintain their identity and their self-renewal capacity. As a result, the number of SCs within a particular tissue can be regulated by controlling the number or size of available niches.

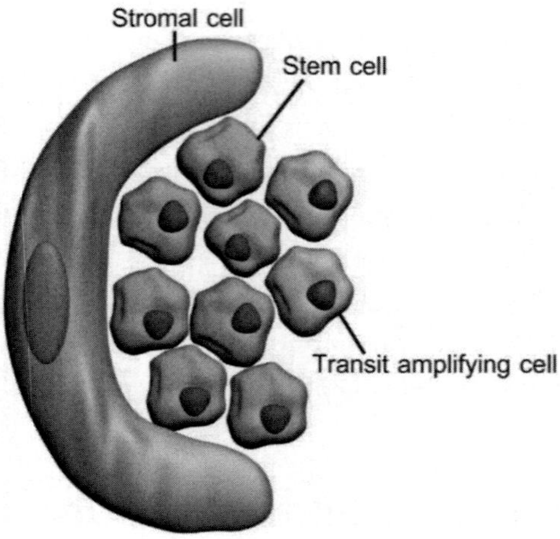

Figure 2. General structure of niche: a stem cell is supported by stromal cells.

There is ample evidence that the maintenance of a functional tissue depends on extensive regulation of SC interaction with components of the ECM. Retention and loss of SCs from the niche may be best achieved by regulating their adhesion to the ECM. Unequivocal molecular determinants for the SC niche have not been identified yet, although these molecular factors may form part of crosstalk between the niche and the ECM. Mesenchymal matrix, subepithelial fibroblast, and myofibroblast may play a crucial role in defining the SC niche.

The molecular glue that anchors SCs to their niches is, at least in part, E-cadherin, which along with its partner β-catenin, is located in SC-niche contacts. Cadherins and catenins participate in the formation of specialized intercellular junctions, called adherent junctions, which can be remodelled through their association with the actin cytoskeleton.

The N-cadherin is expressed by putative stem/progenitor cells in the epithelial SC niche. The N-cadherin is a member of the classic cadherin family that mediates cell adhesion [28]. The N-cadherin may be a critical adhesion molecule between epithelial stem/progenitor cells and their corresponding niches.

Other possible players in establishing SC niches are the integrins, which mediate adhesion of cells to a basal lamina composed of ECM. Elevated levels of integrins are often characteristic of SCs. Studies in mice reveal that both integrins and adherent junctions play critical roles in maintaining the location, adhesiveness, and proliferative status of epithelial cells within tissues [29]. β_1-integrins, specially $\alpha_4\beta_1$, $\alpha_5\beta_1$, play a vital role in the early interaction of hematopoietic progenitor cells with the bone marrow niche [30,31].

Adhesion between SCs and the surrounding support cells is important for the maintenance of SCs within the niche, close to self-renewal signals and away from differentiation cues. Thus, clusters of adherent junctions are observed between SCs and adjacent cells.

Intercellular communication via gap junctions may also be involved in the survival and differentiation of early SCs. The presence of gap junctions between SCs and adjacent support cells suggests that signaling via gap junctions may play a role in SC maintenance. Additionally, gap junctions may help to maintain physically SCs in their niche.

NICHE FUNCTION AND TYPES

Niches are composed not only of SCs but also of distinct types of surrounding differentiated cells, which secrete and organize a rich milieu of ECM and molecular factors. This microenvironment allows to SCs to manifest their unique intrinsic properties, including the ability to self-renew, while keeping their repertoire of differentiation programs.

Without the microenvironment formed by intercellular interactions and the appropriate cellular organization, SCs can become an undesirable beast. The properties and potencial of stem cells are determined by the combination of their intrinsic characteristics and their environment. The direct physical interactions between SCs and their neighbours in the niche are critical in keeping SCs in this specialized compartment and in maintaining SC properties.

The regulation of SC self-renewal is an essential feature of the niche. In the niche, the regulation of balance between symmetric and asymmetric divisions of SCs is critical to maintain the proper SC number within the niche and to meet the demand for differentiated cells within surrounding tissue.

For a daughter cell can be a SC, it must retain self-renewal capacity and inhibitory factors of differentiation. For a daughter cell can proliferate and differentiate along a particular lineage, it must receive scarce stemness factors and/or inherit differentiation factors.

Table I. Markers for the isolation of some adult stem cells in distinct tissues

Tissue/organ	Markers
Hematopoietic	$CD34^+$, $CD45^+$ [37]
Skin	Rac1, Lhx2, CD200, CD59, CD24, CD34, CD71, CD146, β4 and β1 integrins, keratin 15 (K15) [38]
Intestinal	Telomerase, Txf4, EphB3, P-PTEN, P-Akt, 14-3-3c, Noggin, Musashi-1 [39]
Brain	Nestin [40]
Liver	c-kit, CD34, Sca1, CD90 [41]
Pancreas	Homeobox gene-1 (PDX-1) [42]
Lung	CD45-, CD31-, CD34-, Sca-1(low), AF(low) (bronchiolar stem cells) [43]
Breast	K14, K19 [44]
Prostate	**α2β1, α3β1, and α6β4 integrins** [45] CD34+, CD24+CD49f-, CD45+, Ter119+ [46] Sca1[47]
Heart	$Sca1^+$, $CD117^+$ (c-Kit) [48]
Cornea	Hoechst 33342 and propidium iodide staining [49]
Dental/Tooth	Bone sialoprotein, alkaline phosphatase, Type I collagen, osteocalcin [50]

To maintain tissue homeostasis, the number of daughter cells that retain SC identity must be strictly controlled, such that differentiated cells can be generated in response to, for example, a wounding. The SCs are physically attached to the niche. When SCs divide, they orient their mitotic spindles with respect to the niche, so that one daughter cell stays in the niche, whereas another is displaced away from the niche and activates the differentiation [32-35].

The regulatory mechanisms of SC division within the niche are yet unknown, although there is many potential models [36]. When a SC divides, the possible outcomes are the production of two SCs (A), the production of two daughter cells destined to the differentiation (B), or an asymmetric division resulting in one A cell and in one B cell.

Adult SCs from distinct tissues are characterized by the expression of a number of molecular markers, proteins involved in fate determination of each lineage (Table I) [37-50].

Hematopoietic

Hematopoietic stem cells, perhaps the SC population better studied in mammals, arise from a transiently-occurring mesodermal progenitor which is not classified as a SC [51]. In the bone marrow, multiple SC niches have been proposed including, osteoblastic, vascular and CXCL12-abundant reticular (CAR) niches [52]. Many studies strongly suggest that the hematopoietic SC niche is located within the endosteal region of the bone marrow, at the interface between the bone and bone marrow [53] (Figure 3).

Skin and Hair Follicle

Epidermal SCs reside in the basal layer of the epidermis, while hair follicle SCs reside in the bulge, at the base of the permanent portion of the follicle [54]. Each hair follicle is composed of a permanent portion which includes sebaceous glands and the underlying bulge area, and a portion that is dynamically renewed. This part undergoes cycles through a number of stages: anagen phase (a period of active growth), catagen phase (apoptosis-driven retraction), and telogen phase (a short period of rest). Hair stem cells are located in the bulge area that operates as niche. During the early anagen phase, the dermal papilla region may provide the dynamic signals to activate hair SCs [39] (Figure 4).

Intestinal

The epithelium of the gastrointestinal tract has a well-defined organizational structure. This epithelium can be divided into two regions: a functional region that contains differentiated cells (villi) and a proliferative region (crypts of Lieberkühn) that represents the epithelial SC niche. Multipotent epithelial SCs reside in the crypts and give rise to four principal epithelial lineages: absorptive enterocytes, mucin-secreting goblet cells, peptide hormone-secreting enteroendocrine cells, and Paneth cells [39]. Although the location of active SCs in each crypt is debated, they are thought to reside approximately four cell strata above the base of the crypt [55] (Figure 5).

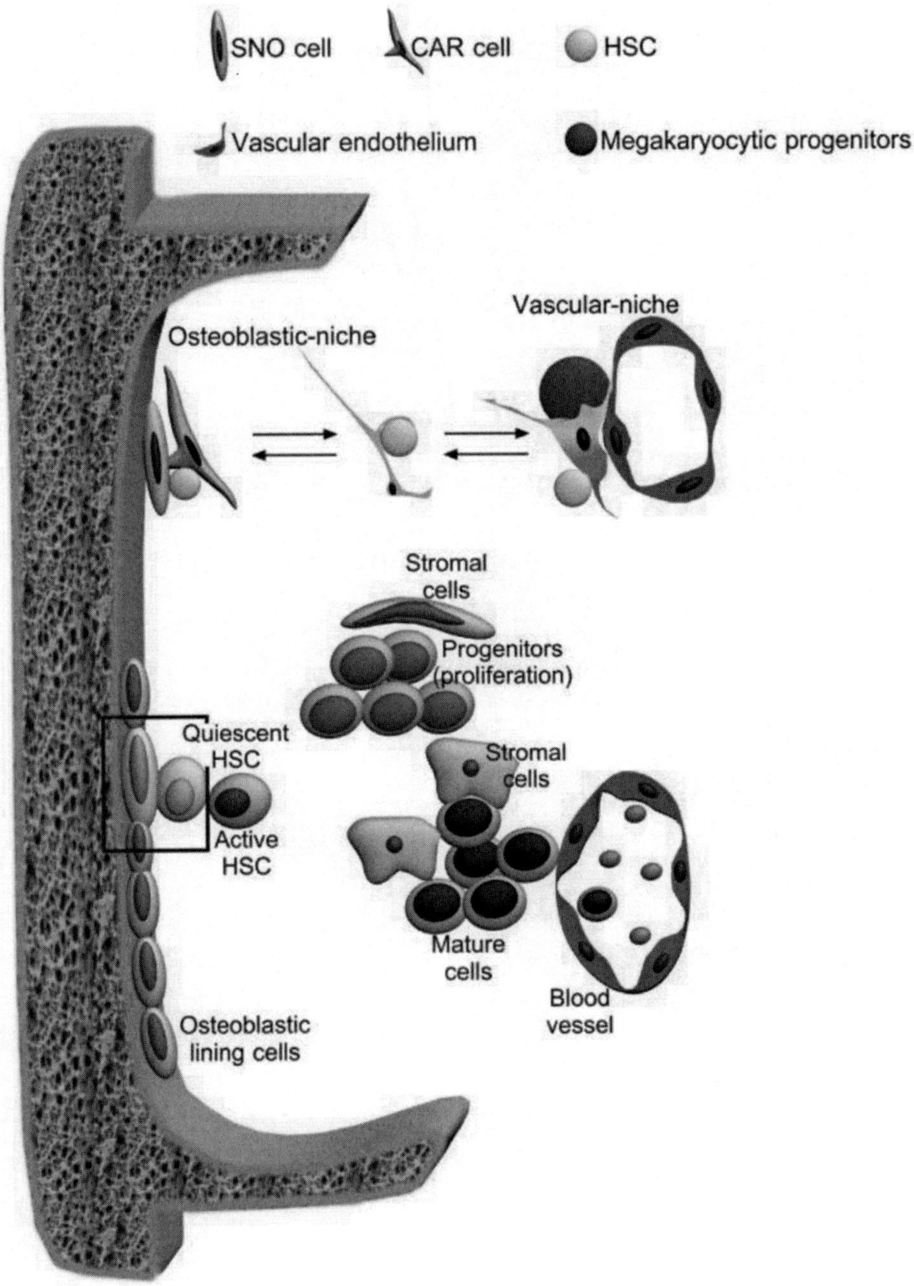

Figure 3. Scheme of hematopoietic niche in bone. HSC: hematopoietic stem cell; CAR cell: CXCL12-abundant reticular cell; SNO: spindle-shaped N-cadherin positive osteoblast.

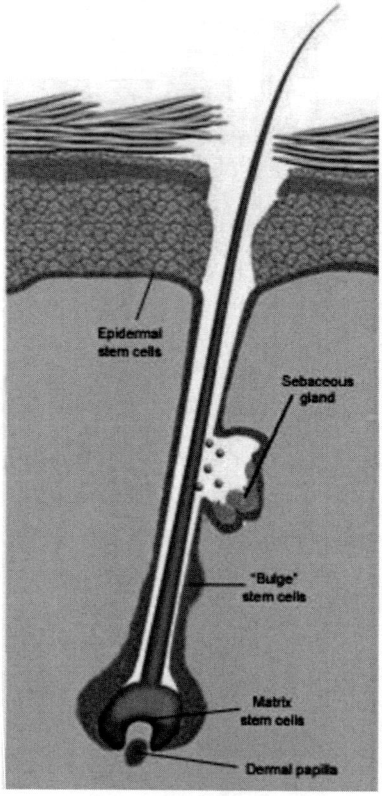

Figure 4. Skin and hair stem cell niche. In hair, stem cells and their niche are located in the area of bulge. On the other hand, epidermal stem cells are located on basal layer of epidermis.

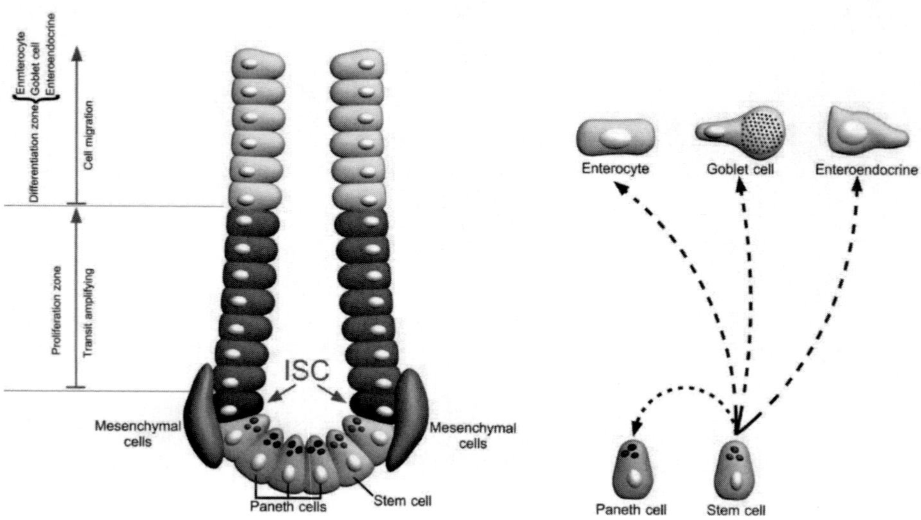

Figure 5. Intestinal stem cell niche. Stem cells are located in the base of crypts of Lieberkühn and are able to generate four mature lineages of intestinal cells: goblet cells, Paneth cells, enteroendocrine cells and enterocytes.

Brain

Neural stem cells can be isolated from various regions in the adult brain and peripheral nervous system (Figure 6). However, the subventricular zone (SVZ) and the subgranular zone (SGZ) of the hippocampus region are the primary regions in which reside neural stem cells, to support neurogenesis in the adult brain. Subventricular astrocytes that express the astrocyte marker, glyal fibrillary acidic protein (GFAP), have SC features. These astrocytes undergo self-renewal and give rise to transit-amplifying precursor C cells, which give rise later to neuroblasts. Neuroblasts differentiate toward the olfactory bulb and other regions. Both in the SVZ and SGZ, endothelial cells and the specialized basal lamina are an essential component of the neural stem cell niche. Endothelial cells provide a substrate for the subventricular and subgranular astrocytes. Additionally, they generate a variety of signals to control SC self-renewal and lineage commitment [39].

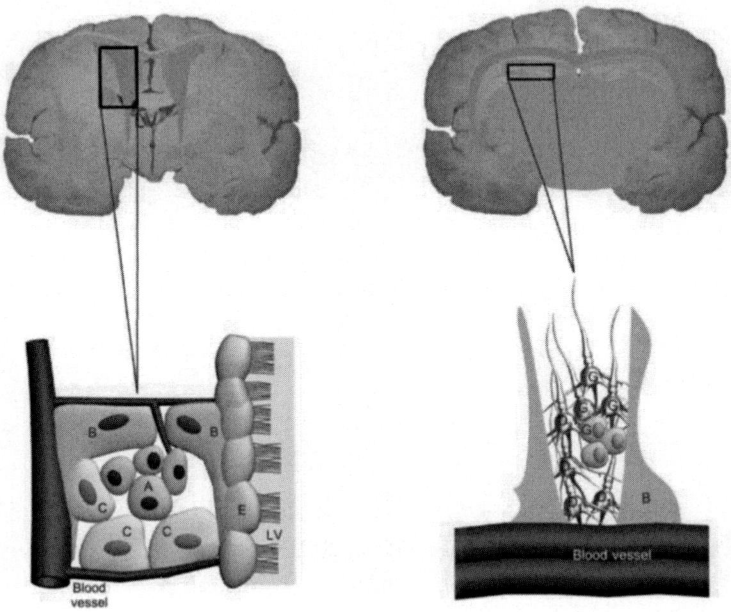

Figure 6. Brain stem cell niche. Stem cells in brain reside primarily into hippocampus, in subventricular zone (SVZ) and subgranular zone (SGZ). Left panel: in box is marked and amplified the SVZ, where the astrocytes (B) are stem cells which generate neuroblasts (A), destined for the olfactory bulb via a rapidly dividing transit-amplifying cell (C). E: Multi-ciliated ependymal cells line the coats of the lateral ventricle. LV: lateral ventricle. Right panel: in box is marked and amplified the SGZ, where the astrocytes (B) divide to generate intermediate precursors (D) which progressively generate more differentiated progeny (D1/D2/D3), to give rise to mature granule neurons (G).

Liver

Different studies indicate the presence of resident SCs in the adult liver. Results, based primarily on models of liver injury and carcinogenesis in experimental animals, led to the concept that a possible SC compartment is located within the smallest branches of the intra-hepatic biliary tree. Proliferation of cells located in this area, defined as **"oval cell response"**, has showed to contribute to regeneration of biliary cells and hepatocytes. Oval cells express markers both of fetal hepatocytes and biliary cells. Additionally, they are bipotential progenitors capable of generating hepatocytes and bile duct cells. The most likely location for the oval cell niche is the canal of Hering (Figure 7). However, recent studies suggest that oval cells are a resident population rather than bone marrow-derived population [41].

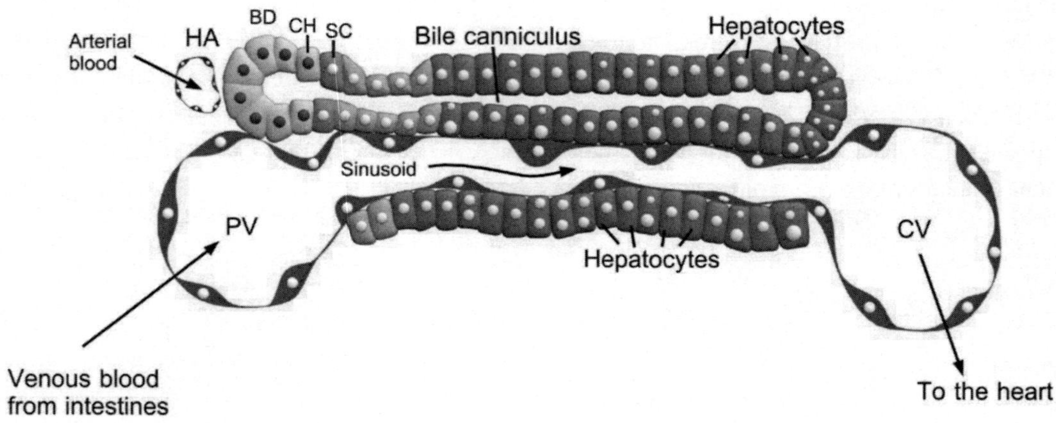

Figure 7. Liver stem cell niche. Stem cells in liver are located into intra-hepatic biliary tract. HA: hepatic artery; BD: bile ducts; CH: canal of Hering; SC: stem cell; PV: portal vein; CV: central vein. Adapated from Grompe M. Adult liver stem cells. In: Robert Lanza, Helen Blau, Douglas Melton, Malcolm Moore, E. Donnall Thomas, Catherine Verfaillie, Irving Weissman, Michael West, editors. *Handbook of Stem Cells*. Volume 2: Adult and fetal stem cells. San Diego, California, USA. Elsevier Academic Press, 2004. 483 p.

Pancreas

The presence of adult SCs or pancreatic-specific progenitor cells in human pancreas is controversial (Figure 8). Islet cells can be generated from adult pancreatic ductal cells of mice and humans, providing evidences for the presence of SC-like progenitor cells in adult pancreas [56]. The better candidates for adult pancreatic stem or progenitor cells are: (1) duct cells; (2) exocrine tissue; (3) nestin-positive islet-derived progenitor cells; (4) neurogenin-3-positive cells; (5) pancreas-derived multipotent precursors; and (6) mature β-cells [45]. Pancreas duodenum homeobox gene-1 (PDX-1), a homeodomain protein, is the first molecular marker expressed by islet SCs during development [42].

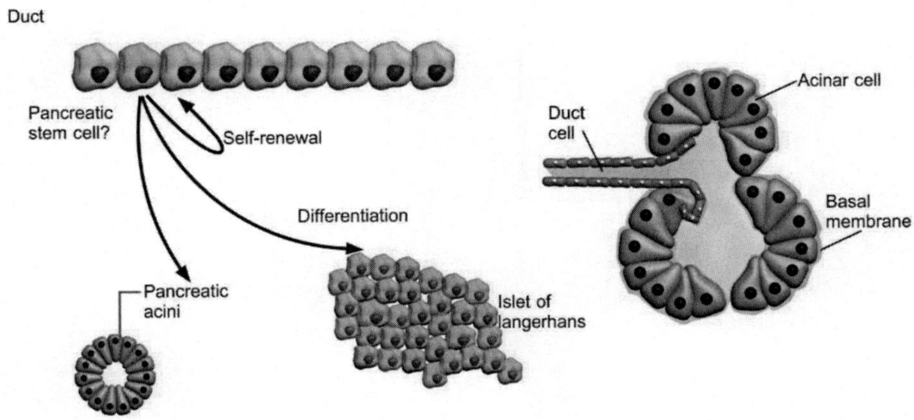

Figure 8. Pancreas stem cells. There is a number of hypothetical cells that can work as stem or progenitor cells in pancreas among which are cells contained in islets, exocrine cells, or duct cells.

Lung

In the lung, epithelial SCs include the type II pneumocyte in the alveolus and the Clara cell in the bronchiole. Recently, a more primitive putative population of adult SCs has been identified in the lung, the lung side population, which seems to have both mesenchymal and epithelial potential [58]. Airway submucosal glands are one proximal tracheobronchial airway component that has been proposed as an important airway SC niche (Figure 9) [59]. Recently, it has been identified a putative phenotype for the stem cell bronchiolar population: CD45-, CD31-, CD34-, Sca-1(low), AF(low) [43].

Breast

The adult mammary gland has a lobulo-alveolar structure, composed of three cell lineages: myoepithelial cells which form the basal layer of ducts and alveoli, ductal epithelia cells, and alveolar epithelial cells. The SCs of the human mammary gland may be contained within the luminal epithelial lineage (Figure 10). However, detailed characterization of human breast gland has led to the widely discussed hypothesis that breast stem cell is a 'basal cell' with clear cytoplasm. On the other hand, it has been established the existence of multipotent progenitor cells in human breast which can be immortalized without loss of SC-like features [60]. Recentley, it has been demonstrated the existence of a breast stem cell population defined by immuno-staining for keratins K14 and K19 [44].

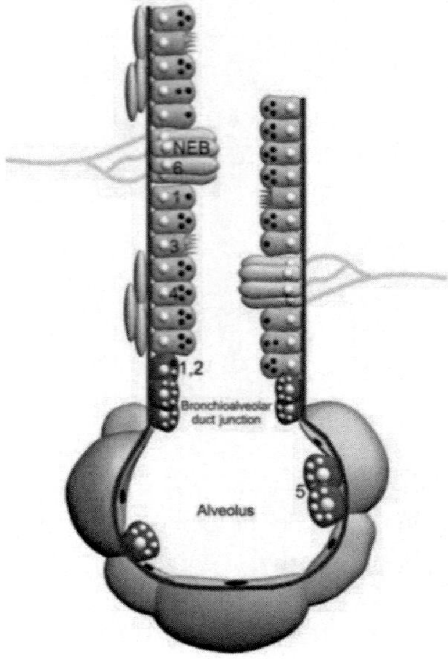

Figure 9. Location of airway stem cells. NEB: neuroendocrine bodies. 1: variant Clara cells. 2: putative bronchioalveolar stem cells (BASCs). 3: ciliated cells. 4: clara cells. 5: Type II cells wich give rise to type I cells. 6: Neuroendocrine cells.

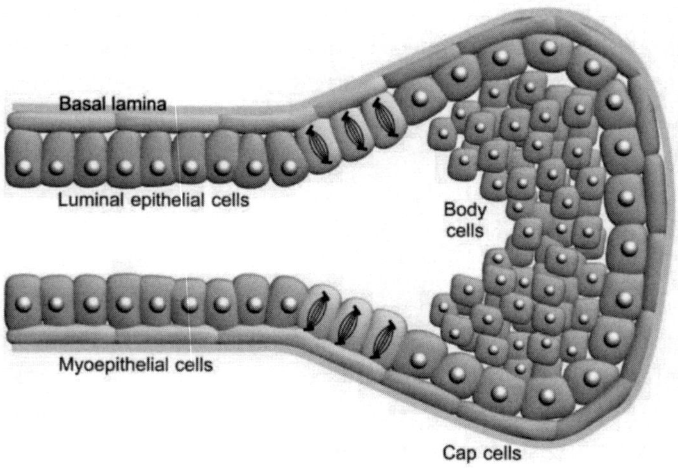

Figure 10. Breast stem cells. Stem cells in breast are hypothetically contained within luminal cells.

Prostate

Several putative SC populations have been identified in the prostate gland and have been characterized by the cell surface markers CD44, CD133, and $\alpha2\beta1^{high}$ integrin [45,46,61]. The proximal region of mouse prostatic ducts is enriched in cells exhibiting SC-like properties, i.e. in quiescence, with a high proliferative potential, and with the ability to generate progeny of different lineages [47]. A number of evidences suggest that the prostatic stem/progenitor cells may also persist in the mature prostatic epithelium, within the niches localized in the basal compartment (Figure 11) [62].

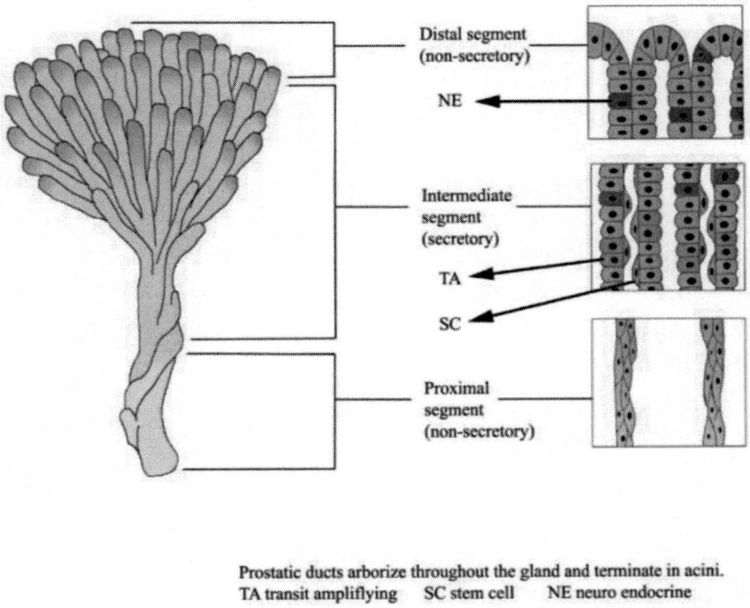

Figure 11. Stem cells in prostate. Hypothetically, stem cells in adult prostate may be located at the intermediate secretory segment of ducts, between non-secretory proximal region and distal region.

Heart/Muscle

The recent identification of different classes of cardiac progenitor cells has suggested that the heart is not a terminally differentiated postmitotic organ, but an organ regulated by a SC compartment (Figure 12). However, the existence of a human cardiac stem cell remains to be demonstrated [63].

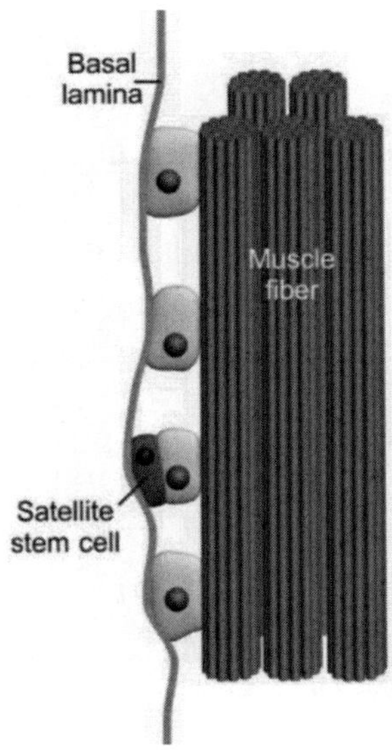

Figure 12. Stem cells in muscle. Satellite stem cells in muscle are located adjacent to basal lamina.

Cornea

Limbal epithelial stem cells are found at the corneal limbus (the peripheral edge of the cornea), and are anatomically segregated from their transit-amplifying cell progeny, which migrate centrally to cover the paracentral and central cornea (Figure 13). It has been shown in histological tissue sections that limbal epithelial stem cells are discretely located in the basal layer of the corneal limbal epithelium, at the junction between the cornea and the sclera. However, the total number and the distribution of limbal epithelial stem cells within the niche are unknown [64]. Surface markers for isolation of limbal epithelial stem cells do not have been yet identified [65].

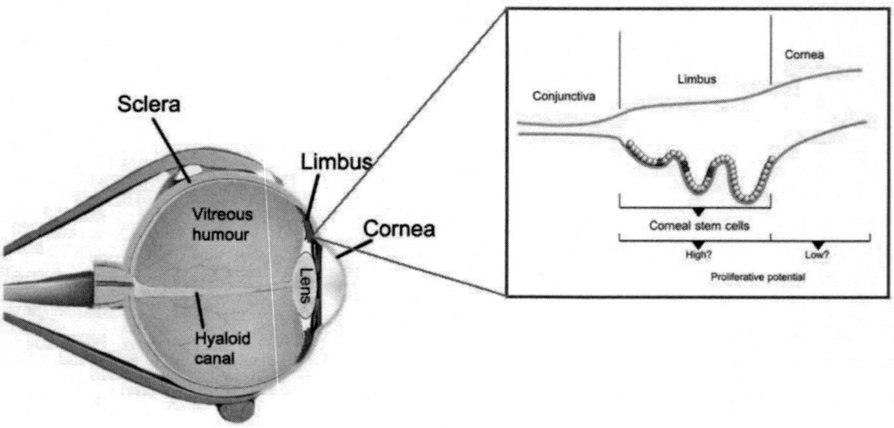

Figure 13. Limbal epithelial stem cells (LESCs) are responsible for regenerating cornea and are discretely located in the basal layer of limbal epithelium, at the junction between cornea and sclera.

Dental/Tooth

Human adult dental pulp stem cells reside within the perivascular niche of dental pulp and are thought to originate from migrating cranial neural crest cells. It has been showed that the dental pulp tissue contains different populations of multipotential stem/progenitor cells [66,50]. Development analysis of the rodent incisor has established that the dental SC niche is located at the apical end of the tooth, in the stellate reticulum, and that its progeny gives rise to ameloblastas [67].

Germinal

To our knowledge, there isn't a specific niche defined for germinal stem cells (GSC) in mammalians. However, it has been established and deeply studied GSC niche in model organisms *Caenorhabditis elegans* and *Drosphila* [63,64]. In *C. elegans*, GSC niche is formed by a single cell called distal tip cell (DTC) whose role is maintain stem cell identity and to regulate the balance between mitotic (self-renewal) and meiotic (differentiation) division in GSCs. The lack of DTC triggers that GSCs enter meiosis to give germinal cells and loss of the feature of mitotic self-renewal. DTC establishes intimate contacts with GSC, even enveloping it, so that it is easy deduce the importance of physical contact between stem and supporting cell. It is through these physical contacts by which DTC, the supporting cell, provides signals to maintain GSC properties while at the same time the proximity of both cells facilitates the action on GSC of diffusible factors generated by DTC. The deletion of glp-1 gene (homologous of Notch), which results in behavior of GSC similar to when DTC is removed, highlights the importance of physical contacts between DTC and GSC. The ligand for glp-1 is lag-2 (Delta homolog) which is expressed by DTC whereas glp-1 is expressed by GSC. The activation of glp-1 is able to inhibit the accumulation of gld-1 which allows GSC self-renewal by mitotic proliferation. On the other hand, the major diffusible factors involved in the maintaining of GSC niche are the belonging to Wnt signaling pathway. The deletion of pop-1 (homologue of Tcf) and sys-1 (β-catenin homologue) triggers loss of DTC and subsequently loss of GSC niche [68].

A similar structure of GSC niche is found in *Drosophila.* This niche is composed by a cluster of stromal cells located on the apex of *Drosophila* gonads. These cells are named hub cells in the testis, and cap cells in the ovary. Both stromal cell types are in intimate contact with GSCs to provide critical physical and diffusible factor-mediated interactions. The signaling by physical interaction involves shotgun (E-cadherin homologue) located in adherent junctions between stromal and germinal stem cells. Shotgun is able to activate Armadillo (β-catenin homologue) and the Wnt pathway which is essential in GSC self-renewal. In addition, there are also a number of diffusible signals which are involved in the maintaining of GSC population as well as in their differentiation, such as decapentaplegic (BMP homologue) and hedgehog, respectively [69].

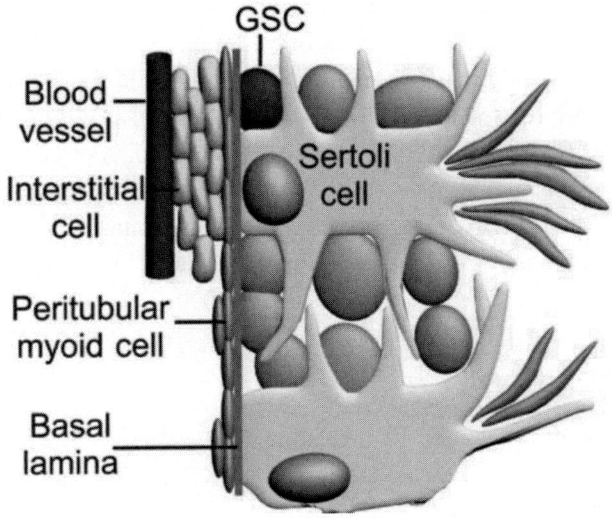

Figure 14. Male germinal stem cells (GSC) are located adjacent to basal lamina of epithelium that coats the seminiferous tubules in testes. Germinal stem cells in males are looked after Sertoli cells. Sertoli cells nurture the developing spermatozoids and phagocyte the residual cytoplasm of germinal cells. Morphological and physiological changes of germinal cells occur as germ cells move from basal lamina to lumen of seminiferous through margins of Sertoli cells.

In mammals, the niche for GSCs is not well established. However, in males, it is possible that GSCs are located adjacent to basal lamina of epithelium that coats the seminiferous tubules. In this way, a good candidate for niche supporting cell might be Sertoli cell which is responsible for nurturing the developing spermatozoids and to phagocyte the residual cytoplasm of germinal cells in differentiation (Figure 14).

Stem Cell Division

SCs have a finite replicative life span, defined as the number of daughter cells that can produce before they cease to divide and ultimately die.

Cell cycle includes a number of processes which make possible that a cell can give rise to two daughter cells. These processes involve specific regulatory networks so that they are strictly controlled both in time and space. Progression through the cell cycle requires duplication of the genetic material and its delivery of the newly duplicated genomes to the two daughter cells during mitosis, which represent one of the key processes in living organisms. This genetic duplication occurs in coordination with an increase in cellular components and changes in cell architecture. In fact, it is more appropriate to consider the concept of cell proliferation instead of cell division, since it includes cell cycle control, its arrest and reactivation, the endocycle, and the cell differentiation and death. In addition, these processes, considered at the cellular level, must be connected to the particular ontogenic program. While organogenesis in mammals occurs during embryogenesis, the organ maintenance and reparation are post-embryonic and continuous processes that occur over the entire lifespan of the organism: this remarkable fact relies on the existence of SC niches.

The balance between SC division and differentiation implies an accurate joint regulation of cell division, cell cycle arrest and its reactivation, the endoreplication, and differentiation. The emerging view is that regulatory components of cell cycle, in addition to controlling cell cycle progression, also have some roles in the coordination of cell division in the context of a developing organism: cell cycle regulators, in addition to controlling cell division, also function as targets for maintaining cell homeostasis during development.

The balance between cell division and differentiation during SC life span is provided by a number of mechanisms and interactions among different regulatory pathways, as it is discussed below.

Symmetric versus Asymmetric Division

Each cell splits in two at the final step of division cycle. This division generates normally two equal daughters. When SCs divide, their daughters may maintain SC identity, or may initiate differentiation. Conceptually, there are only three potential outcomes for SCs after division: (1) a symmetrical division leading to net expansion of SC pool; (2) a symmetrical division that leads to the production of differentiated cells; and (3) an asymmetrical division leading to the maintenance of the SC population while they are generated differentiated cells [70-72] (Figure 15).

Some mammalian SC populations may undergo both asymmetric and symmetric divisions depending on their circumstances [73]. The relative proportion of symmetric divisions appears to change over time, with symmetric divisions predominating at early time points when the SC pool is in expansion [73,74]. However, it is not clear whether a single cell can switch from a symmetric to an asymmetric mode of cell division.

One of the challenges in biology of development is to understand how a SC generates the cell diversity necessary to produce a mature, multicellular organism. Although it is well established that asymmetric cell divisions play a crucial role during development, their exact mechanism is yet to be fully understood. Over the past years, studies in model organisms [75-78] have shown that establishment of an axis of cell polarity, the translocation of the nucleus to specific cellular locations, the orientation of the mitotic spindle along the axis of polarized cell, and a precise spatio-temporal execution of cytokinesis to achieve preferential segregation of cell fate determinants, constitute the basis of SC asymmetric divisions.

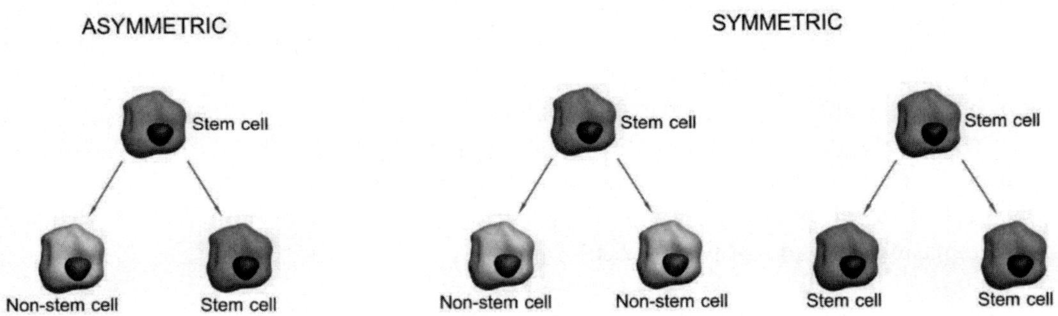

Figure 15. Asymmetric and symmetric division in stem cells.

In addition, developmental signals can change the plane of cell division to facilitate the asymmetric segregation of fate determinants and to control the position and the relative sizes of daughter cells. The anaphase spindle is the milestone in the regulation of symmetric versus asymmetric cell division.

To produce new tissue cells, SCs undergo repeated, intrinsically determined asymmetric cell divisions to produce one differentiated (progenitor) daughter cell and another daughter cell that is still a SC [79] (Figure 15). The choice between these alternative fates is critical to maintain the SC number and their potentially.

Asymmetrical divisions may have extrinsic or intrinsic regulators. Intrinsically, asymmetrical divisions can be the result of a differential distribution of determinants during cellular division. Asymmetrically distributed factors are differentially located in daughter cells by: (1) a selective localization of the mRNA; (2) a preferential transport of the protein; or (3) distinct mechanisms of degradation [80]. The segregation of cell fate determinants, which leads to the production of two cells with different properties, is closely linked to cell cycle machinery and to mitotic cues.

Among extrinsic determinants, positional information in the niche may represent a general and evolutionary conserved regulatory mechanism for cellular assymmetry [72]. Cell in close contact with the niche maintain its SC identity, while separated cells enter differentiation. This system is regulated by direct interaction between SCs and niche surrounding cells, as well as by the spindle orientation during mitosis.

Symmetric divisions allow the maintaining of SC pool by factors that control the probability of self-renewal versus differentiation [81]. Contrary to asymmetrical divisions, symmetrical divisions are an important mechanism for SC expansion during embryonic development and for their replacement after an injury in adult tissues [82].

Polarity

Cell polarity is defined as asymmetry in cell shape, its functions, and protein distribution [83]. Polarity is a property of virtually all eukaryotic cells and is linked to other aspects of developmental biology, such as differentiation, signaling, cytoskeletal organization, and division [84]. In eukaryotic cells, the polarity axis is established using two general mechanisms: (1) cytoskeleton orientation to direct polar intracellular transport; and (2) establishment of stable marks within the actin rich cortex which serve as reference points to capture and concentrate specific markers in cell poles [85].

They are required three steps for successful execution of asymmetric cell division: (1) a polarity cue for the determination of cell axis; (2) subsequent formation of cortical domains which establish cell polarity; and (3) communication of this polarity to the cytoskeleton so that cytoplasm can divide [86] (Figure 16).

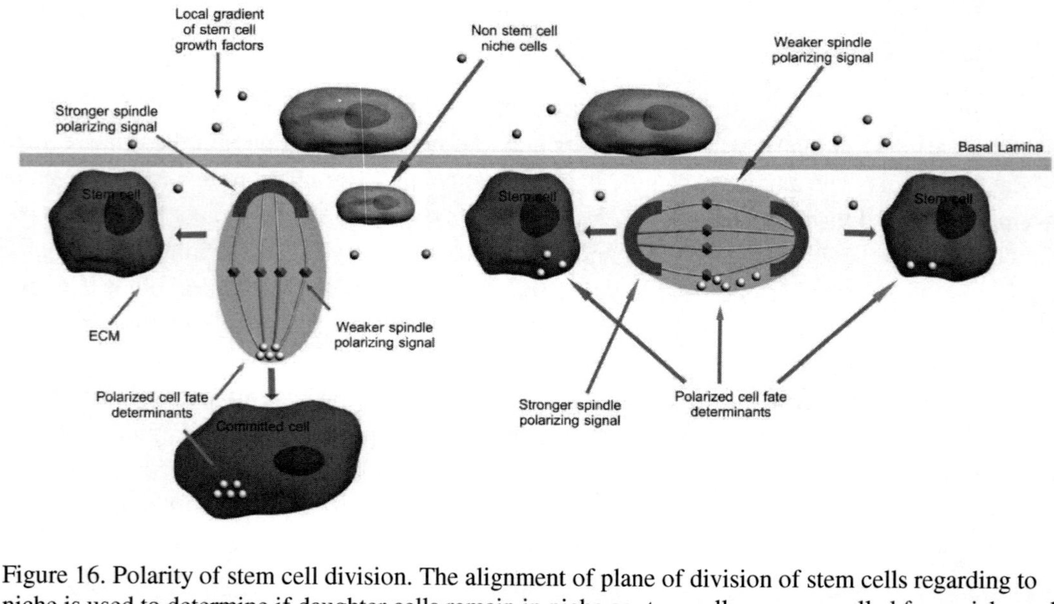

Figure 16. Polarity of stem cell division. The alignment of plane of division of stem cells regarding to niche is used to determine if daughter cells remain in niche as stem cells or are expelled from niche and begin their walk toward differentiation. ECM: extracellular matrix.

In most instances, an external asymmetrical signal is necessary to generate polarity. This signal comes generally from asymmetries present within the niche microenvironment.

In general, cell polarity is a succession of events initiated by extrinsic or intrinsic signals which are interpreted by receptors at the cell surface and transduced to cytoskeleton. The cytoskeleton establishes cellular asymmetry in terms of positioning and orientation of the mitotic spindle, the allocation of cell fate determinants, and their stabilization by cytokinesis [87].

There are two general mechanisms by which cell polarity determines the division plane: (1) the orientation and positioning of mitotic spindle (or the nucleus) in a precise cellular location; and (2) determination of the division plane in accordance with localized cortical cues.

With regard to the mitotic spindle role during SC division, the nucleus, initially central, is moved to an asymmetric position before division. This movement is mediated by a specialized microtubule system that appears at the generative cell pole. Both microtubules and actin filaments are involved in maintenance of the excentric position of nucleus. Regarding to cortical cues, cells have an internal microenvironment which is asymmetrically distributed and provides characteristics of each cell type.

Cell Fate Determinants' Allocation

Cellular differentiation has been depicted in terms of localization of cell fate determinants in daughter cells. In the knowledge of SC division, a central issue is the specific segregation of cell fate determinants (signaling molecules and transcription factors). Cell fate determinants' segregations are also coordinated with nuclear/spindle translocation/orientation and cell plate formation. In animals, the involvement of the spindle in cell fate determinants' allocation has been established in relation to proteins/genes identified and their cell-specific interactions. In most cell types, the relative size of the two daughter cells is determined by the position of the mitotic spindle within the geometry of the mother cell. This is generally accomplished by the cortical forces generated by the microtubules, linked to specific sites in the cortex. In the asymmetric SC division, cytokinesis is coordinated by the spindle orientation along the polar axis. As result, the mother cell is partitioned into two daughter cells of different fates, depending on the allocation of cell fate determinants.

The influence of cytokinesis [88] on division appears to be critical. Cytoplasmic constituents, as water and soluble proteins, are inherited by daughter cells in equal quantities by a passive process during equatorial cytokinesis [89]. Most models for cytokynesis include the generation of a differential cortical tension to facilitate the flow of components (such as membrane receptors), which provide regional specificity for cytokynesis.

In division process, certain organelles and cell fate determinants may be allocated in a cell-specific manner. The mRNA localization and its post-transcriptional regulation may also accomplish the same results. The intracellular asymmetries in mRNA localization are used to modulate the local concentration of proteins and their assembly, and can profoundly influence cellular identity and morphogenesis [90,91]. Recently, it has been shown that microRNAs are actively transcribed during the process of cell division. MicroRNAs may play a significant role in cellular fate assignation, as regulators of gene expression [90,92], by regulating mRNA cellular fate determinants, during stem cell division.

SIGNALING FROM A SUPPORT CELL NICHE

One of the critical questions in the adult SC field is the mechanism that regulates the decision between self-renewal and differentiation. Adult SCs have two fundamental properties: a long-term capacity to divide, and the ability to produce daughter cells. The balance between self-renewal and initiation of differentiation is crucial. If too many daughter cells begin differentiation, the SC population may be depleted. Conversely, if too many daughter cells maintain SC identity, the SC population may expand out of proportion, giving rise to a pool of proliferative, incompletely differentiated cells that may mutate and become tumorigenic.

For cells, signals are crucial in the decision of division, specialization, death, or metabolism control [93]. In general, the outcome of a signal depends on the developmental status of the reciving cell [94].

Multicellular organisms depend on molecular signaling pathways designed to receive and interpret signals from their environment and to respond appropriately. In the context of the study of asymmetric cell divisions, it is necessary to understand the integration of distinct signaling pathways and their relation with SC fate and differentiation.

Cell division and differentiation share a reciprocal relationship: cells in active division do not differentiate, while differentiated cells do not divide.

A dynamic coordination is seen during many developmental situations, where the cell cycle transitions of G1-S-G2-M, followed by cytokinesis, are shuffled, altered, or removed to facilitate a concrete developmental output.

The cell cycle is an orderly and unidirectional sequence of events driven by a common class of proteins: the Cdk-cyclin complexes [95,96]. The decision of divisionor differentiation is normally made in the G1 phase. Then, cells undergo G1-S transition and enter the S phase for DNA/chromatin replication.

The entry and exit from mitosis are governed by the activation/inactivation of Cdk-activating kinases, Cdk inhibitors, and phosphorylation/dephosphorylation events [97]. Several Cdks and kinases, involved in signaling, play important roles in formation of the cell plate. The pRb-cyclinD-E2F pathway regulates the G1 to S phase transition. E2F pathway and phosphorylation of pRb, with the involving of cyclinD-Cdk and cyclinE-Cdk complexes, pave the way for S phase entry. pRb, together its partners p107/p130, bind to the E2F familiy transcription factors to block entry to S phase (Figure 17).

The physical adhesion to the niche, through adherent junctions, may be a feature of many adult SCs. The mechanisms by which the SC niche might to control the SC number and maintain the correct balance between self-renewal and differentiation are complex. These processes involve a complex interplay between intercellular and intracellular mechanisms. First, the number of SCs in niches is defined by short-range signals coming from the neighbouring cells. Second, cell-cell interactions enable SCs to remain tightly associated with the niche. Third, SCs are polarized with respect to the niche. Finally, SCs can orient their mitotic spindles to ensure the outcome of asymmetric cell divisions.

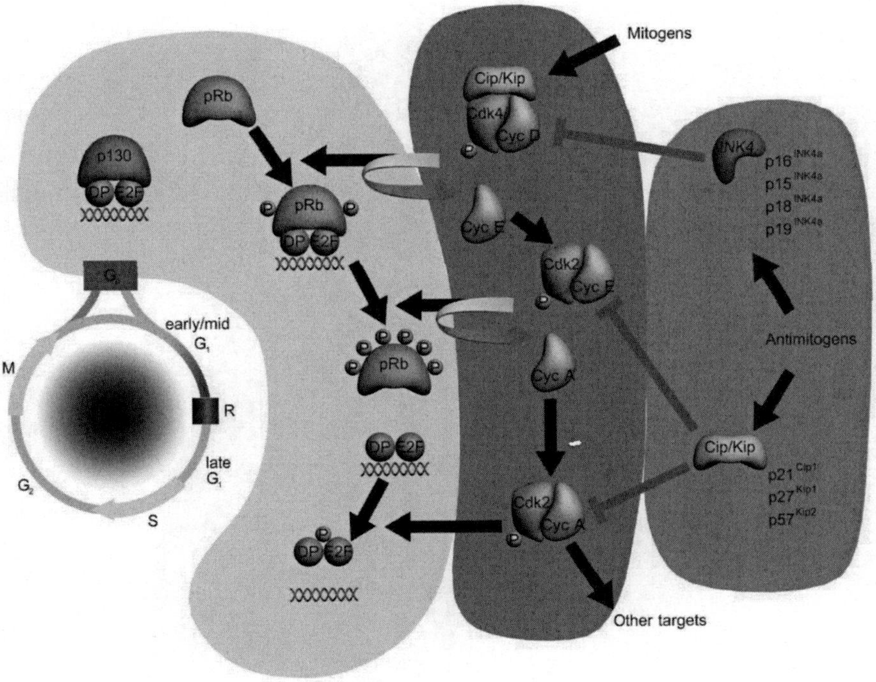

Figure 17. Overview of molecular regulation of cell cycle.

SCs are located usually adjacent to niche cells that secrete factors, required for the maintaining of SC identity. In addition, niche-SC interactions are required for the maintenance of SCs within the niche and to ensure that SCs are close to self-renewal signals.

SCs reside in niches with little vasculature, so they are in a low-oxygen environment [98-100]. Therefore, the SCs potential for self-renewal may be strictly linked to the capacity of SCs to survive in a hypoxic environment. The control of SC survival and the regulation of hypoxic response may be intimately coupled and to share common genes and pathways [101]. Recent data indicate that the SC regulation through Notch pathway is linked to the hypoxia modulator, HIF-1α, to promote stem/undifferentiated phenotype [102]. There are evidences that the hypoxia affects to function and survival of SCs [103,104]. *In vitro*, hypoxia maintains SCs in immaturity state, induces a loss of differentiation markers, and inhibits differentiation. *In vivo*, SCs express high levels of glycolitic enzymes. In addition, SCs express higher levels of hypoxia-regulated proteins than do their mature progeny.

They are thought that various transcription factors, growth factors, and matrix molecules act as guidance cues to help orchestrate homeostasis and dynamic of different SC niches [105].

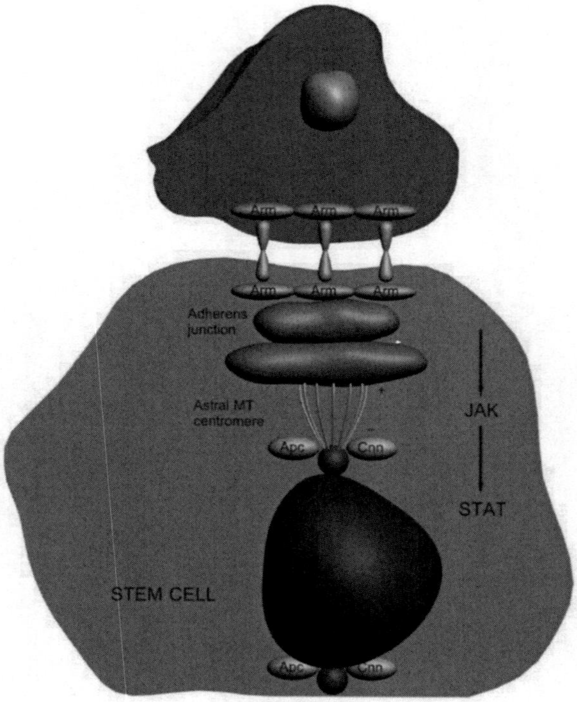

Figure 18. Intercellular communication is used by niche stromal cells to maintain the self-renewal features of their SCs. The signal flows from stromal cell to SC through adherent junctions established between both and this signal triggers the activation of JAK-STAT signaling into SC. Here, this is exemplified with germinal SC and its niche partner cell in *Drosophila*. Apc: adenomatous polyposis coli gene. Cnn: centrosomin, integral centrosome component in *Drosophila*. Arm: armadillo (β-catenin counterpart in fly).

A number of papers have revealed that niche is able to promote SC self-renewal of SCs through the activation of Janus-kinase (JAK) and signal transducer and activator of transcription (STAT) pathways (Figure 18) [25,33]. JAKs are non-receptor tyrosine kinases that mediate signalling downstream of many cytokines and receptors in mammalians, due in part to the phosphorylation and activation of STAT [106].

Wnt

Wnt has vital roles in adult tissues—for example, in the regulation of cell proliferation and motility, generation of cell polarity, and specification of cell fate. Recently, Wnt signaling has been involved in the control of differentiation in hematopoietic and skin SCs, as well as in regulating myogenesis, chondrogenesis, and adipogenesis [107]. Wnt signaling is involved in a variety of cellular processes, including proliferation, differentiation, survival, apoptosis, and cell motility [108]. Wnt signals are transduced to the canonical pathway for cell fate determination, and to the noncanonical pathway for controlling cellular movement and polarity [109]. A surprising number of Wnt cytoplasmic regulators (e.g., β-catenin, Bcl-9/Lgs, APC, Axin) appear often, also in the nucleus [108]. In the absence of Wnt, a β-catenin-Axin-GSK3β-APC complex is located in the cytoplasm. GSK3β phosphorylates β-catenin, and phosphorylated β-catenin is degraded. When Wnt binds to its Frizzled receptor, Axin is recruited to the cell membrane where joins to LRP5 and, subsequently, is subjected to degradation. As a result, GSK3β no longer phosphorylates β-catenin. β-catenin goes to the nucleus and forms complexes with TCF/LEF transcription factors to activate the transcription of Wnt target genes [110].

Nanog

Nanog, a homebox factor, is expressed *in vivo* in the internal cells of compacted morulae, in the inner cell mass of perimplantation blastocyst, and in post-migratory germ cells. Nanog is a marker of all pluripotent cell lines and is essential for maintenance of a pluripotent phenotype both *in vivo* and *in vitro*. Nanog is considered as master transcriptional regulator of self-renewal and pluripotency in embryonic stem cells. Nanog is expressed in embryonic stem cells and is rapidly downregulated during in vitro differentiation. Upon aggregation of embryonic stem cells, Nanog downregulation occurs at the embryonal outer layer, even in the presence of Leukemia inhibitory factor (LIF). Interestingly, Nanog overexpression maintains embryonic stem cells in the absence of LIF, and inhibits primitive endoderm differentiation. These data suggest that primary function for Nanog is the repression of differentiation to primitive endoderm. The mechanism by which Nanog is able to inhibit primitive endoderm differentiation is probably multifaceted. In fact, Nanog, in combination with Oct4 and Sox2, can work both as activator and repressor of target genes responsible for self-renewal and differentiation [111].

Notch

Notch proteins are involved in distinct processes related with the cellular fate determination in vertebrates, including lateral inhibition [55]. Canonical Notch signaling is mediated by a family of cell surface transmembrane receptors. Upon interaction with cell surface ligands of the DELTA/JAGGED family, expressed on neighboring cells, the intracellular domain of the NOTCH receptor is cleaved and translocated to nucleus, where it associates with the transcription factor CBF-1 (RBPJK, RBPSUH) [112]. In mammals, four Notch receptors (Notch1–Notch4) and five Notch ligands (Delta-like1, Delta-like3, Delta-like4, Jagged1, and Jagged2) have been identified. However, there are few evidences about the ability of Delta-like3 for the binding to Notch receptors, and about its role as Notch ligand [113].

Hedgehog

The Hedgehog (Hh) gene is involved in various aspects of embryonic development, such as bilateral symmetry, the anterior-posterior patterning of limb bud, and neural tube formation [55]. In higher vertebrates, the Hh gene family consists of at least three members: Sonic,

Indian, and Desert hedgehog (Shh, Ihh, and Dhh, respectively). Among these, Shh and Ihh are involved in the skeletal formation during development and in skeletal repair [114]. The Hh pathway has been closely associated with cell cycle regulation. In addition, Hh signaling was involved both in the specification of hematopoiesis during embryogenesis and proliferation of hematopoietic stem cells in adulthood [115]. All Hh molecules act through Patched (PTCH1) and Smoothened (SMO) receptors, to lead to the activation of the Gli transcription factors [116].

BMP

BMPs are members of the TGF-β superfamily of secreted signaling molecules. BMPs have important functions in many biological processes, as embryogenesis. BMPs join to specific receptors serine/theronine kinase, to send signals to cell nucleus through Smad proteins [55].

Plasticity / Trans-differentiation / Cell fusion

The capacity of stem cells from a certain tissue to give rise differentiated cells from another tissue is called plasticity or trans-differentiation [117,118]. The differentiated cells resulting from plasticity have usually the same morphological characteristics of the differentiated cells, and display their characteristic surface markers.

The SC plasticity has been a subject of major debate since its discovery, when it was demonstrated that adult brain cells can give rise to blood cells *in vitro* [119]. In papers where transplanted SCs show plasticity *in vivo*, these integrate into a different tissue from its original tissue, and assume at least some of its characteristics [120-124]. However, to date, there are limited evidences that adult SCs can generate mature, fully functional cells belonging to other tissues distinct to its original tissue [118].

Collectively, studies about plasticity suggest that SC populations in adult mammals are not stable entities, so that, after exposure to a new environment, they may populate other tissues, and possibly differentiate into other cell types. These results imply that somatic SCs can switch fates across the barrier of different germ layers to generate ectoderm, mesoderm, and endoderm.

Trans-differentiation implies the loss of the tissue-specific markers and function of original cell, and the acquisition of markers and functions of the trans-differentiated cell. Trans-differentiation is a mechanism by which SCs could contribute to different cellular lineages.

Studies about SC trans-differentiation have been highly controversial [117,125,126], insofar as they have been invoked as possible mechanisms to explain SC plasticity and lineage conversion [127] (Figure 3). First, lineage conversion may occur by activation of a differentiation program, otherwise dormant, to alter the lineage specificity of cells. Second, lineage conversion could also occur via dedifferentiation of a tissue-specific cell to a more primitive, multipotent cell, and subsequent re-differentiation along a new lineage. Third, to demonstrate trans-differentiation, it is essential ensuring the purity or homogeneity of SC population, to exclude the possibility that different stem or progenitor may be contributing to these results. Fourth, lineage conversion may also arise from a single and rare, pluripotent adult progenitor cell, present in many tissues, which is able to generate tissues from multiple germ layers. Finally, it is possible consider cellular fusion as a mechanism to explain SC plasticity. Molecular mediators for heterotypic fusion among cells of different lineages have been already elucidated, and represent a physiological mechansim designed to repair a wide variety of tissues [128]. In fact, cell fusion data put in question the validity of many trans-differentiation observations [129,130]. In cell fusion, generated cells inherit molecular markers and properties from both SC types. Thus, fused cells can contribute, as chimeras, to somatic tissues.

Cell fusion is an important issue in the debate about SC plasticity. The question is especially relevant in cell types that normally fuse, such as skeletal and cardiac muscle cells.

Conclusions

Conceptually, a niche is a physical place that provides protection and nourishment to SCs, away from molecules that may cause differentiation, proliferation, or distinct alterations. At the niche, SCs are virtually enveloped by differentiated cells, specialized to accomodate and interact with SCs.

Adult SCs are often localized in specific niches. SCs reside, in many tissues, within a niche formed by a group of surrounding cells and ECM, which provide an optimal microenvironment for SC function. Distinct evidences suggest that, when a SC compartment is formed in a tissue, SCs live there for long time.

Some SCs from adult mammalians don't seem to have a specified niche within their respective tissue (e.g., skeletal muscle). In other cases, however, an SC niche is established in a developing tissue. Cells within this niche are then activated in response to specific environmental cues (e.g., skin, hair, follicles, epidermis, mammary gland, lung, brain, etc.).

In general, the niche consists in a highly organized microenvironment in which distinct factors, such as secreted cytokines, interactions with ECM, and intercellular adhesion operate cooperatively.

References

[1] Xie, T; Spradling, AC. A niche maintaining germ line stem cells in the Drosophila ovary. *Science*, 2000 290, 328-330.

[2] Kiger, AA; Jones, DL; Schulz, C; Rogers, MB; Fuller, MT. Stem cell self-renewal specified by JAK-STAT activation in response to a support cell cue. *Science*, 2001 294, 2542-2545.

[3] Crittenden, SL; Bernstein, DS; Bachorik, JL; Thompson, BE; Gallegos, M; Petcherski, AG; Moulder, G; Barstead, R; Wickens, M; Kimble, J. A conserved RNA-binding

protein controls germline stem cells in Caenorhabditis elegans. *Nature*, 2002 417, 660-663.

[4] Sprandling, A; Dummond-Barbosa, D; Kai, T. Stem cells find their niche. *Nature*, 2001 414, 98-104.

[5] Potten, CS. The epidermal proliferative unit: the possible role of the central basal cell. *Cell Tissue Kinet*, 1974 7, 77-88.

[6] Furstenberger, G; Gross, M; Schweizer, J; Vogt, I; Marks, F. Isolation, characterization and in vitro cultivation of subfractions of neonatal mouse keratinocytes: effectss of phorbol esters. *Carcinogenesis*, 1986 7, 1745-1753.

[7] Yuspa, SH; Ben, T; Hennings, H; Lichti, U. Divergent response in epidermal basal cells exposed to the tumor promoter 12 O-tetradecanoylphorbol-13-acetate. *Cancer Res*, 1982 42, 2344-2349.

[8] Smith, GH; Strickland, P; Daniel, CW. Putative epithelial stem cells loss corresponding with mammary growth senescence. *Cell Tissue Res*, 2002 310, 313-320.

[9] Hayflick, L. The limited in vitro lifetime of human diploid cell strains. *Exp Cell Res*, 1965 37, 614-636.

[10] Hammond, SL; Ham, RG; Stampper, MR. Serum-free growth of human mammary epithelial cells: rapid clonal growth in defined medium and extended serial passage with pituitary extract. *Proc Natl Acad Sci USA*, 1984 81, 5435-5439.

[11] Edwards, PA; Brooks, IM; Bunnage, HJ; Foster, AV; Ellison, MC; O´Hare, MJ. Clonal analysis of expression of epithelial antigens in cultures of normal human breast. *J cell Sci*, 1986 80, 91-101.

[12] O´Hare, MJ; Ormerod, MG; Monaghasa, P; Lane, EB; Gusterson, BA. Characterization in vitro of luminal and myoepithelial cells isolated from the human mammary gland by cell sorting. *Differentiation*, 1991 46, 209-221.

[13] Clayton, H; Titley, I; Vivanco, M. Growth and differentiation of progenitor/stem cells derived from the human mammary gland. *Exp Cell Res*, 2004 297, 444-460.

[14] Stingl, J; Eanes, CJ; Zandieh, I; Emerman, JT. Characterization of bipotent mammary epithelial progenitor cells in normal adult human breast tissue. *Breast Cancer res Treat*, 2001 67, 93-109.

[15] Dontu, G; Abdallah, WM; Foley, JM; Jackson, KW; Clarke, MF; Kawamura, MJ; Wicha, MS. In vitro propagation and transcriptional profiling of human mammary stem/progenitor cells. Genes Dev, 2003 17, 1253-1270.

[16] Alvi, AJ; Clayton, H; Joshi, C; Enver, T; Ashworth, A; Vivanco, MM. Functional and molecular characterization of mammary side population cells. *Breast Cancer Res*, 2003 5, R1-R8.

[17] Clarke, RB; Spence, K; Andersonm, E; Howell, A; Okano, H; Potten, CS. A putative human breast stem cell population is enriched for steroid receptor-positive cells. *Dev Biol*, 2005 277, 443-456.

[18] Slack, JM. Stem cells in epithelial tissues. *Science*, 2000 287, 1431-1433.

[19] Smith, GA. Label-retaining epithelial cells in mouse mammary gland divide asymmetrically and retain their template DNA strands. *Development*, 2005 132, 681-687.

[20] Sujata, L; Chaudhuri, S. Stem cell niche, the microenvironment and immunological crosstalk. *Cellular & Molecular Immunology*, 2008 5, 107-112.

[21] Li, L; Neaves, WB. Normal stem cells and cancer stem cells: the niche matters. *Cancer Res*, 2006 66, 4553-4557.

[22] Schofield, R. The relationship between the spleen colony-forming cell and the haemopoietic stem cell. *Blood Cells*, 1978 4, 7-25.

[23] Conti, L; Pollard, SM; Gorba, T; et al. Niche-independent symmetrical self-renewal of a mammalian tissue stem cell. *PLos Biol*, 2005 3, e283.

[24] Sprandling, A; Drummond-Barbosa, D; Kai, T. Stem cells find their niche. *Nature*, 2001 414, 98-104.

[25] Tulina, N; Matrevis, E. Control of stem cell self-renewal in Drosophila spermatogenesis by JAK-STAT signalling. *Science*, 2001 294, 2546-4549.

[26] Morrison, SJ; Shah, NM; Anderson, DJ. Regulatory mechanisms in stem cell biology. *Cell*, 1997 88, 287-298.

[27] Kiger, AA; Jones, DL; Schulz, C; Rogers, MB; Fuller, MT. Stem cell self-renewal specified by JAK-STAT activation in respond to a support cell cue. *Science*, 2001 294, 2542-2545.

[28] Takeichi, M. Cadherin cell adhesion receptors as a morphogenetic regulator. *Science*, 1991 251, 1451-1455.

[29] Watt, FM; Hohar, BL. Out of Eden: stem cells and their niches. *Science*, 2000 287, 1427-1430.

[30] Voura, EB; Billia, F; Iscove, NN; et al. Expression mapping of adhesion receptor genes during differentiation of individual hematopoietic progenitors. *Exp Hematol*, 1997 25, 1172-1179.

[31] Papayannopoulou, T; Priestley, GV; Nakamoto, B; et al. Molecular pathways in bone marrow homing: Dominant role of $\alpha_4\beta_1$ over β_2-integrins and selectins. *Blood*, 2001. 98, 2403-2411.

[32] Chen, D; McKearin, D. Dpp signaling silences bam transcription directly to establish asymmetric divisions of germline stem cells. *Curr Biol*, 2003 13, 1786-1791.

[33] Kiger, AA; White-Cooper, H; Fuller, MT. Somatic support cells restrict germline stem cell self-renewal and promote differentiation. *Nature*, 2000 407, 750-754.

[34] Xie, T; Spradling, AC. A niche maintaining germ line stem cells in the Drosophila ovary. *Science*, 2000 290, 328-330.

[35] Yamashita, YM; Jones, DL; Fuller, MT. Orientation of asymmetric stem cell division by the APC tumor suppressor and centrosome. *Science*, 2003 301, 1547-1550.

[36] Loeffler, M; Roeder, I. Tissue stem cells: definition, plasticity, heterogeneity, self-organization, and models a conceptual approach. *Cells Tiss Organs*, 2002 171, 8-26.

[37] Svensson, M. Isolation and culture of human hematopoietic progenitors for studies of dendritic cell biology. *Methods Mol Biol*, 2009 531, 187-202.

[38] Watt FM. Human epidermal stem cells. In: Robert Lanza, Helen Blau, Douglas Melton, Malcolm Moore, E. Donnall Thomas, Catherine Verfaillie, Irving Weissman, Michael West editors. *Handbook of stem cells. Volumen 2: Adult and fetal stem cells.* San Diego, California, USA. Elsevier Academic Press, 2004 245 pp.

[39] Linheng Li; Ting Xie. Stem cell niche: structure and function. *Annu Rev Cell Dev Biol*, 2005 21, 605-631.

[40] Gilyarov, AV. Nestin in central nervous system cells. *Neurosci Behav Physiol*, 2008 38: 165-169.

[41] Herrera, MB; Bruno, S; Buttiglieri, S; Tetta, C; Gatti, S; Deregibus, MC; Bussolati, B; Camussi, G. Isolation and characterization of a stem cell population from adult human liver. *Stem Cells*, 2006 24, 2840–2850.

[42] Fernandes, A; King, C; Guz, Y; Stein, R; Wright, CVE; Teitelman, G. Differentiation of new insulin-producing cells is induced by injury in adult pancreatic islets. *Endocrinology*, 1997 138, 1750-1762.

[43] Teisanu RM, Lagasse E, Whitesides JF, Stripp BR. Prospective isolation of bronchiolar stem cells based upon immunophenotypic and autofluorescence characteristics. *Stem Cells*, 2009 27, 612-22.

[44] Villadsen R, Fridriksdottir AJ, Rønnov-Jessen L, Gudjonsson T, Rank F, LaBarge MA, Bissell MJ, Petersen OW. Evidence for a stem cell hierarchy in the adult human breast. *J Cell Biol*, 2007 177, 87-101.

[45] Collins, AT; Habib, FK; Maitland, NJ; Neal, DE. Identification and isolation of human prostate epithelial stem cells based on α2β1-integrin expression. Journal of Cell *Science*, 2001 114, 3865-3872.

[46] Lawson, DA; Xin, L; Lukacs, RU; Cheng, D; Witte, ON. Isolation and functional characterization of murine prostate stem cells. *Proc Natl Acad Sci USA*, 2007 104, 181-186.

[47] Goto, K; Salm, SN; Coetzee, S; Xiong, X; Burger, PE; Shapiro, E; Lepor, H; Moscatelli, D; Wilson, EL. Proximal Prostatic Stem Cells Are Programmed to Regenerate a Proximal-Distal Ductal Axis. *Stem Cells*, 2006 24, 1859–1868.

[48] Barile, L; Messina, E; Giacomello, A; Marbán, E. Endogenous cardiac stem cells. *Prog Cardiovasc Dis*, 2007 50: 31-48.

[49] Umemoto, T; Yamato, M; Nishida, K; Yang, J; Tano, Y; Okano, T. Limbal epithelial side-population cells have stem cell-like properties, including quiescent state. *Stem Cells*, 2006 24: 86-94.

[50] Gronthos, S; Brahim, J; Li, W; Fisher, LW; Cherman, N; Boyde, A; DenBesten, P; Gehron Robey, P; Shi, S. Stem cell properties of human dental pulp stem cells. *J Dent Res*, 2002 81, 531-535.

[51] Martinez-Agosto, JA; Mikkola, HKA; Hartenstein, V; Banerjee, U. The hematopoietic stem cell and its niche: a comparative view. *Genes Dev*, 2007 21, 3044–3060.

[52] Ting Xie; Linheng Li. Stem cells and their niche: an inseparable relationship. *Development*, 2007 134, 2001-2006.

[53] Haylock, DN; Williams,B; Johnston, HM; Liu, MCP; Rutherford, CE; Whitty, GA; Simmons, PJ; Bertoncello, I; Nilsson, SK. Hemopoietic stem cells with higher hemopoietic potential reside at the bone marrow endosteum. *Stem Cells*, 2007 25, 1062–1069.

[54] Drummond-Barbosa, D. Stem cells, their niches and the systemic environment: an aging network. *Genetics*, 2008 180: 1787–1797.

[55] Rizvi, AZ; Wong, MH. Epithelial stem cells and their niche: there's no place like home. *Stem Cells*, 2005 23, 150-165.

[56] Zhao, M; Amiel, SA; Christie, MR; Muiesan, P; Srinivasan, P; Littlejohn, W; Rela, M; Arno, M; Heaton, N; Huang, GC. Evidence for the presence of stem cell-like progenitor cells in human adult pancreas. *J Endocrinol*, 2007 195, 407–414.

[57] Soria, B; Bedoya, FJ; Martin, F. Gastrointestinal stem cells I. Pancreatic stem cells. *Am J Physiol Gastrointest Liver Physiol*, 2005 289, G177–G180.

[58] Majka, SM; Beutz, MA; Hagen, M; Izzo, AA; Voelkel, N; Helm, KM. Identification of novel resident pulmonary stem cells: form and function of the lung side population. *Stem Cells*, 2005 23, 1073–1081.

[59] Liu, X; Engelhardt, JF. The glandular stem/progenitor cell niche in airway development and repair. *Proc Am Thorac Soc*, 2008 5, 682–688.

[60] Gudjonsson,T; Villadsen, R; Nielsen, HL; Rønnov-Jessen, L; Bissell, MJ; Petersen, OW. Isolation, immortalization, and characterization of a human breast epithelial cell line with stem cell properties. *Genes Dev*, 2002 16, 693–706.

[61] Dubrovskaa, A; Kimb, S; Salamoneb, RJ; Walkerb, JR; Mairac, SM; Garcia-Echeverria, C; Schultza, PG; Reddy, WA. The role of PTEN/Akt/PI3K signaling in the maintenance and viability of prostate cancer stem-like cell populations. *Proc Natl Acad Sci USA*, 2009 106, 268-273.

[62] Mimeault, M; Mehta, PP; Hauke, R; Batra, SK. Functions of normal and malignant prostatic stem/progenitor cells in tissue regeneration and cancer progression and novel targeting therapies. *Endocr Rev*, 2008 29, 234–252.

[63] Bearzi, C; Rota, M; Hosoda, T; Tillmanns, J; Nascimbene, A; De Angelis, A; Yasuzawa-Amano, S; Trofimova, I; Siggins, RW; LeCapitaine, N; Cascapera, S; Beltrami, AP; D'Alessandro, DA; Zias, E; Quaini, F; Urbanek, K; Michler, RE; Bolli, R; Kajstura, J; Leri, A; Anversa, P. Human cardiac stem cells. *Proc Natl Acad Sci USA*, 2007 104, 14068-14073.

[64] Shortt, AJ; Secker, GA; Munro, PM; Khaw, PT; Tuft, SJ; Daniels, JT. Characterization of the limbal epithelial stem cell niche: novel imaging techniques permit in vivo observation and targeted biopsy of limbal epithelial stem cells. *Stem Cells*, 2007 25, 1402-1409.

[65] Takács L, Tóth E, Berta A, Vereb G. Stem cells of the adult cornea: from cytometric markers to therapeutic applications. *Cytometry A*, 2009 75, 54-66.

[66] Arthur, A; Rychkov, G; Shi, S; Koblar, SA; Gronthos, S. (2008). Adult human dental pulp stem cells differentiate toward functionally active neurons under appropriate environmental cues. *Stem Cells*, 2008 26, 1787–1795.

[67] Tummers, M; Thesleff, I. Root or crown: a developmental choice orchestrated by the differential regulation of the epithelial stem cell niche in the tooth of two rodent species. *Development*, 2003 130, 1049-1057.

[68] Walker, MR; Patel, KK; Stappenbeck, TS. The stem cell niche. *J Pathol*, 2009 217, 169-180.

[69] Naveiras, O; Daley, GQ. Stem cells and their niche: a matter of fate. *Cell Mol Life Sci*, 2006 63, 760-766.

[70] Morrison, SDJ; Kimble, J. Asymmetric and symmetric stem-cell divisions in development and cancer. *Nature*, 2006 441, 1068-1074.
[71] Knoblich, JA. Mechanisms of asymmetric stem cell division. *Cell*, 2008 132,583-597.
[72] Gonczy, P; DiNardo, S. The germ line regulates somatic cyst cell proliferation and fate during Drosophila spermatogenesis. *Development*, 1996 122, 2437-2447.
[73] Chenn, A; McConnell, SK. Cleavage orientation and the asymmetric inheritance of Notch1 immunoreactivity in mammalian neurogenesis. *Cell*, 1995 82, 631-641.
[74] Takahashi, T; Nowakowski, RS; Caviness, VS. The leaving or Q fraction of the murine cerebral proliferative epithelium: a general model of neocortical neuronogenesis. *J Neurosci*, 1996 16, 6183-6196.
[75] Knoblich, JA. Asymmetric divisions during animal development. *Nat Rev Mol Cell Biol*, 2001 2, 11-20.
[76] Bardin, AJ; Borgne, RL; Schweisguth, F. Asymmetric localization and functionm of cell fate determinants: a fly´s view. *Curr Opin Neurobiol*, 2004 14, 6-14.
[77] Betschinmger, J; Knoblich, J. Dare to be different: asymmetric cell division in Drosophila, C. elegans and vertebrates. *Curr Biol*, 2004 14, R674-R685.
[78] Rogiers, F; Jan, YN. Asymmetric cell division. *Curr Opin Cell Biol*, 2004 16, 195-205.
[79] Horvitz, HR; Herskowitz, I. Mechanims of asymmetric division: two Bs are not two Bs, that is the question. *Cell*, 1992 68, 237-255.
[80] Jan, YN; Jan, LY. Asymmetric cell division. *Nature*, 1998 392, 775-778.
[81] Potten, CS; Loeffler, M. Stem cells: attributes, cycles, spirals, pitfalls and uncertainties. Lessons for and from the crypt. *Development*, 1990 110, 1001-1020.
[82] Morrison, SJ; Kimble, J. Asymmetric and symmetric stem-cell divisions in development and cancer. *Nature*, 2006 441, 1068-1074.
[83] Nelson, WJ. Adaptation of core mechanisms to generate cell polarity. *Nature*, 2003 422, 766-774.
[84] Mostov, K; Su, T; Bees, M-T. Polarized epithelial membrane traffic: conservation and plasticity. *Nat Cell Biol*, 2003 5, 287-293.
[85] Baum, B. Drosophila oogenesis: generating an axis of polarity. *Curr Biol*, 2002 12, R835-R837.
[86] Cowan, CR; Hyman, AA. Asymmetric cell division in C. elegans: Cortical polarity and spindle positioning. *Ann Rev Cell Dev Biol*, 2004 20, 427-453.
[87] Drubin, DG. Cell polarity. Series: *Frontiers in Molecular Biology*. Oxford: Oxford University Press, 2000.
[88] Hales, KG; Bi, E; Wu, J-Q; Adam, JC; Yu, I-C; Pringle, JR. Cytokinesis: an emerging unified theory for eukaryotes? *Curr Opin Cell Biol*, 1999 11, 717-725.
[89] Shima, DT; Warren, G. Inheritance of the cytoplasm during cell division. In: *Dynamics of Cell Division.* Endow SA & Glover DM, editorss. New York:Oxford University Press, 1998; pp 248-269.
[90] He, L; Hannon, GJ. MicroRNAs: small RNAs with a big role in gene regulation. *Nat Rev Genet*, 2004 5, 522-531.
[91] Hahe, S. Mobile protein signals cell fate. *Nature*, 2001 413, 261-262.
[92] Hannon, GJ. RNA interference. *Nature*, 2002 418, 244-251.
[93] Downward, J. The ins and outs of signaling. *Nature*, 2001 11, 759-762.

[94] Freeman, M. Feedback control of intracellular signaling in development. *Nature*, 2000 408, 313-319.

[95] Amon, A. Controlling cell cycle and cell fate: common strategies in prokaryotes and eukaryotes. *Proc Nat Acad Sci USA*, 1998 95, 85-86.

[96] Nigg, EA. Cyclin-dependent protein kinases: key regulators of the eukaryotic cell cycle. *BioEssays*, 1995 17, 759-768.

[97] Morgan, DO. Cyclin-dependent kinases: Engines, clocks and microprocessors. *Ann Rev Cell Dev Biol*, 1997 18, 261-284.

[98] Cipolleschi, MG; Dello Sbarba, P; Olivotto, M. The role of hypoxia in the maintenance of hematopoietic stem cells. *Blood*, 1993 82, 2031-2037.

[99] Suda, T; Raia, F; Hirao, A. Hematopoietic stem cells and their niche. *Trends Immunol*, 2005 26, 426-433.

[100] Nilsson, SK; Johnston, HM. Coverdale JA. Spatial localization of transplanted hemopoietic stem cells: Interferences for the localization of stem cell niches. *Blood*, 2001 97, 2293-2299.

[101] Sansone, P; Storci, G; Giovannini, C; Pandolfi, S; et al. p66Shc/Notch-3 interplay controls self-renewal and hypoxia survival in human stem/progenitor cells in mammary gland expanded in vitro as mammospheres. Stem Cells, 2007 25, 807-815.

[102] Gustafson, MV; Zheng, X; Pereira, T; et al. Hypoxia requires Notch signaling to maintain the undifferentiated cell state. *Dev Cell*, 2005 9, 617-628.

[103] Cejudo-Martin, P; Johnson, RS. A new Notch in the HIF belt: How hypoxia impacts differentiation. *Dev Cell*, 2005 9, 575-576.

[104] Covello, KL; Kehler, J; Yu, H; et al. HIF-2 alpha regulates Oct-4: Effects of hypoxia on stem cell function, embryonic development, and tumor growth. *Genes Dev*, 2006 20, 557-570.

[105] Fuchs, E; Segre, JA. Stem cells: A new lease on life. *Cell*, 2000 200, 143-155.

[106] Ihle, JN; Stravapodis, D; Parganas, E; Thierfelder, W; Feng, J; Wang, D; Teglund, S. The roles of Jaks and Stats in cytokine signaling. *Cancer J Sci Am*, 1998 4(suppl), 84-91.

[107] Etheridge, SL; Spencer, GJ; Heath, DJ; Genever, PG. Expression profiling and functional analysis of Wnt signaling mechanisms in mesenchymal stem cells. *Stem Cells*, 2004 22, 849-860.

[108] Willert, K; Jones, KA. Wnt signaling: is the party in the nucleus? *Genes Dev*, 2006 20, 1394-1404.

[109] Katoh, M; Katoh, M. Wnt signaling pathway and stem cell signaling network. *Clin Cancer Res*, 2007 13, 4042-4045.

[110] Chen, HJ; Lin, CM; Lin, CS; Perez-Olle, R; Leung, CL; Liem, RKH. The role of microtubule actin cross-linking factor 1 (MACF1) in the Wnt signaling pathway. *Genes Dev*, 2006 20, 1933-1945.

[111] Singh, AM; Hamazaki, T; Hankowski, KE; Terada, N. A heterogeneous expression pattern for Nanog in embryonic stem cells. *Stem Cells*, 2007 25, 2534-2542.

[112] Fox, V; Gokhale, PJ; Walsh, JR; Matin, M; Jones, M; Andrews, PW. Cell-cell signaling through Notch regulates human embryonic stem cell proliferation. *Stem Cells*, 2008 26, 715-723.

[113] Chiba, S. Concise review: Notch signaling in stem cell systems. *Stem Cells*, 2006 24, 2437-2447.
[114] Yamaguchi, A; Komori, T; Suda, T. Regulation of osteoblast differentiation mediated by bone morphogenetic proteins, Hedgehogs, and Cbfa1. *Endocr Rev*, 2000 21, 393-411.
[115] Trowbridge, JJ; Scott, MP; Bhatia, M. Hedgehog modulates cell cycle regulators in stem cells to control hematopoietic regeneration. *Proc Natl Acad Sci USA*, 2006 103, 14134-14139.
[116] Cai, C; Thorne, J; Grabel, L. Hedgehog serves as a mitogen and survival factor during embryonic stem cell neurogenesis. *Stem Cells*, 2008 26, 1097-1108.
[117] Anderson, DJ; Gage, FH; Weissman, IL. Can stem cell cross lineage boundaries? *Nat Med*, 2001 7, 393-395.
[118] Lagasse, E; Connors, H; Al Dhalims, M; Reitsma, M. Purified hematopoietic stem cells can differentiate into hepatocytes in vivo. *Nat Med*, 2000 6, 1229-1234.
[119] Bjornson, CR; Rietze, RL; Reynolds, BA; Magli, MC; Vescovi, AL. Turning brain into blood: a hematopoietic fate adopted by adult neural stem cells in vivo. *Science*, 1999 283, 534-537.
[120] Brazelton, TR; Rossi, FM; Keshet, GI; Blau, HM. From marrow to brain: expression of neurona phenotypes in adult mice. *Science*, 2000 290, 1775-1779.
[121] Ferrari, G; Cusella-De Angelis, G; Coletta, M; Paolucci, E; Stornainolo, P; Cossu, G; Mavilio, F. Muscle regeneration by bone marrow-derived myogenic progenitors. *Science*, 1998 279, 1528-1530.
[122] Kocher, AA; Schuster, MD; Szaboles, MJ; Takuma, S. Neovascularization of ischemic myocardium by human bone-marrow-derived angioblasts prevents cardiomyocyte apoptosis, reduces remodelling and improves cardiac function. *Nat Med*, 2001 7, 430-436.
[123] Mezey, E; Chandross, KJ; Harta, G; Maki, RA; McKercher, SR. Turning blood into brain: cells bearing neuronal antigens generated in vivo from bone marrow. *Science*, 2000 290, 1779-1782.
[124] Orlic, D; Kajstura, J; Chimenti, S; Jakoniuk, I. Bone marrow cells regenerate infracted myocardium. *Nature*, 2001 410, 701-705.
[125] Goodell, MA. Stem-cell "plasticity": befuddled by the muddle. *Curr Opin Hematol*, 2003, 10 208-213.
[126] Blau, HM; Brazelton, TR; Weimann, JM. The evolving concept of a stem cell: entity or function? *Cell*, 2001 105, 829-841.
[127] Wagers, AJ; Weisman, IL. Plasticity of adult stem cells. *Cell*, 2004 116, 639-648.
[128] Blau, HM. A twist of fate. *Nature*, 2002 419, 437.
[129] Ying, QL; Nichols, J; Evans, EP; Smith, AG. Changing potency by spontaneous fusion. *Nature*, 2002 416, 545-548.
[130] Terada, N; Hamazaki, T; Oka, M; Hoki, M; Mastalerz, DM; Nakano, Y; Meyer, EM; Morel, L; Petersen, BE; Scott, EW. Bone marrow cells adopt the fenotype of other cells by spontaneous cell fusion. *Nature*, 2002 416, 542-545.

In: Daughter Cells
Editors: A. Hitomi et al., pp. 197-220

ISBN: 978-1-60876-790-8

Chapter V

Regeneration of the Kidney: Involvement of Intra- and Extra-Renal Stem Cells

Akito Maeshima
Gunma University, Graduate School of Medicine, Japan.

Abstract

Kidney has a capacity to regenerate itself after a variety of insults. For the past few decades, the search for the factor(s) that can promote regeneration of the injured kidney has been a principal and fundamental theme in nephrology. Using various kidney injury models, the role of growth factors in kidney regeneration has been extensively examined and several factors such as hepatocyte growth factor (HGF) have been demonstrated to act as renotropic factor(s). Lately, along with other organs, the identification of adult tissue stem cell is one of the most exciting subjects in this area. A number of cell populations expressing putative stem cell markers or possessing stem cell property has been found within the kidney. Using BrdU labeling methods, it was shown that a slow-cycling tubular cell population was actively engaged in the regeneration processes of the kidney after ischemia and function as an origin of regenerating cells. Like in other tissues, side population cells have been identified in the kidney. Several groups have reported that cell therapies with bone marrow-derived cells (hematopoietic stem cell or mesenchymal stem cell) are highly effective for the treatment of acute or chronic renal failure in animals, which seems to be via the production of regenerative factor(s) rather than direct differentiation into a component of renal cells. In the near future, therapeutic strategy with stem cells might be possible for the treatment of various kidney diseases.

In this chapter, we summarize and discuss the current understanding of kidney regeneration after injury from the perspective of renotropic factors, renal stem/progenitor cells, and stem cell therapies.

INTRODUCTION

Epithelial tissues are thought to constantly replace damaged or dead cells throughout life. The process of continual cell replacement is an essential mechanism of the maintenance of adult tissues and is typically maintained through the presence of stem cells. Adult tissue stem cells are functionally defined by their ability to self-renew and to differentiate into the cell lineages of their tissue of origin. Once activated, epithelial stem cells can generate proliferating progeny, which are often referred to as transiently amplifying cells. In their normal environment, transiently amplifying cells will divide actively for a restricted period, expanding the cellular pool that will then differentiate along a particular cell lineage to make the tissue [1].

In the kidney, renal tubules are composed of polarized mature epithelial cells indispensable for tissue homeostasis as well as regeneration. The degree of tubular damage is directly related to the impairment of renal function. Understanding the tubular repair process is required for developing new strategies for the treatment of kidney disease. It is thought that, in acute kidney injury (AKI), the surrounding tubular cells that border the region of injury mediate tubular recovery. After the insult occurs, these cells rapidly lose their brush border and dedifferentiate into a more mesenchymal phenotype. The dedifferentiated cells migrate into the regions where cell necrosis, apoptosis, or detachment has resulted in denudation of the tubular basement membrane. They proliferate and eventually redifferentiate into an epithelial phenotype, completing the repair process [2].

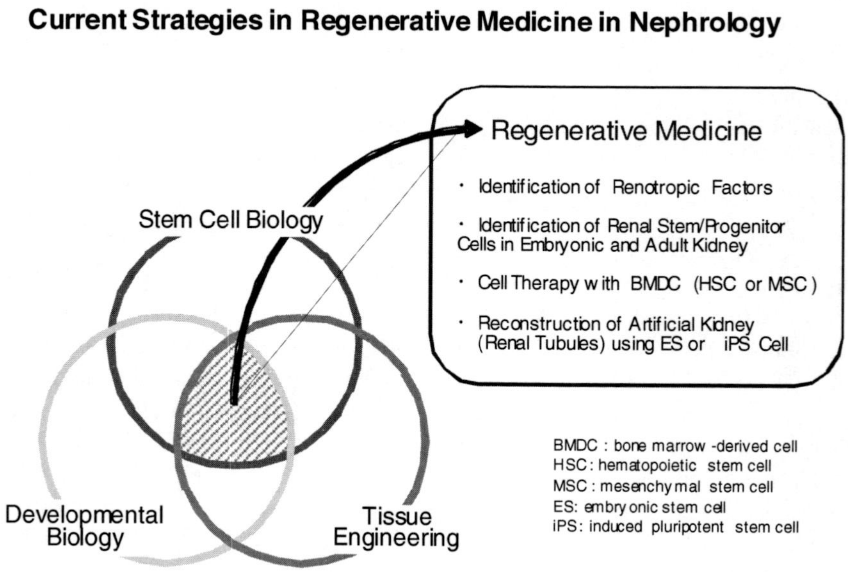

Figure 1. Current Strategies in Regenerative Medicine in Nephrology.

The kidney is the target of several injurious stimuli such as sepsis-associated cytokines, endogenous or exogenous toxins and ischemia. The susceptibility to develop AKI and the recovery of renal function following AKI depend on the ability of renal tubules to regenerate and to replace necrotic tubular cells with functional tubular epithelium. The absence or reduction of epithelial regeneration may predispose to tubulointerstitial scarring and chronic kidney disease. Although the clinical management of AKI patients has significantly improved in recent years, we still lack specific therapies to enhance kidney repair. Recovery after acute injury is critical for patient morbidity and mortality in the hospital setting. To overcome these problems, regenerative therapies have been extensively investigated based on the evidence originated from stem cell biology, developmental biology and tissue engineering (Figure 1). In this chapter, recent advances regarding renotropic factors, renal stem/progenitor cells, and stem cell therapies will be summarized.

RENOTROPIC FACTORS

The regeneration process is thought to resemble its developmental paradigm. The remodeling and maturation of restored epithelium after renal injury have many parallels with the growth and maturation that occur during kidney organogenesis. Development of the kidney depends on a series of reciprocal inductive interactions between the epithelial ureteric bud and metanephric mesenchyme [3]. The signals from metanephric mesenchyme initiate kidney development by inducing formation of the ureteric bud from the Wolffian duct. Subsequently, the growing ureteric bud also starts to branch as a response to the mesenchymal signals and in return secretes signals that induce mesenchymal cells to condensate and generate pretubular aggregates at the tips of ureteric bud branches. Aggregates undergo tubulogenesis via comma-shaped and S-shaped bodies. These tubular structures will finally fuse to the forming collecting duct. Forming nephrons also attract endothelial cells to make a functional glomerulus [4] (Figure 2). A large number of soluble factors have been reported to be involved in kidney development by using gene targeting, in vitro tubulogenesis model, and organ culture system. Most of these factors including hepatocyte growth factor (HGF), epidermal growth factor (EGF), insulin-like growth factor (IGF)-I, heparin-binding EGF-like growth factor (HB-EGF), platelet-derived growth factor (PDGF) and bone morphogenetic protein (BMP)-7, have been shown to be involved in kidney repair after injury.

HGF possess mitogenic, motogenic, morphogenic, and anti-apoptotic activities on renal tubular epithelial cells and is a potential renotropic factor for renal protection and repair [5]. EGF [6] as well as IGF-I [7,8] accelerates the repair process of the kidney after an injurious insult. HB-EGF is considered an important growth factor involved in renal epithelial cell repair, proliferation, and regeneration in the early stages of recovery after acute renal injury, as well as in nephrogenesis [9,10]. PDGF signaling pathway plays an important role in the regeneration of tubular cells from acute ischemic injury. The expression of PDGF-B, which is expressed in immature tubules of the developing kidney, was enhanced in the S3 segment of the proximal tubule after injury. The inhibition of the PDGF signaling with selective inhibitor of PDGF receptor phosphorylation resulted in a rise of serum creatinine, higher mortality rate, abnormal regenerating process, and suppressed proliferation of tubular epithelial cells [11]. BMPs play a key role in kidney development and postnatal function [12]. The kidney is a major site of BMP-7 synthesis during embryonic and postnatal development, which mediates differentiation and maintenance of metanephric mesenchyme. Targeted disruption of BMP-7 gene expression in mice resulted in dysgenic kidneys with hydroureters. Several experimental animal models of acute and chronic renal injury have all unequivocally shown beneficial effect of BMP-7 in ameliorating the severity of damage by preventing inflammation and fibrosis [13,14]. Consistent with this notion, mice lacking uterine sensitization-associated gene-1 (USAG1), a novel BMP antagonist abundantly expressed in the kidney, were reported to be resistant to renal injury [15]. These factors are considered to mediate the migration, dedifferentiation, and proliferation of surviving cells at sites of tubular injury, and also to regulate cell maturation and restoration of polarity, allowing the organ to return to its basal physiological role as regulator of the internal milieu [16].

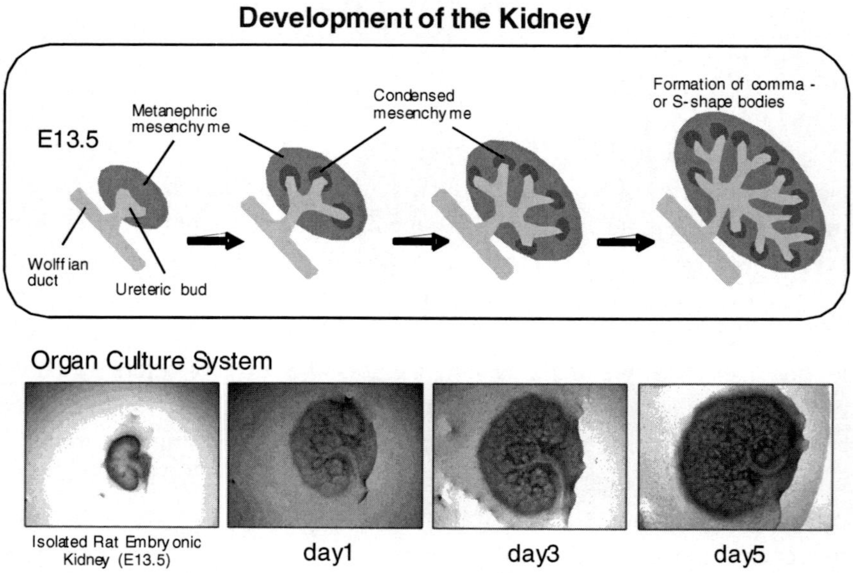

Figure 2. Development of the Kidney.

Utilizing the data obtained from developmental biology of the kidney, several molecules critical for kidney repair have been identified. Activin A, a member of the transforming growth factor (TGF)-beta superfamily, is known to regulate cell growth and differentiation in various tissues. In an in vitro Wolffian duct (WD) culture system, activin A was found to negatively regulate ureteric bud outgrowth from the Wolffian duct [17,18]. Activin A also inhibits ureteric bud branching [19,20] and is required for metanephric mesenchyme differentiation during kidney development [19]. In the kidneys of transgenic mice expressing truncated activin type II receptor, the number of glomeruli was significantly increased when compared to that in wild-type mice [21]. Collectively, it is considered that activin A serves as a negative regulator of renal organogenesis [22,23]. This notion was supported by the results obtained from an in vitro tubulogenesis model. HGF induced branching tubulogenesis in Madin-Darby canine kidney (MDCK) cells, which was inhibited by activin A. Blockade of activin action by follistatin (activin antagonist) or overexpression of activin mutant receptor induces tubulogenesis, even in the absence of HGF [24]. Recently, the role of the activin-follistatin system in the tissue repair and regeneration processes of the kidney was investigated [25]. Expression of activin increased significantly in the kidneys after ischemic injury. In contrast, expression of follistatin, abundant in tubular cells of the outer medulla of normal kidneys, decreased significantly after renal ischemia. Expression of the activin receptors was also detected in tubules of normal kidneys. Exogenous follistatin was found to alleviate renal damage after ischemic injury by promoting proliferation and reducing apoptosis in tubular cells. The blockade of endogenous growth inhibitors may be an alternative method for accelerating renal regeneration, instead of growth-promoting factor infusion.

Developmental paradigms are considered to be recapitulated in order to restore mature kidney function during the regeneration processes. Consistent with this idea, reactivation of Pax-2, a key regulatory molecule during renal organogenesis [26], was observed in the nuclei of regenerating tubular cells after ischemic injury [27,28]. During mouse kidney development, Pax-2 is expressed in the undifferentiated mesenchyme in response to ureter induction and continues to be expressed in the developing Comma- and S-shaped bodies. These structures harbor the immediate precursors of the proximal tubular epithelial cells. Pax-2 expression is down-regulated as the differentiation of the functional units of the nephron proceeds. In normal kidney, Pax-2 protein is detectable exclusively in the epithelium of the collecting ducts. A transient, temporally and locally restricted re-expression of Pax-2 was observed in regenerating proximal tubular epithelial cells following kidney damage. Interestingly, the number of Pax-2-positive cells in ischemic kidneys was increased after the administration of follistatin and, conversely, was reduced after the administration of activin A. Inactivation of activin signaling pathways also induced the upregulation of Pax-2 expression in a renal tubular epithelial cell line, suggesting that the activin-follistatin system may modulate renal regeneration after injury by controlling Pax-2 expression [29].

Similar to Pax-2, it has been demonstrated that leukemia inhibitory factor (LIF), known to play a crucial role in the conversion of mesenchyme into epithelium during nephrogenesis [30], participates in the regeneration process after tubular injury [31]. In developing kidney, LIF was expressed in the ureteric buds and LIF receptor was located in nephrogenic mesenchyme and the ureteric buds. In adult kidney, LIF and LIF receptor expression was confined to the collecting ducts. LIF as well as LIFR mRNA expression was significantly increased after renal ischemia. The increased expression of LIF was most marked in the outer medulla, especially in the S3 segment of the proximal tubules. In an in vitro model of acute renal failure, blockade of endogenous LIF with a neutralizing antibody significantly reduced the cell number and DNA synthesis during the recovery period. The Wnt-beta-catenin pathway, which plays key roles in embryogenesis, has also been demonstrated to regulate the transcription of cyclin D1 and control the regeneration of renal tubules in acute kidney injury [32]. Wnt-4 is known to be expressed in the mesonephric duct in embryonic development. Wnt-4 mRNA and protein expression were strongly increased after ischemia. The overexpression of Wnt-4 and beta-catenin promoted the cell cycle and increased the promoter activity and protein expression of cyclin D1 in renal tubular epithelial cell line, suggesting that the Wnt-4-beta-catenin pathway plays a key role in the cell cycle progression of renal tubules after injury.

Although many renotropic factors have been identified so far, the mechanism by which these growth factors mediate recovery from renal injury is not totally understood. Most of these factors are epithelial cell mitogens in vitro and exogenously administration of growth factors increase DNA synthesis in tubular cells after injury. However, these factors may not only act as mitogens, but also are involved in cell spreading, motility, and re-establishment of cell polarity. These factors might serve as cytoprotective agents or may exert biological effects independent of their mitogenic effect, such as the modulation of renal blood flow or neutrophil infiltration. It also remains unknown whether these renotropic factors promote renal regeneration via the activation of renal stem cell. Further study will be needed to clarify this issue.

Renal Stem/Progenitor Cells

The adult kidney contains more than 24 mature cell types arranged in distinct vascular, interstitial, glomerular and tubular compartments [4]. There is a considerable capacity in the kidney for morphological restoration of tubules and recovery of function post-injury. This structural complexity of the adult kidney has presented many challenges to the identification and characterization of kidney stem cells [33]. Although many attempts have been made to identify such a population, no definitive data to date establish the existence of a long-term, self-renewing cell population with the capacity to generate distinct daughter cells with renal potential in the adult kidney. However, many approaches have been taken to look for such a population. Attempts to identify adult kidney stem cells were made based on the broad principles of stem cell biology, such as prolonged cell-cycling time (label-retaining cells), ability to extrude Hoechst dye (side population cells), by restrictive cell culture conditions, or by using markers expressed by other stem cells or embryonic kidney.

Identification of Renal Stem/Progenitor Cells based on Cell Behavior

1. Label-Retaining Cell (LRC)

Stem cells generally possess an inexhaustible capacity for self-renewal and differentiation to ensure the lifelong maintenance of tissue homeostasis. Stem cells have a slow turnover and display minimal physiological differentiation. To conserve growth potential and prevent genetic injury during mitosis, stem cells are thought to cycle slowly and are recruited only as demanded by tissue turnover. As early descendents of stem cells, transit-amplifying (TA) cells retain significant growth capacity while acquiring differentiated functions. TA cells eventually become incapable of proliferation and enter a terminally differentiated state. One of the most common methods to identify stem cells is to search for slow-cycling cells by labeling their DNA. A pulse of 5-bromo-2-deoxyuridine (BrdU) mostly labels TA cells. Long-term BrdU labeling is thought to mark stem cells that retain the label for an extended period because of slow turnover. Adequate labeling intensity and a suitable washout period for the TA and terminal differentiation compartments will thus result in so-called LRCs, which are believed to represent the stem cell compartment. Utilizing this design, LRCs have now been reported in many adult organs. These include the lung [34,35], pancreas [36], conjunctiva [37], hair follicles [38], liver [39], heart [40], skin [41,42], nasopharynx [43], esophagus [44], endometrium [45], mammary glands [46], bladder [47] and muscle [48]. Some of these populations display pluripotency or marked degrees of plasticity.

Several groups have used this approach to identify slow-cycling cells in the kidney, which may represent renal stem cells. Maeshima et al. studied the existence of LRCs in normal rat kidneys by in vivo BrdU labeling [49]. LRCs were present in renal epithelial tubular cells of normal rat kidneys. After renal ischemia, LRCs underwent cell division and most became positive for proliferating-cell nuclear antigen (PCNA). Descendants of LRCs expressed a mesenchymal marker, vimentin, at an early phase of tubular regeneration, and eventually became positive for an epithelial marker, E-cadherin, after multiple cell divisions, suggesting that regenerating cells during tubular repair are essentially derived from LRCs. During tissue regeneration, stem cells are thought to divide asymmetrically into stem cells and TA cells, the latter of which vigorously proliferate, differentiate, and finally reconstitute part of the tissue [50]. At each asymmetric cell division, stem cells are postulated to selectively retain a set of chromosomes that contain old template DNA strands [51]. In this study, a difference in the BrdU content between daughter cells was found. Among each pair of LRCs, one expressed vimentin whereas the other did not, suggesting that LRCs undergo asymmetric cell division. Their descendents may acquire an immature mesenchymal phenotype, as well as having a high potential to proliferate and differentiate into epithelial tubular cells and function as TA cells during tubular regeneration [49]. Interestingly, tubular LRCs were found to be involved in epithelial to mesenchymal transition (EMT) in a rat renal fibrosis model with unilateral ureteral obstruction (UUO) [52]. Yamashita et al have demonstrated that LRCs were detected not only in tubules but also in the interstitium after UUO. A number of LRCs were adjacent to the destroyed tubular basement membrane and some tubules containing LRCs lost the ability to express E-cadherin after UUO. In vitro analysis demonstrated that, when renal tubules isolated from BrdU-treated rats were cultured in gel, some cells protruded from the periphery of the tubules and migrated into the gel, and most of these protruding cells were BrdU-positive [52]. Maeshima et al. have characterized tubular LRCs in vitro [53]. The fluorescence intensity of membrane-permeable Hoechst 33342 was reduced by the incorporation of BrdU into DNA, allowing the separation of viable BrdU-positive cells from viable BrdU-negative cells [54]. Using this evidence, the LRC population from BrdU-treated rat kidney was isolated as Hoechst-negative cells by FACS. LRCs formed tubule-like or tubulocystic structures in response to HGF when cultured in gel. When transplanted into a metanephric kidney, LRCs, but not non-LRCs, were integrated into epithelial components of the nephron, including the proximal tubular cells and the UB, suggesting that LRCs are a multipotent cell population with tubulogenic capacity [53].

Other groups have utilized similar methods to identify slow-cycling cells in the kidney [55,57]. Oliver et al. discovered a population of BrdU label–retaining cells within the papilla of the kidney [57]. BrdU was pulsed during the first postnatal week, during which nephrogenesis is continuing in the rodent and the chase endured for 2 months, indicating long-term label retention. These cells can be cultured under hypoxic conditions to form aggregates of nestin-positive cells, a marker in other stem cells types. In response to renal ischemia, these cells seemed to reenter mitosis rapidly, particularly within the outer medulla. The above research justifiably has generated a great deal of excitement but it is unknown whether these cells possess renal stem cell properties such as long-term self-renewal capacity and clonogenicity. Definitive markers for the isolation of these cell types will be required to address this issue. Vogetseder et al also defined a LRC population in the kidney, using BrdU pulse-chase in the rat [55]. This population appeared to be morphologically mature and fully differentiated epithelial cells. The authors suggest that renal LRCs are simply cells that have ceased to cycle because they are fully differentiated, and question whether this approach can be used to identify a stem cell in this organ.

As described above, the location, properties, and behavior of LRCs after injury differ or remain controversial. This is probably due to differences in timing of the pulse, duration of the pulse, and length of the chase. There is no consensus, at present, as to whether these LRCs with the capacity to regenerate should be referred to as unipotent stem cells or simply as cells that have retained proliferative potential. It is undeniable that every tubular cell shares the capacity to dedifferentiate and proliferate after injury. Nonetheless, the patterns of growth and differentiation of LRCs should be clarified in detail, because factors that can activate LRCs may possess renoprotective effects.

2. Side Population (SP) Cell

The ability of hematopoietic stem cells to efflux dyes such as Hoechst 33342 and Rhodamine 123 has been used as the basis of a single-step hematopoietic stem cell (HSC) isolation protocol [58]. The term side population (SP) is used to describe HSC that are isolated in this way. The observation of cells with the same efflux profile in solid organs has raised the possibility of organ-based SP, which also may represent stem cells. Several groups have reported the existence of an SP in the adult rodent kidney. These data remain contradictory in terms of the relative size, origin, and lineage capacity of the renal SP. Iwatani et al. showed no evidence for a capacity to transdifferentiate into renal cells in vivo [59]. Hishikawa et al. reported that these cells possess renal and multilineage capacity in culture [60]. When assessing the effect of renal SP cell introduction into a model of renal damage, they observed little evidence for transdifferentiation, with SP cells being located in the interstitium of the recipient kidney. Challen et al., using more stringent isolation procedures, demonstrated that the renal SP represented 0.1% of the kidney [61]. These cells showed an immunophenotype that was distinct from that of bone marrow SP, and these cells demonstrated evidence for multilineage potential in vitro and in vivo. However, despite stringent selection, this population remains heterogeneous with evidence of a monocytic fraction, which may result in the apparent phenotypic plasticity. In both studies [60,61], the introduction of SP cells into a model of acute experimental renal damage was therapeutically beneficial, suggesting a paracrine role for this cell type. While SP cells from the bone marrow do represent HSC, dye efflux activity alone is not enough to define as stem cells. There are no studies proving self-renewal potential of the renal SP cells. The definition of a marker phenotype that allows isolation without the assessment of dye efflux will be needed. The problems of SP homogeneity and the reproducibility of current isolation techniques remain unsolved [61].

Identification of Renal Stem/Progenitor Cells based on Specific Marker Expression

1. CD 133-Positive Cell

The CD133 is mainly known as a marker of HSC and endothelial progenitors [62], but recent reports indicate its expression in adult tissue stem cells such as those present in the prostate and brain [63,64]. Bussolati et al. found a rare population of CD133-positive cells in the interstitium close to the proximal tubules and glomeruli, or within tubules [65]. Once isolated, these cells lacked the expression of hematopoietic markers (CD34 and CD45), whereas they expressed some mesenchymal stem cell markers, such as CD29, CD90, CD44 and CD73. Moreover, they expressed Pax2, a kidney developmental gene, suggesting their renal origin. These cells were found to undergo epithelial and endothelial differentiation both in vitro and in vivo. When injected in vivo subcutaneously in Matrigel, CD133-positive cells spontaneously differentiated in tubular structures expressing proximal and distal tubular epithelial markers and endothelial differentiated CD133-positive cells formed vessels connected with the mouse vasculature. Moreover, when injected into mice with glycerol-induced acute renal injury, CD133-positive renal progenitors homed to the kidney and integrated into proximal and distal tubules during the repair [65].

2. CD24/CD133 Double-Positive Cell

A subset of parietal epithelial cells localized to the urinary pole of the Bowman's capsule was identified in human adult kidneys, based on co-expression of CD24 and CD133, both of which are used as a marker of several types of adult tissue stem cells [66]. These cells purified from cultured capsulated glomeruli exhibited multidifferentiation potential and long-term proliferative capacity in vitro. Injection of these CD24/CD133 double-positive cells into SCID mice with acute renal failure induced a complete recovery of renal function and restoration of tubular structures compared to vehicle (saline)-treated mice. It is unknown whether these cells elicit repair via functional integration or humoral induction when delivered into the recipient animal. Same group have identified CD24/CD133 double-positive multipotent progenitors in human embryonic kidneys as well [67]. CD24 was expressed in both metanephric mesenchyme- and ureteric bud-derived epithelial structures in human embryonic kidney, while CD133 selectively marked a subset of CD24-positive cells ultimately restricted to the parietal epithelial cells of the Bowman's capsule. This subpopulation strongly expressed stem cell-specific transcription factors BmI-1, Oct-4 and Nanog and demonstrated proliferation and multidifferentiation potential in vitro. Engraftment of CD24/CD133 double-positive renal progenitor cells into glycerol-induced rhabdomyolysis in SCID mice significantly improved renal function [67]. Considering the similarity of its location, it is possible that CD24/CD133 double-positive cells in the adult human kidney represent a residual population directly derived from this embryonic kidney progenitor population.

3. SCA-1-Positive Cell

A non-tubular multipotent stem/progenitor cell population was isolated from the adult mouse kidney and characterized as being Sca-1-positive cells [68]. These cells were capable of differentiation into myogenic, adipogenic and neural lineages. When injected directly into the renal parenchyma, shortly after ischemic/reperfusion injury, renal Sca-1-positive cells derived from ROSA26 reporter mice adopt a tubular phenotype and potentially could contribute to kidney repair. Unfortunately, the results were assessed by LacZ staining only and detection of specific tubular markers was not performed. It is not deniable the possibility that Sca-1-positive cell population is heterogeneous.

Identification of Renal Stem/Progenitor Cells based on selective Culture Conditions

1. OCT4-Positive Cell

A unique population of cells that behave in a manner that is consistent with a renal stem cell were isolated from rat kidneys. Features of these cells include spindle-shaped morphology, self-renewal for more than 200 population doublings without evidence for senescence, normal karyotype and DNA analysis. These cells express vimentin, CD90 (thy1.1), Pax-2, and Oct4 but not cytokeratin, MHC class I or II, or other markers of more differentiated cells. The ability of the cells to be induced to express endothelial, hepatocyte, and neural markers is demonstrated, suggesting the plasticity of these cells. The cells can differentiate into renal tubules when injected under the capsule of an uninjured kidney or intra-arterially after renal ischemia-reperfusion injury. As Oct4 expression was seen in some tubular cells in the adult kidney, the authors consider that these cells may be candidate renal stem cells [69].

2. Cell Line from S3 Segment

Kitamura et al. screened for stem cell potential in various regions of the postnatal murine kidney *via* dissection of various nephron segments and culture after dissociation to single cell [70]. In this way, the authors defined a cell line that was derived from the S3 segment of the proximal tubules. The authors showed that these cells that expressed Pax2, Wnt4, and WT1 could be maintained long term without transformation and replaced partly in injured tubules when engrafted to kidney in rat ischemic reperfusion model. Although the improvement in renal function was not assessed and potential of clonogenic self-renewal was not analyzed, these cells seemed to contribute to the regeneration of renal tubules after injury.

Stem Cell Therapy

Several studies have suggested that cells from bone marrow (BM) might possess a surprising degree of plasticity and could differentiate into cell types of multiple organs of the body [71,72]. BM transplantation is commonly used to study BM cell plasticity. The host BM is replaced by donor BM, and after BM chimerism is established, donor cells are tracking in the target organs. The donor BM cells are distinguished from host cells by virtue of their chromosome content (male Y chromosome-positive cells in a female host), the expression of a reporter molecule (beta-galactosidase, luciferase, enhanced green fluorescent protein), or the performance of a function (re-establishment of a function in a knockout mouse model). The ability of cells that originate from BM to move into distant sites within the body, including the kidney, is now well recognized. As in most organs, bone marrow–derived cells (BMDCs) appear in the kidney in response to damage. Recent reports have suggested that BMDCs can transdifferentiate into renal tubular epithelial cells [73-75], mesangial cells [76-79], glomerular endothelial cells [80,81], and even podocytes [82,83]. According to this evidence, cell therapy with BMDCs has been extensively examined in recent years and has been reported to be successful. In light of their ease of accessibility, BMDCs seem to be strong candidate for the treatment of renal diseases. In this chapter, the involvement of BMDCs (hematopoietic stem cell: HSC and mesenchymal stem cell: MSC) in the regeneration process of renal tubules after injury will be focused.

1. Hematopoietic Stem cell (RSC)

HSC are undifferentiated cells capable of self-renewal and stepwise differentiation into fully specialized cells of the blood such as erythrocytes, thrombocytes and leukocytes. Using whole-BM transplantation in the mouse, Poulsom et al. have demonstrated that BMDCs could contribute to regeneration of the renal tubular epithelium [74]. Kale S et al. have also shown that HSC from BM are mobilized into the circulation by transient renal ischemia and home specifically to injured regions of the renal tubules. HSC were found to differentiate into renal tubular epithelial cells and appear to constitute the majority of the cells present in necrotic tubules. Loss of HSC following BM ablation results in a greater rise in BUN after renal ischemia. In contrast, HSC infusion after BM ablation reverses this effect [73]. The capacity of HSC to contribute to the regeneration of renal tubular epithelial cells after ischemia/reperfusion (I/R) injury has been revealed by Lin F et al. HSC isolated from male Rosa26 mice expressing beta-galactosidase constitutively were transplanted into female nontransgenic mice after unilateral renal I/R injury. Four weeks after HSC transplantation, beta-galactosidase-positive cells were detected in renal tubules of the recipients by X-Gal staining. PCR analysis of the male-specific Sry gene and Y chromosome fluorescence in situ hybridization confirmed the presence of male-derived cells in the kidneys of female recipients [75].

These data suggest that the enhancement of the mobilization, propagation, and delivery of BM stem cells to the kidney hold potential as entirely new approaches for the treatment of acute kidney injury. If this process could be recapitulated pharmacologically, tubular repair may occur without the need for cell recruitment. Stockman et al. investigated whether the enhanced availability of HSC, induced by stem cell factor and granulocyte colony-stimulating factor (G-CSF), to the injured kidney provides an adequate strategy for stem cell-based therapy to counteract renal I/R injury. The authors found that cytokine treatment before injury resulted in significant enhancement of functional recovery of the kidney. However, this was not due to increased incorporation of tubular epithelial cells from bone marrow origin. Since cytokine treatment resulted in impaired influx of granulocytes into the injured kidney, it is possible that the underlying mechanism is not based on stem cell transdifferentiation but rather on altered inflammatory kinetics [84]. Iwasaki et al. also have demonstrated that

BMDCs mobilized by G-CSF accelerate the improvement in renal functions and prevent the renal tubular injury induced by cisplatin and that M-CSF enhances the effects of G-CSF [85]. In contrast, the data against the use of granulocytosis-inducing hematopoietic stem cell mobilization protocols for the prevention or treatment of ischemic ARF was also reported. Togel et al. have tested whether a further increase in circulating stem cell numbers, induced by their mobilization from the bone marrow, would improve renal function and outcome in mice with ischemic ARF. Unlike the reports above, it was found that the boosting of peripheral stem cell numbers failed to exert any renoprotective effects but rather was associated with greatly increased severity of renal failure and mortality. Because ischemic injury in neutropenic mice resulted in milder renal insufficiency and significantly reduced mortality, it was deduced that the adverse effects of pharmacologic stem cell mobilization are primarily mediated by the concomitant induction of marked granulocytosis. High numbers of activated granulocytes seem to obscure the potential renoprotective and positive survival effects of HSC via their injurious renal and systemic actions [86].

In terms of potential of BMDCs to transdifferentiate into tubular cells, there are several reports against this possibility. It was suggested that while BMDCs recruitment occurs, repair is predominantly elicited via proliferation of endogenous renal cells [87,88]. Duffield et al has reported that in chimeric mice expressing GFP in BMDCs, some interstitial cells but not tubular cells expressed GFP after ischemic injury. More than 99% of those GFP interstitial cells were leukocytes. Whereas there was an absence of BM-derived tubular cells, many tubular cells expressed proliferating cell nuclear antigen (PCNA), which is reflective of a high proliferative rate of endogenous surviving tubular cells. The authors conclude that BMDCs do not make a significant contribution to the restoration of epithelial integrity after an ischemic insult. [88]. Lin F et al also elegantly analyzed the relative contribution of intrarenal cells and extrarenal cells to kidney regeneration. In transgenic mice expressing enhanced GFP (EGFP) specifically and permanently in mature renal tubular epithelial cells, EGFP-positive cells showed incorporation of BrdU after I/R injury, indicating that regenerating tubules are derived from renal tubular epithelial cells. As a single injection of BMDCs did not make a significant contribution to renal functional or structural recovery, it is likely that intrarenal cells are the main source of renal repair [87]. Similar results were obtained in human BMDCs. Dekel et al. have investigated the potential of human adult CD34-positive progenitor cells to undergo renal differentiation after transplantation into ischemic kidneys. Although kidney ischemia enhanced engraftment of human cells, they were mostly hematopoietic CD45-positive cells residing in the interstitial spaces. Few other engrafted cells demonstrated an endothelial phenotype, but no evidence of tubular cells of human origin was found. The authors conclude that human CD34-positive stem cell do not acquire a tubular phenotype and are restricted mainly to hematopoietic and, to a lesser extent, to endothelial lineages [89].

The contradictory results in the localization of BMDCs and the degree of BMDCs contribution to kidney regeneration after injury may be due to methodological limitation in tracking BMDCs, particularly in injured tissues. There is the difference of protocols used in these studies (species, type of injury, transplantation methods, type of cells used for transplantation, specificity and sensitivity of the detection methods for BM cell origin). Cell fusion may also explain discrepancy in the different studies. It has been shown that 10-20% of

BM-derived stem cells engrafted in the kidney during regeneration after folic acid injury were dependent on cell fusion with resident epithelial cells [90]. Moreover, in a model of renal injury due to a toxic intermediate of tyrosine metabolism in a fumarylacetoacetate-hydrolase (Fah)-deficient mouse, Fah-positive transplanted BM-derived cells, which are resistant to this metabolite, produced a significant replacement of damaged proximal tubular epithelium by fusion with tubular cells reaching up to 50%, and not by transdifferentiation [91].

In all of these reports of bone marrow recruitment to damaged kidneys, the lineage of the BMDCs that were recruited has not been established. It remains unknown whether the recruitment of BMDCs has beneficial effect on chronic renal damage.

2. Mesenchymal Stem Cell (MSC)

The other possible candidate for the BM cell responsible for ameliorating renal damage is the MSC. The MSC are undifferentiated adult stem cell of mesodermal origin that have the capacity to differentiate into a wide range of mesenchymal tissue types, including cartilage, bone, muscle, stroma, fat, tendon, and other connective tissues. MSC represent a very small fraction of the total population of nucleated cells in the BM, but they can be isolated and expanded with high efficiency in culture as plastic adherent cells. The characterization of MSC includes their multipotency, the expression of typical surface markers such as CD90, CD73, CD105, CD44 and CD29, and the absence of hematopoietic lineage markers such as CD34 and CD45. Unlike HSC, once isolated, these MSC can be grown in culture for many population doublings and now have been shown to have a much broader potential, including neural differentiation. Several groups have now investigated the effect of delivery of MSC in models of acute renal damage [92].

Morigi et al. have reported that MSC markedly accelerated tubular proliferation in response to cisplatin-induced damage, while HSC failed to exert beneficial effects. Injection of MSC of male BM origin remarkably protected cisplatin-treated syngeneic female mice from renal function impairment and severe tubular injury. Y chromosome-containing cells localized in the context of the tubular epithelial lining, indicating that MSC engraft the damaged kidney and differentiate into tubular epithelial cells, thereby restoring renal structure and function. [93]. A greater understanding of the factors that regulate their homing and renoprotective activity may help to establish delivering a cell-free approach to renal disease. Same group examined the process of kidney cell repair in co-culture experiments with MSC and cisplatin-injured proximal tubular epithelial cells (PTEC). Exposure of PTEC to cisplatin markedly reduced cell viability, but co-culture with MSC provided a protective effect by promoting tubular cell proliferation. This effect was mediated by IGF-I, highly expressed by MSC. Knocking down IGF-I expression in MSC by small interfering-RNA also resulted in a significant decrease in PTEC proliferation and increased apoptosis. Furthermore, administration of IGF-I gene-silenced MSC limited their protective effect on renal function and tubular structure in cisplatin-induced kidney injury model. The authors conclude that MSC exert beneficial effects on tubular cell repair in acute kidney injury by producing the mitogenic and pro-survival factor IGF-I [94]. This group also investigated whether human MSC treatment could prevent AKI induced by cisplatin and prolong survival in an immunodeficient mouse model. Results showed that infusion of human MSC decreased proximal tubular epithelial cell injury and ameliorated the deficit in renal function, resulting in reduced recipient mortality. Infused MSC became localized predominantly in peritubular areas and acted to reduce renal cell apoptosis and to increase proliferation. MSC were protective against AKI-related peritubular capillary changes consisting of endothelial cell abnormalities, leukocyte infiltration, and low endothelial cell and lumen volume density. These findings indicate that human MSC of BM origin hold potential to prolong survival in AKI [95].

The other groups have also shown that in acute kidney injury model induced by intramuscular injection of glycerol, which is associated with an extensive necrosis of tubular epithelial cells due to myoglobin- and hemoglobin-induced injury, GFP-positive MSC injected intravenously homed to the kidney of mice with glycerol-induced injury but not to the kidney of normal mice. GFP-positive MSC localized in the context of the tubular epithelial lining, differentiated into tubular epithelial cells and promoted the recovery of morphological and functional alterations [96]. To examine the mechanisms underlying the migration and homing of these cells, Herrera et al have examined the role of CD44 and its major ligand, hyaluronic acid, in the trafficking of intravenously injected MSC. In vitro, hyaluronic acid promoted a dose-dependent migration of the stem cells that was inhibited by an anti-CD44 blocking monoclonal antibody. In vivo, stem cells injected into mice with acute kidney injury migrated to the injured kidney where hyaluronic acid expression was increased. Renal localization of the MSC was blocked by pre-incubation with the CD44 blocking antibody or by soluble hyaluronic acid. Stem cells derived from CD44 knockout mice did not localize to the injured kidney and did not accelerate morphological or functional recovery. Together, it is likely that CD44 and hyaluronic acid interactions recruit exogenous MSC to injured renal tissue and enhance renal regeneration [97].

Lange et al. have evaluated the therapeutic effectiveness of BM-derived MSC in a rat model of I/R injury. They showed that infusion of MSC enhances recovery of renal function. MSC were found to be located in the kidney cortex after injection, as demonstrated by MRI. MSC-treated animals had both significantly better renal function and better injury scores after renal ischemia. Histologically, MSC were predominantly located in glomerular capillaries and there was no tubular transdifferentiation [98]. It is possible that MSC possess an anti-inflammatory, antiapoptotic and immunosuppressive potential that may be involved in the protection of tissue injury. In this regard, Togel et al. have proposed a different mechanism by which transplanted MSC improve renal dysfunction after injury. Intracarotid administration of MSC after renal ischemia resulted in significantly improved renal function, higher proliferative and lower apoptotic indexes, as well as lower renal injury and unchanged leukocyte infiltration. Using in vivo two-photon laser confocal microscopy, fluorescence-labeled MSC were detected in glomeruli early after injection, and low numbers attached at microvasculature sites. However, none of the administered MSC had differentiated into a tubular or endothelial cell phenotype within 3 days of administration. At 24 h after injury, expression of proinflammatory cytokines such as IL-1beta, TNF-alpha, IFN-gamma, and inducible nitric oxide synthase was significantly reduced and that of anti-inflammatory IL-10 and basic FGF, TGF-alpha, and Bcl-2 was highly upregulated in treated kidneys. Based on these results, the authors considered that the beneficial effects of MSC are primarily mediated via complex paracrine actions and not by their differentiation into target cells [99]. The authors have also investigated the possibility of vasculoprotective activity in MSC. MSC-conditioned medium contains VEGF, HGF, and IGF-1 and augments aortic endothelial cell (EC) growth and survival. MSC and EC share vasculotropic gene expression profiles, as both form capillary tubes in vitro on Matrigel alone or in cooperation without fusion. MSC differentiate into an endothelial-like cell phenotype in vitro and develop into vascular structures in vivo. Infused MSC were readily detected in the kidney early after reflow but were only rarely engrafted at 1 week post-injury. MSC attached in the renal microvascular circulation significantly decreased apoptosis of adjacent cells, suggesting the significance of vasculotropic, paracrine actions of infused MSC [100].

Recently, it has been suggested that homing is not an absolute need for the beneficial effect of MSC-based therapy. The systemic concentration of critical factors derived from MSC might be sufficient to stimulate renal resident cells. Bi B et al. have reported that MSC express erythropoietin (EPO) receptor and that EPO can protect MSC from serum deprivation-induced cell death and can stimulate MSC proliferation in vitro. The administration of EPO to mice resulted in the expansion of MSCs in the BM as well as in the spleen and mobilized these cells into the peripheral circulation. Consistent with previous reports, the administration of EPO diminished the decline in renal function associated with cisplatin administration, suggesting that EPO protects renal tubular epithelial cells from toxic injury through the in vivo expansion and/or activation of MSC. The authors conclude that endocrine action of factors secreted by MSC is responsible for the renoprotective effect of MSC-based therapy [101].

CONCLUSION

In this chapter, the involvement of intra- and extra- renal stem cells in kidney regeneration after injury was summarized (Figure 3). Compared to other organs, data regarding renal stem/progenitor cells remain at a preliminary stage. The precise location, size of the pool and cellular morphology are either unknown or controversial. Recent studies have demonstrated that BMDCs contribute to the regenerative process by producing protective and regenerative factors, rather than by differentiating to directly replace damaged cells. To support this idea, the hypothesis that extra-renal cell contribute significantly to renal regeneration after injury has recently been questioned by a study by Humphreys et al. who have developed a method for distinguishing the source of kidney tubular regeneration using transgenic mice [102]. In this model, the promoter of homeodomain transcriptional regulator Six2 drives a fusion protein of GFP and Cre recombinase, which is expressed transiently in renal epithelial precursors during nephrogenesis. GFPCre expression is not present in the adult kidney even after injury. When Six2–GFPCre mice are crossed with a floxed STOP reporter strain, Cre-dependent removal of the stop sequence in progeny leads to constitutive expression of a marker gene such that all mesenchyme-derived renal epithelial cells are heritably labeled. In contrast, the entire interstitial compartment is unlabelled. Thus, the maintenance of labeled tubules post-injury would support a model of epithelial tubule repair due to surviving tubular epithelial cells or stem/progenitor cells localized within the labeled nephron, while label dilution would implicate an unlabelled, interstitial stem cell in the repair process. This study indicates that papillary interstitial cells [57] or other types of interstitial stem/progenitor cells [60-68] do not directly contribute to renal epithelial cells regeneration and clearly show that repair of injured nephrons is principally accomplished by intrinsic, surviving tubular epithelial cells or a subset of stem/progenitor cells localized within the nephron [102].

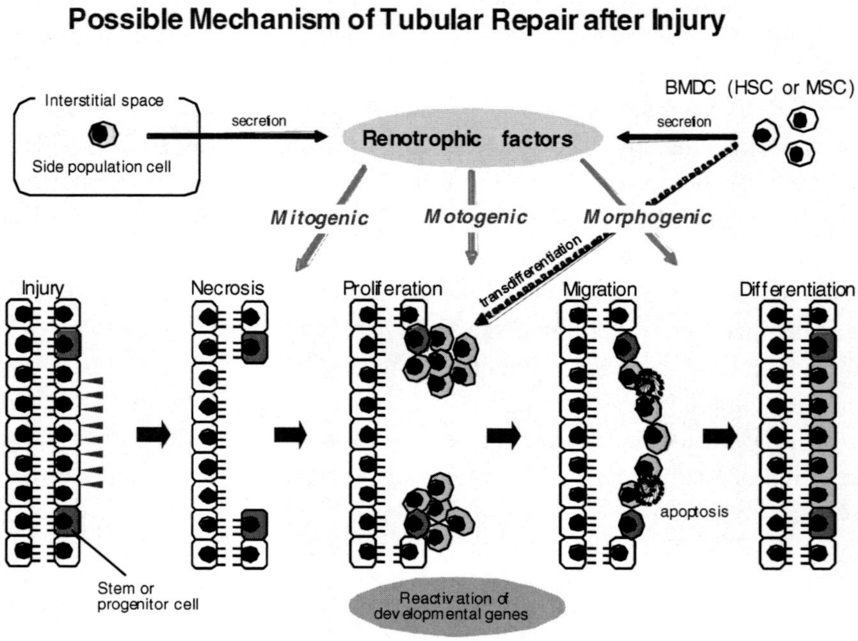

Figure 3. Possible Mechanism of Tubular Repair after Injury.

The delivery of soluble factors with the potential to improve the ability of the tissue to repair itself is the most pharmacologically attractive strategy for organ repair in situ. In this regard, clarification of the factors or the signaling pathways that enhance the regenerative capacity of stem/progenitor cells will lead to a better understanding of the mechanisms of tubular regeneration, as well as to the identification of novel therapeutic strategies to facilitate renal repair after acute kidney injury in humans. Considering the recapitulation of the developmental process in kidney regeneration, such factors may be produced by the embryonic kidney. Understanding the molecular basis of kidney development will thus be the key to further development of regenerative therapies for various kidney diseases.

References

[1] Blanpain, C., Horsley, V. & Fuchs, E. (2007). Epithelial stem cells: turning over new leaves. *Cell 128*; 445-58.

[2] Bonventre, J.V. (2003). Dedifferentiation and proliferation of surviving epithelial cells in acute renal failure. *J Am Soc Nephrol 14 Suppl 1*; S55-61.

[3] Saxén, L. (1987). *Organogenesis of the kidney*. Cambridge; New York: Cambridge University PRess.

[4] Dressler, G.R. (2006). The cellular basis of kidney development. *Annu Rev Cell Dev Biol 22*; 509-29.

[5] Matsumoto, K., Mizuno, S. & Nakamura, T. (2000). Hepatocyte growth factor in renal regeneration, renal disease and potential therapeutics. *Curr Opin Nephrol Hypertens 9*; 395-402.

[6] Humes, H.D., Cieslinski, D.A., Coimbra, T.M., Messana, J.M. & Galvao, C. (1989). Epidermal growth factor enhances renal tubule cell regeneration and repair and accelerates the recovery of renal function in postischemic acute renal failure. *J Clin Invest 84*; 1757-61.

[7] Ding, H., Kopple, J.D., Cohen, A. & Hirschberg, R. (1993). Recombinant human insulin-like growth factor-I accelerates recovery and reduces catabolism in rats with ischemic acute renal failure. *J Clin Invest 91*; 2281-7.

[8] Miller, S.B., Martin, D.R., Kissane, J. & Hammerman, M.R. (1992). Insulin-like growth factor I accelerates recovery from ischemic acute tubular necrosis in the rat. *Proc Natl Acad Sci U S A 89*; 11876-80.

[9] Sakai, M., Zhang, M., Homma, T., Garrick, B., Abraham, J.A., McKanna, J.A. & Harris, R.C. (1997). Production of heparin binding epidermal growth factor-like growth factor in the early phase of regeneration after acute renal injury. Isolation and localization of bioactive molecules. *J Clin Invest 99*; 2128-38.

[10] Homma, T., Sakai, M., Cheng, H.F., Yasuda, T., Coffey, R.J., Jr. & Harris, R.C. (1995). Induction of heparin-binding epidermal growth factor-like growth factor mRNA in rat kidney after acute injury. *J Clin Invest 96*; 1018-25.

[11] Nakagawa, T., Sasahara, M., Haneda, M., Kataoka, H., Nakagawa, H., Yagi, M., Kikkawa, R. & Hazama, F. (1999). Role of PDGF B-chain and PDGF receptors in rat tubular regeneration after acute injury. *Am J Pathol 155*; 1689-99.

[12] Simic, P. & Vukicevic, S. (2005). Bone morphogenetic proteins in development and homeostasis of kidney. *Cytokine Growth Factor Rev 16*; 299-308.

[13] Zeisberg, M., Hanai, J., Sugimoto, H., Mammoto, T., Charytan, D., Strutz, F. & Kalluri, R. (2003). BMP-7 counteracts TGF-beta1-induced epithelial-to-mesenchymal transition and reverses chronic renal injury. *Nat Med 9*; 964-8.

[14] Vukicevic, S., Basic, V., Rogic, D., Basic, N., Shih, M.S., Shepard, A., Jin, D., Dattatreyamurty, B., Jones, W., Dorai, H., Ryan, S., Griffiths, D., Maliakal, J., Jelic, M., Pastorcic, M., Stavljenic, A. & Sampath, T.K. (1998). Osteogenic protein-1 (bone morphogenetic protein-7) reduces severity of injury after ischemic acute renal failure in rat. *J Clin Invest 102*; 202-14.

[15] Yanagita, M., Okuda, T., Endo, S., Tanaka, M., Takahashi, K., Sugiyama, F., Kunita, S., Takahashi, S., Fukatsu, A., Yanagisawa, M., Kita, T. & Sakurai, T. (2006). Uterine sensitization-associated gene-1 (USAG-1), a novel BMP antagonist expressed in the kidney, accelerates tubular injury. *J Clin Invest 116*; 70-9.

[16] Nigam, S. & Lieberthal, W. (2000). Acute renal failure. III. The role of growth factors in the process of renal regeneration and repair. *Am J Physiol Renal Physiol 279*; F3-F11.

[17] Maeshima, A., Vaughn, D.A., Choi, Y. & Nigam, S.K. (2006). Activin A is an endogenous inhibitor of ureteric bud outgrowth from the Wolffian duct. *Dev Biol 295*; 473-85.

[18] Maeshima, A., Sakurai, H., Choi, Y., Kitamura, S., Vaughn, D.A., Tee, J.B. & Nigam, S.K. (2007). Glial cell-derived neurotrophic factor independent ureteric bud outgrowth from the Wolffian duct. *J Am Soc Nephrol 18*; 3147-55.

[19] Maeshima, A., Yamashita, S., Maeshima, K., Kojima, I. & Nojima, Y. (2003). Activin a produced by ureteric bud is a differentiation factor for metanephric mesenchyme. *J Am Soc Nephrol 14*; 1523-34.

[20] Ritvos, O., Tuuri, T., Eramaa, M., Sainio, K., Hilden, K., Saxen, L. & Gilbert, S.F. (1995). Activin disrupts epithelial branching morphogenesis in developing glandular organs of the mouse. *Mech Dev 50*; 229-45.

[21] Maeshima, A., Shiozaki, S., Tajima, T., Nakazato, Y., Naruse, T. & Kojima, I. (2000). Number of glomeruli is increased in the kidney of transgenic mice expressing the truncated type II activin receptor. *Biochem Biophys Res Commun 268*; 445-9.

[22] Maeshima, A., Nojima, Y. & Kojima, I. (2001). The role of the activin-follistatin system in the developmental and regeneration processes of the kidney. *Cytokine Growth Factor Rev 12*; 289-98.

[23] Maeshima, A., Miya, M., Mishima, K., Yamashita, S., Kojima, I. & Nojima, Y. (2008). Activin A: autocrine regulator of kidney development and repair. *Endocr J 55*; 1-9.

[24] Maeshima, A., Zhang, Y.Q., Furukawa, M., Naruse, T. & Kojima, I. (2000). Hepatocyte growth factor induces branching tubulogenesis in MDCK cells by modulating the activin-follistatin system. *Kidney Int 58*; 1511-22.

[25] Maeshima, A., Zhang, Y.Q., Nojima, Y., Naruse, T. & Kojima, I. (2001). Involvement of the activin-follistatin system in tubular regeneration after renal ischemia in rats. *J Am Soc Nephrol 12*; 1685-95.

[26] Dressler, G.R. (1996). Pax-2, kidney development, and oncogenesis. *Med Pediatr Oncol 27*; 440-4.

[27] Maeshima, A., Maeshima, K., Nojima, Y. & Kojima, I. (2002). Involvement of Pax-2 in the action of activin A on tubular cell regeneration. *J Am Soc Nephrol 13*; 2850-9.

[28] Imgrund, M., Grone, E., Grone, H.J., Kretzler, M., Holzman, L., Schlondorff, D. & Rothenpieler, U.W. (1999). Re-expression of the developmental gene Pax-2 during experimental acute tubular necrosis in mice 1. *Kidney Int 56*; 1423-31.

[29] Maeshima, A., Nojima, Y. & Kojima, I. (2002). Activin A: an autocrine regulator of cell growth and differentiation in renal proximal tubular cells. *Kidney Int 62*; 446-54.

[30] Barasch, J., Yang, J., Ware, C.B., Taga, T., Yoshida, K., Erdjument-Bromage, H., Tempst, P., Parravicini, E., Malach, S., Aranoff, T. & Oliver, J.A. (1999). Mesenchymal to epithelial conversion in rat metanephros is induced by LIF. *Cell 99*; 377-86.

[31] Yoshino, J., Monkawa, T., Tsuji, M., Hayashi, M. & Saruta, T. (2003). Leukemia inhibitory factor is involved in tubular regeneration after experimental acute renal failure. *J Am Soc Nephrol 14*; 3090-101.

[32] Terada, Y., Tanaka, H., Okado, T., Shimamura, H., Inoshita, S., Kuwahara, M. & Sasaki, S. (2003). Expression and function of the developmental gene Wnt-4 during experimental acute renal failure in rats. *J Am Soc Nephrol 14*; 1223-33.

[33] Little, M.H. (2006). Regrow or repair: potential regenerative therapies for the kidney. *J Am Soc Nephrol 17*; 2390-401.

[34] Borthwick, D.W., Shahbazian, M., Krantz, Q.T., Dorin, J.R. & Randell, S.H. (2001). Evidence for stem-cell niches in the tracheal epithelium. *Am J Respir Cell Mol Biol 24*; 662-70.

[35] Hong, K.U., Reynolds, S.D., Giangreco, A., Hurley, C.M. & Stripp, B.R. (2001). Clara cell secretory protein-expressing cells of the airway neuroepithelial body microenvironment include a label-retaining subset and are critical for epithelial renewal after progenitor cell depletion. *Am J Respir Cell Mol Biol 24*; 671-81.

[36] Duvillie, B., Attali, M., Aiello, V., Quemeneur, E. & Scharfmann, R. (2003). Label-retaining cells in the rat pancreas: location and differentiation potential in vitro. *Diabetes 52*; 2035-42.

[37] Wei, Z.G., Cotsarelis, G., Sun, T.T. & Lavker, R.M. (1995). Label-retaining cells are preferentially located in fornical epithelium: implications on conjunctival epithelial homeostasis. *Invest Ophthalmol Vis Sci 36*; 236-46.

[38] Cotsarelis, G., Sun, T.T. & Lavker, R.M. (1990). Label-retaining cells reside in the bulge area of pilosebaceous unit: implications for follicular stem cells, hair cycle, and skin carcinogenesis. *Cell 61*; 1329-37.

[39] Kuwahara, R., Kofman, A.V., Landis, C.S., Swenson, E.S., Barendswaard, E. & Theise, N.D. (2008). The hepatic stem cell niche: identification by label-retaining cell assay. *Hepatology 47*; 1994-2002.

[40] Urbanek, K., Cesselli, D., Rota, M., Nascimbene, A., De Angelis, A., Hosoda, T., Bearzi, C., Boni, A., Bolli, R., Kajstura, J., Anversa, P. & Leri, A. (2006). Stem cell niches in the adult mouse heart. *Proc Natl Acad Sci U S A 103*; 9226-31.

[41] Tumbar, T., Guasch, G., Greco, V., Blanpain, C., Lowry, W.E., Rendl, M. & Fuchs, E. (2004). Defining the epithelial stem cell niche in skin. *Science 303*; 359-63.

[42] Tani, H., Morris, R.J. & Kaur, P. (2000). Enrichment for murine keratinocyte stem cells based on cell surface phenotype. *Proc Natl Acad Sci U S A 97*; 10960-5.

[43] Zhang, H.B., Ren, C.P., Yang, X.Y., Wang, L., Li, H., Zhao, M., Yang, H. & Yao, K.T. (2007). Identification of label-retaining cells in nasopharyngeal epithelia and nasopharyngeal carcinoma tissues. *Histochem Cell Biol 127*; 347-54.

[44] Kalabis, J., Oyama, K., Okawa, T., Nakagawa, H., Michaylira, C.Z., Stairs, D.B., Figueiredo, J.L., Mahmood, U., Diehl, J.A., Herlyn, M. & Rustgi, A.K. (2008). A subpopulation of mouse esophageal basal cells has properties of stem cells with the capacity for self-renewal and lineage specification. *J Clin Invest 118*; 3860-9.

[45] Chan, R.W. & Gargett, C.E. (2006). Identification of label-retaining cells in mouse endometrium. *Stem Cells 24*; 1529-38.

[46] Smith, G.H. (2005). Label-retaining epithelial cells in mouse mammary gland divide asymmetrically and retain their template DNA strands. *Development 132*; 681-7.

[47] Kurzrock, E.A., Lieu, D.K., Degraffenried, L.A., Chan, C.W. & Isseroff, R.R. (2008). Label-retaining cells of the bladder: candidate urothelial stem cells. *Am J Physiol Renal Physiol 294*; F1415-21.

[48] Shinin, V., Gayraud-Morel, B., Gomes, D. & Tajbakhsh, S. (2006). Asymmetric division and cosegregation of template DNA strands in adult muscle satellite cells. *Nat Cell Biol 8*; 677-87.

[49] Maeshima, A., Yamashita, S. & Nojima, Y. (2003). Identification of renal progenitor-like tubular cells that participate in the regeneration processes of the kidney. *J Am Soc Nephrol 14*; 3138-46.

[50] Alison, M.R., Poulsom, R., Forbes, S. & Wright, N.A. (2002). An introduction to stem cells. *J Pathol 197*; 419-23.

[51] Knoblich, J.A. (2001). Asymmetric cell division during animal development. *Nat Rev Mol Cell Biol 2*; 11-20.

[52] Yamashita, S., Maeshima, A. & Nojima, Y. (2005). Involvement of renal progenitor tubular cells in epithelial-to-mesenchymal transition in fibrotic rat kidneys. *J Am Soc Nephrol 16*; 2044-51.

[53] Maeshima, A., Sakurai, H. & Nigam, S.K. (2006). Adult kidney tubular cell population showing phenotypic plasticity, tubulogenic capacity, and integration capability into developing kidney. *J Am Soc Nephrol 17*; 188-98.

[54] Mozdziak, P.E., Pulvermacher, P.M., Schultz, E. & Schell, K. (2000). Hoechst fluorescence intensity can be used to separate viable bromodeoxyuridine-labeled cells from viable non-bromodeoxyuridine-labeled cells. *Cytometry 41*; 89-95.

[55] Vogetseder, A., Karadeniz, A., Kaissling, B. & Le Hir, M. (2005). Tubular cell proliferation in the healthy rat kidney. *Histochem Cell Biol 124*; 97-104.

[56] Fujigaki, Y., Goto, T., Sakakima, M., Fukasawa, H., Miyaji, T., Yamamoto, T. & Hishida, A. (2006). Kinetics and characterization of initially regenerating proximal tubules in S3 segment in response to various degrees of acute tubular injury. *Nephrol Dial Transplant 21*; 41-50.

[57] Oliver, J.A., Maarouf, O., Cheema, F.H., Martens, T.P. & Al-Awqati, Q. (2004). The renal papilla is a niche for adult kidney stem cells. *J Clin Invest 114*; 795-804.

[58] Goodell, M.A., Brose, K., Paradis, G., Conner, A.S. & Mulligan, R.C. (1996). Isolation and functional properties of murine hematopoietic stem cells that are replicating in vivo. *J Exp Med 183*; 1797-806.

[59] Iwatani, H., Ito, T., Imai, E., Matsuzaki, Y., Suzuki, A., Yamato, M., Okabe, M. & Hori, M. (2004). Hematopoietic and nonhematopoietic potentials of Hoechst(low)/side population cells isolated from adult rat kidney. *Kidney Int 65*; 1604-14.

[60] Hishikawa, K., Marumo, T., Miura, S., Nakanishi, A., Matsuzaki, Y., Shibata, K., Ichiyanagi, T., Kohike, H., Komori, T., Takahashi, I., Takase, O., Imai, N., Yoshikawa, M., Inowa, T., Hayashi, M., Nakaki, T., Nakauchi, H., Okano, H. & Fujita, T. (2005). Musculin/MyoR is expressed in kidney side population cells and can regulate their function. *J Cell Biol 169*; 921-8.

[61] Challen, G.A., Bertoncello, I., Deane, J.A., Ricardo, S.D. & Little, M.H. (2006). Kidney side population reveals multilineage potential and renal functional capacity but also cellular heterogeneity. *J Am Soc Nephrol 17*; 1896-912.

[62] Miraglia, S., Godfrey, W., Yin, A.H., Atkins, K., Warnke, R., Holden, J.T., Bray, R.A., Waller, E.K. & Buck, D.W. (1997). A novel five-transmembrane hematopoietic stem cell antigen: isolation, characterization, and molecular cloning. *Blood 90*; 5013-21.

[63] Richardson, G.D., Robson, C.N., Lang, S.H., Neal, D.E., Maitland, N.J. & Collins, A.T. (2004). CD133, a novel marker for human prostatic epithelial stem cells. *J Cell Sci 117*; 3539-45.

[64] Uchida, N., Buck, D.W., He, D., Reitsma, M.J., Masek, M., Phan, T.V., Tsukamoto, A.S., Gage, F.H. & Weissman, I.L. (2000). Direct isolation of human central nervous system stem cells. *Proc Natl Acad Sci U S A 97*; 14720-5.

[65] Bussolati, B., Bruno, S., Grange, C., Buttiglieri, S., Deregibus, M.C., Cantino, D. & Camussi, G. (2005). Isolation of renal progenitor cells from adult human kidney. *Am J Pathol 166*; 545-55.

[66] Sagrinati, C., Netti, G.S., Mazzinghi, B., Lazzeri, E., Liotta, F., Frosali, F., Ronconi, E., Meini, C., Gacci, M., Squecco, R., Carini, M., Gesualdo, L., Francini, F., Maggi, E., Annunziato, F., Lasagni, L., Serio, M., Romagnani, S. & Romagnani, P. (2006). Isolation and characterization of multipotent progenitor cells from the Bowman's capsule of adult human kidneys. *J Am Soc Nephrol 17*; 2443-56.

[67] Lazzeri, E., Crescioli, C., Ronconi, E., Mazzinghi, B., Sagrinati, C., Netti, G.S., Angelotti, M.L., Parente, E., Ballerini, L., Cosmi, L., Maggi, L., Gesualdo, L., Rotondi, M., Annunziato, F., Maggi, E., Lasagni, L., Serio, M., Romagnani, S., Vannelli, G.B. & Romagnani, P. (2007). Regenerative potential of embryonic renal multipotent progenitors in acute renal failure. *J Am Soc Nephrol 18*; 3128-38.

[68] Dekel, B., Zangi, L., Shezen, E., Reich-Zeliger, S., Eventov-Friedman, S., Katchman, H., Jacob-Hirsch, J., Amariglio, N., Rechavi, G., Margalit, R. & Reisner, Y. (2006). Isolation and characterization of nontubular sca-1+lin- multipotent stem/progenitor cells from adult mouse kidney. *J Am Soc Nephrol 17*; 3300-14.

[69] Gupta, S., Verfaillie, C., Chmielewski, D., Kren, S., Eidman, K., Connaire, J., Heremans, Y., Lund, T., Blackstad, M., Jiang, Y., Luttun, A. & Rosenberg, M.E. (2006). Isolation and characterization of kidney-derived stem cells. *J Am Soc Nephrol 17*; 3028-40.

[70] Kitamura, S., Yamasaki, Y., Kinomura, M., Sugaya, T., Sugiyama, H., Maeshima, Y. & Makino, H. (2005). Establishment and characterization of renal progenitor like cells from S3 segment of nephron in rat adult kidney. *Faseb J 19*; 1789-97.

[71] Krause, D.S., Theise, N.D., Collector, M.I., Henegariu, O., Hwang, S., Gardner, R., Neutzel, S. & Sharkis, S.J. (2001). Multi-organ, multi-lineage engraftment by a single bone marrow-derived stem cell. *Cell 105*; 369-77.

[72] Herzog, E.L., Chai, L. & Krause, D.S. (2003). Plasticity of marrow-derived stem cells. *Blood 102*; 3483-93.

[73] Kale, S., Karihaloo, A., Clark, P.R., Kashgarian, M., Krause, D.S. & Cantley, L.G. (2003). Bone marrow stem cells contribute to repair of the ischemically injured renal tubule. *J Clin Invest 112*; 42-9.

[74] Poulsom, R., Forbes, S.J., Hodivala-Dilke, K., Ryan, E., Wyles, S., Navaratnarasah, S., Jeffery, R., Hunt, T., Alison, M., Cook, T., Pusey, C. & Wright, N.A. (2001). Bone

marrow contributes to renal parenchymal turnover and regeneration. *J Pathol 195*; 229-35.

[75] Lin, F., Cordes, K., Li, L., Hood, L., Couser, W.G., Shankland, S.J. & Igarashi, P. (2003). Hematopoietic stem cells contribute to the regeneration of renal tubules after renal ischemia-reperfusion injury in mice. *J Am Soc Nephrol 14*; 1188-99.

[76] Imasawa, T., Utsunomiya, Y., Kawamura, T., Zhong, Y., Nagasawa, R., Okabe, M., Maruyama, N., Hosoya, T. & Ohno, T. (2001). The potential of bone marrow-derived cells to differentiate to glomerular mesangial cells. *J Am Soc Nephrol 12*; 1401-9.

[77] Ito, T., Suzuki, A., Imai, E., Okabe, M. & Hori, M. (2001). Bone marrow is a reservoir of repopulating mesangial cells during glomerular remodeling. *J Am Soc Nephrol 12*; 2625-35.

[78] Masuya, M., Drake, C.J., Fleming, P.A., Reilly, C.M., Zeng, H., Hill, W.D., Martin-Studdard, A., Hess, D.C. & Ogawa, M. (2003). Hematopoietic origin of glomerular mesangial cells. *Blood 101*; 2215-8.

[79] Cornacchia, F., Fornoni, A., Plati, A.R., Thomas, A., Wang, Y., Inverardi, L., Striker, L.J. & Striker, G.E. (2001). Glomerulosclerosis is transmitted by bone marrow-derived mesangial cell progenitors. *J Clin Invest 108*; 1649-56.

[80] Ikarashi, K., Li, B., Suwa, M., Kawamura, K., Morioka, T., Yao, J., Khan, F., Uchiyama, M. & Oite, T. (2005). Bone marrow cells contribute to regeneration of damaged glomerular endothelial cells. *Kidney Int 67*; 1925-33.

[81] Uchimura, H., Marumo, T., Takase, O., Kawachi, H., Shimizu, F., Hayashi, M., Saruta, T., Hishikawa, K. & Fujita, T. (2005). Intrarenal injection of bone marrow-derived angiogenic cells reduces endothelial injury and mesangial cell activation in experimental glomerulonephritis. *J Am Soc Nephrol 16*; 997-1004.

[82] Prodromidi, E.I., Poulsom, R., Jeffery, R., Roufosse, C.A., Pollard, P.J., Pusey, C.D. & Cook, H.T. (2006). Bone marrow-derived cells contribute to podocyte regeneration and amelioration of renal disease in a mouse model of Alport syndrome. *Stem Cells 24*; 2448-55.

[83] Sugimoto, H., Mundel, T.M., Sund, M., Xie, L., Cosgrove, D. & Kalluri, R. (2006). Bone-marrow-derived stem cells repair basement membrane collagen defects and reverse genetic kidney disease. *Proc Natl Acad Sci U S A 103*; 7321-6.

[84] Stokman, G., Leemans, J.C., Claessen, N., Weening, J.J. & Florquin, S. (2005). Hematopoietic stem cell mobilization therapy accelerates recovery of renal function independent of stem cell contribution. *J Am Soc Nephrol 16*; 1684-92.

[85] Iwasaki, M., Adachi, Y., Minamino, K., Suzuki, Y., Zhang, Y., Okigaki, M., Nakano, K., Koike, Y., Wang, J., Mukaide, H., Taketani, S., Mori, Y., Takahashi, H., Iwasaka, T. & Ikehara, S. (2005). Mobilization of bone marrow cells by G-CSF rescues mice from cisplatin-induced renal failure, and M-CSF enhances the effects of G-CSF. *J Am Soc Nephrol 16*; 658-66.

[86] Togel, F., Isaac, J. & Westenfelder, C. (2004). Hematopoietic stem cell mobilization-associated granulocytosis severely worsens acute renal failure. *J Am Soc Nephrol 15*; 1261-7.

[87] Lin, F., Moran, A. & Igarashi, P. (2005). Intrarenal cells, not bone marrow-derived cells, are the major source for regeneration in postischemic kidney. *J Clin Invest 115*; 1756-64.

[88] Duffield, J.S., Park, K.M., Hsiao, L.L., Kelley, V.R., Scadden, D.T., Ichimura, T. & Bonventre, J.V. (2005). Restoration of tubular epithelial cells during repair of the postischemic kidney occurs independently of bone marrow-derived stem cells. *J Clin Invest 115*; 1743-55.

[89] Dekel, B., Shezen, E., Even-Tov-Friedman, S., Katchman, H., Margalit, R., Nagler, A. & Reisner, Y. (2006). Transplantation of human hematopoietic stem cells into ischemic and growing kidneys suggests a role in vasculogenesis but not tubulogenesis. *Stem Cells 24*; 1185-93.

[90] Fang, T.C., Alison, M.R., Cook, H.T., Jeffery, R., Wright, N.A. & Poulsom, R. (2005). Proliferation of bone marrow-derived cells contributes to regeneration after folic acid-induced acute tubular injury. *J Am Soc Nephrol 16*; 1723-32.

[91] Held, P.K., Al-Dhalimy, M., Willenbring, H., Akkari, Y., Jiang, S., Torimaru, Y., Olson, S., Fleming, W.H., Finegold, M. & Grompe, M. (2006). In vivo genetic selection of renal proximal tubules. *Mol Ther 13*; 49-58.

[92] Humphreys, B.D. & Bonventre, J.V. (2008). Mesenchymal stem cells in acute kidney injury. *Annu Rev Med 59*; 311-25.

[93] Morigi, M., Imberti, B., Zoja, C., Corna, D., Tomasoni, S., Abbate, M., Rottoli, D., Angioletti, S., Benigni, A., Perico, N., Alison, M. & Remuzzi, G. (2004). Mesenchymal stem cells are renotropic, helping to repair the kidney and improve function in acute renal failure. *J Am Soc Nephrol 15*; 1794-804.

[94] Imberti, B., Morigi, M., Tomasoni, S., Rota, C., Corna, D., Longaretti, L., Rottoli, D., Valsecchi, F., Benigni, A., Wang, J., Abbate, M., Zoja, C. & Remuzzi, G. (2007). Insulin-like growth factor-1 sustains stem cell mediated renal repair. *J Am Soc Nephrol 18*; 2921-8.

[95] Morigi, M., Introna, M., Imberti, B., Corna, D., Abbate, M., Rota, C., Rottoli, D., Benigni, A., Perico, N., Zoja, C., Rambaldi, A., Remuzzi, A. & Remuzzi, G. (2008). Human bone marrow mesenchymal stem cells accelerate recovery of acute renal injury and prolong survival in mice. *Stem Cells 26*; 2075-82.

[96] Herrera, M.B., Bussolati, B., Bruno, S., Fonsato, V., Romanazzi, G.M. & Camussi, G. (2004). Mesenchymal stem cells contribute to the renal repair of acute tubular epithelial injury. *Int J Mol Med 14*; 1035-41.

[97] Herrera, M.B., Bussolati, B., Bruno, S., Morando, L., Mauriello-Romanazzi, G., Sanavio, F., Stamenkovic, I., Biancone, L. & Camussi, G. (2007). Exogenous mesenchymal stem cells localize to the kidney by means of CD44 following acute tubular injury. *Kidney Int 72*; 430-41.

[98] Lange, C., Togel, F., Ittrich, H., Clayton, F., Nolte-Ernsting, C., Zander, A.R. & Westenfelder, C. (2005). Administered mesenchymal stem cells enhance recovery from ischemia/reperfusion-induced acute renal failure in rats. *Kidney Int 68*; 1613-7.

[99] Togel, F., Hu, Z., Weiss, K., Isaac, J., Lange, C. & Westenfelder, C. (2005). Administered mesenchymal stem cells protect against ischemic acute renal failure

through differentiation-independent mechanisms. *Am J Physiol Renal Physiol 289*; F31-42.

[100] Togel, F., Weiss, K., Yang, Y., Hu, Z., Zhang, P. & Westenfelder, C. (2007). Vasculotropic, paracrine actions of infused mesenchymal stem cells are important to the recovery from acute kidney injury. *Am J Physiol Renal Physiol 292*; F1626-35.

[101] Bi, B., Guo, J., Marlier, A., Lin, S.R. & Cantley, L.G. (2008). Erythropoietin expands a stromal cell population that can mediate renoprotection. *Am J Physiol Renal Physiol 295*; F1017-22.

[102] Humphreys, B.D., Valerius, M.T., Kobayashi, A., Mugford, J.W., Soeung, S., Duffield, J.S., McMahon, A.P. & Bonventre, J.V. (2008). Intrinsic epithelial cells repair the kidney after injury. *Cell Stem Cell 2*; 284-91.

In: Daughter Cells
Editors: A. Hitomi et al., pp. 221-242
ISBN: 978-1-60876-790-8

Chapter VI

EXFOLIATED EPITHELIAL CELLS OF MUCOSA: A REVIEW ON THEIR PHENOTYPES AND THEIR PROPERTIES

Bertrand Kaeffer *

INRA and University of Nantes, (France, EU)

ABSTRACT

Exfoliation is a loss of cellular material retaining the basic cytological features of typical cells (plasma membrane, cytoplasm and nucleus). However, the set of physiological parameters leading to exfoliation is complex and difficult to measure. Exfoliated epithelial cells can be obtained from a wide range of mucosa like mammary glands, oral mucosa by scraping, bronchial mucosa by expectoration and gastric or colonic epithelia by extensive rinsings during routine procedures of endoscopic examination. From an histological point of view, epithelia are organized in functional units containing different cellular compartments (stem, proliferative, mature or functional and senescent cells). Exfoliation is either (1) a natural loss of body cells implying a molecular signal related to the turn-over of terminally differentiated cells and to the progressive mobilization of proliferative as well as stem cells or (2) the result of manual exfoliation by applying mechanical constraints like gentle swabbing or scraping. Exfoliated epithelial cells are believed to be either in apoptosis or in anoïkis. In the latter group, cells may survive by autophagy retaining their growth potential if proper matrix is rapidly provided after exfoliation. Most studies are based on cytological proof using microscopic examination along with a limited number of biomarkers. Only few studies using proteomics or transcriptomics are available and they open discussion about tissue references and normalization. Exfoliated cells may represent valuable source of information on the physiopathological status of the mucosa. In translational research, ethical reasons as well as legal ones are limiting the use of biopsies in favor of measures

* Correspondence concerning this article should be addressed to: Bertrand Kaeffer, UMR 1280 Phan, INRA and University of Nantes, CHU - Hôtel DIEU HNB1, Place Alexis Ricordeau 44093 Nantes Cedex 1, France. Bertrand.Kaeffer@univ-nantes.fr.

realized on epithelial cells isolated from biological fluids (milk, urines, gastric aspirates, stools) collected by way of non-invasive technology. Unambiguous knowledge on the advantages and the limitations of exfoliation methodologies could have far reaching impact on chrononutrition or chronopharmacology.

Keywords: Mammals, gut, adult, neonate, premature, stem, clock.

INTRODUCTION

Exfoliated or sloughed epithelial cells are naturally lost by mammals in their environment. These cells represent valuable tools to study epithelial cell physiology of the mucosa and to predict the outcome of nutritional/clinical intervention studies. Recovery of exfoliated cells from biological fluids is a non-invasive technology which is in high demand in the field of translational research as well as during long-term experiments designed to minimize the sacrifice of long-lived or precious animals. Exfoliated epithelial cells can be used as surrogate for tissue biopsies in predicting changes in gene expression, DNA methylation, DNA damage, protein expression and accumulation of dietary components (Davis, 2003; Kamra, 2005) or to study the expression of circadian clock's components (Kaeffer et al., 2007). In the past, the recovery of cellular material from blood samples or tissue biopsies has been crucial to demonstrate the rhythmic nature of more than a hundred circadian parameters as well as the ultradian or seasonal rhythms of many others (Reinberg, 2003) allowing to sustain the concept of reactive homeostasis (Moore-Ede, 1986). However, the drawbacks of tissue biopsies or explants are well-known. The detection of clock gene products have been realized on biopsies of oral mucosa (Bjarnason et al., 2001), skin (Zanello et al., 2000; Bjarnason et al., 2001) or colon (Pardini et al., 2005) but most biopsies cannot be used in conditions of long-term nutritional/intervention studies. Moreover works in chrononutrition or chronopharmacology have to go beyond the relatively simple detection of molecules coded in presence versus absence of clock components. Most investigators would like to measure, as a minimum, morning versus evening parameters. Preferably, they wish to measure clock gene expression every 1-4 hours over at least three consecutive circadian cycles (72 hours). Recent developments in the concept of epithelial homeostasis and the role of exfoliation in preserving epithelial architecture (Lock & Debnath, 2008) should be reevaluated to foster the use of these cells in chronobiological studies. In addition, methodological improvements have been achieved in the efficientisolation of exfoliated epithelial cells allowing the recording of circadian clock molecular components in situation where biological fluids can be collected with a sufficient density in time.

This review presents the influence of methodology on the isolation of exfoliated epithelial cell phenotypes, new concepts in the understanding of exfoliation and, finally, speculates on the stability of biological information in such cellular material with special emphasis on the regulation of peripheral circadian clocks.

1. Methodology to Study Exfoliated Cells:

A major difficulty in exfoliation is to dispose of a device or a method allowing a reliable and easy recovery of cellular material as well as, for successful chronobiological research, to perform sampling with a sufficient density in time and to take into account legal aspects (Portaluppi et al., 2008).

1.1. Handling of Samples and Cellular Isolation:

Under our hands, exfoliated cells are recovered from gastric fluid aspirates and stools of preterm infants hospitalized in a neonatal intensive care unit. In that case, digestive fluids are not always directly available to scientist and we have designed a technical procedure to circumvent delays. First, digestive fluids have to be kept at room temperature and away from direct sun light to preserve cellular viability preferably no more than 20 min. Then samples should be stored directly at -70°C or snap frozen under liquid nitrogen. Storage at 4°C or at -20°C did not give good results (at -20°C probably due to the formation of ice crystal with deleterious effects at thawing on cellular structures). We have also transported samples during one-hour by immersing gastric aspirates or stools in Dulbecco's growth medium supplemented with D-glucose and L-glutamine (Kaeffer et al., 2007) but the loss of pertinent biological information may be considerable under these conditions of conservation. Anyway, the best strategy is to avoid sample storage by performing cellular isolation immediately on fresh samples. This is highly demanding in time and manpower. A widely-used procedure is to store biological fluids at -70°C (for a maximum of 6 months). But, the gold standard is to prepare unfixed exfoliated cells which have been snap frozen (Mojica et al., 2007). Alternatively, for isolation of freshly exfoliated cells subsequently to exposure to some inducers with the corresponding original epithelium, the cryogenic technique is crucial (Aoyama et al., 2008).

Manual exfoliation of the cells is also accomplished by applying the edge of a glass slide to the exposed surface of the opened mucosa like colonic resection specimen and gently scraping it (Yang et al., 1997; Mojica et al., 2007). Viability of recovered cells by this exfoliation/enrichment method as well as by other similar techniques is on average between 90 to 100% by the Trypan blue exclusion assay. However, centrifugation is the best method to pellet cellular materials from biological fluids devoid of big aggregates or mucus like urine to realize urocytogram (Robine et al., 1988) or recovered by gently swabbing oral mucosa on inner side of cheek with a cytobrush to study cytogenetic damage (Holland et al., 2007 and Figure 1). Some device has also been designed to recover surface exfoliated cells of human rectal mucosa by a minimal invasive scraping (Loktionov, 2006).

The combination of magnetic beads and antibodies constitutes a reliable system to recover low numbers of epithelial cells in biological fluids or to recover highly purified epithelial phenotypes. Antibodies against human cell surface antigens like AntiHep (Bandaletova et al., 2002, Kaeffer et al., 2007 from an original antibody described by Moldenhauer et al., 1987) or anti-Ber-Ep4 (Mojica et al., 2005 and 2007 from an original antibody described by Sheibani et al., 1991) labeled with paramagnetic particles are used to capture and purify epithelial cells. However, in our experience, shieving is necessary to perform immunocapture. With samples containing sheets of 5 to 30 cells, shieving or Percoll gradients are not adapted because this precious cellular material is discarded by the purification process. In addition, microbial contaminations are not easily removed by such density gradient methods as microbes are tightly associated with cells (Figure 1). Even with the manual exfoliation technique described by Mojica et al., 2007; the exfoliated cell populations may contain other cell types, most notably lymphocytes and plasma cells. It should be underlined that the problem of cross-contaminations by other sources of exfoliated epithelial cells like breast cells from the milk with gastric cells of lactating infants or exfoliated cells from manipulators are particularly difficult to ward off calling for the development of biomarkers of tissue origin which can both guarantee the cellular origin and the affordability of testing (Kaeffer & Rozé, 2009). Our current studies demonstrate that few thousands exfoliated cells can be recovered from gastric aspirates in preterm infants to measure the expression of specific genes of interest but the recovery of epithelial cells form stools is very low (Kaeffer et al., 2007). Exfoliated cells are shed in abundance in the stools of adult patients with colorectal cancer. These cells are non-apoptotic, and markers in exfoliated cells originating from the tumor are actively studied as a potential tool for cancer screening (Albaugh et al. 1992; Osborn and Ahlquist, 2005). Exfoliated colonocytes are rare in human adults free of colorectal cancer as these cells are believed to be rapidly removed from colonic epithelium by phagocytosis (Van Lieshout et al., 2004; Loktionov, 2007; Figure 2). Moreover these cells can be mixed up with well-preserved squamous epithelial cells mechanically sloughed in abundance from the surface of the anal canal epithelium during every stool passage. According to Loktionov (2007), few colonocytes can be usually obtained, whereas squamous cells of the anal canal epithelium often dominated the pool of isolated cells. Direct observations of colonic cells recovered from preterm infants did not take this possibility into account (Kaeffer et al., 2007) and more works are needed to design a stratagem allowing a clear-cut identification of colonocytes and squamous cells of the anal canal.

Exfoliated cells are studied by fixed cell immunofluorescent imaging, proteomics or transcriptomics (microarray technology). Methods to recover exfoliated cells are difficult to standardize and rely heavily on cytological approach in order to minimize misinterpretations (due to microbial contaminations or to the presence of cells of other lineages). Single-cell methods provide important information on the mean and variance of cell responses, but allow for far fewer signals to be measured than with proteomics or transcriptomics (Albeck et al., 2006). A direct measurement by fluorospectrometry on living purified cell preparations would be highly informative as normal tissues or diseased tissues (inflammation or cancer) are producing distinctive spectra *in vivo*. Fluorescent profiles are due to NADH, tryptophane, tyrosine and may help to follow epithelial cells' physiology. Profiles of normal gastric mucosa have been reported by Silveira et al (2008) but after sampling tissue biopsies are loosing rapidly coherent fluorescence profiles. Moreover spectral analysis of freshly isolated exfoliated epithelial cells remains unreported, probably due to the low numbers of exfoliated cells and to the sensitivity of instruments.

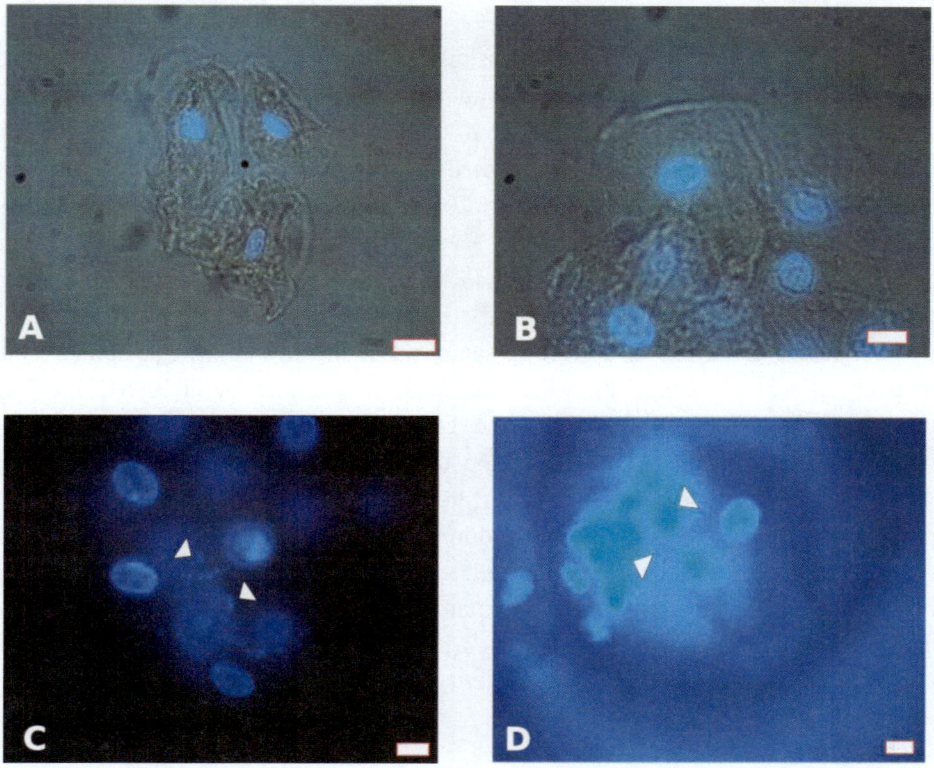

Figure 1. Morphology of exfoliated epithelial cells. Cells were recovered by gentle swabbing of the inner side of cheek of an adult (A) or from gastric fluid aspirates of a preterm infant (B) with associated contaminations by bacteria (C) or yeasts (D) underlined by white arrow heads (Bar stands for 10 µm). Cells labeled by DNA-specific Hoechst fluorochrome are examined under conventional light and ultraviolet light to show cytoplasm and blue nuclei. The recovery of exfoliated cells by gentled swabbing is immediate and can be realized frequently without stress for the patient. However, the period lentgh of human fibroblasts explanted *in vitro* have been shown to vary widely (Brown et al., 2005) indicating that the connection between the state of anoikis of exfoliated epithelial cells retaining a growth potential and circadian clock functioning may open new avenues of research in clock synchronization.

1.2. Microscopic analyses:

Microscopic examinations allow to circumvent a major impediment in working with clinical material, its inherent heterogeneity, by relating all results to the single cell. Exfoliated epithelial cells have been described sometime more than 50 years ago (colonocytes (Bader et al., 1952) or urothelial cells (Felix & Littlefield, 1980; Robine et al., 1988) but such descriptions were based on the microscopic examination of cell morphology and were of little use to characterize the cellular origin of recovered cell phenotypes. In addition, dark material like fecal samples can have adverse effects for direct examination as well as heavy loads in autofluorescent materials like biliary salts. Cell enumeration by microscopic examination is a tedious and slow process but it opens the possibility to easily distinguish quiescent from apoptotic cells (Bandaletova et al., 2002; Kaeffer et al., 2007; Kaeffer et al., 2009) and to evaluate contamination by yeasts and bacteria (Figure 2).

1.3. Transcripts:

The isolation of total RNA and the possibility to use RT-PCR to detect transcripts have been described on exfoliated human colonocytes recovered in exsudates during routine colonoscopy (Davidson et al., 2003) or from stools (Ahmed et al., 2004). Many protocols have been proposed to isolate intact sloughed-off colonocytes from freshly passed stools, most authors propose to limit for less than 20 min the unprocessed time of samples after removal from their normal environment. The methodologies to prepare total RNA are founded on commercially available kits to reduce the chances of contaminating the preparations with degrading nucleases. Briefly, a standard procedure is to extract RNA within a very short period of time after collecting samples (stool or tissue), i. e., within 15 minutes. The recovery of rare RNA molecules has been widely facilitated by the use of kits based on the amplification procedure of Phillips and Erwine, 1996. In parallel, works have been realized to evaluate the RNA quality of postmortem human brain tissue indicating that ribosomal and messenger RNA have different degrees of degradation depending on postmortem conditions and extraction procedures (Miller et al., 2004). Likewise, detection of clock transcripts have been reported on exfoliated gastric and colonic cells of preterm infants (Kaeffer et al., 2007). Future works on human exfoliated cells may take advantage of the strategy used by Loboda et al. (2009) to compare diurnal variation of adipose transcriptome *in silico* through the publicly-available Connectivity Map.

1.4. Proteins:

Comparative proteomic analysis between scrapings of the human intestinal epithelium and colon cancer cell lines have revealed that numerous proteins can be used to profile both scrapings and cell lines (Lenaerts et al., 2007). Vimentin (P08670) has been strongly associated with the differentiation process on 2D gels suggesting that this protein has some value in normalization problems associated with quantification of clock components in epithelial cells of unknown origin. Beta-actin (P60709) is also useful to include in normalization matrix due to its abundance and ubiquity. Among circadian regulated protein, alpha-enolase (P06733) is systematically detected by Lenaerts et al., 2007 on scrapings of large intestine. With exfoliated cells from the cervix, Steinau et al (2005) have found that around 80% of genes are expressed in exfoliated cells comparatively to total cell extract of a single corresponding tissue used as reference. Global proteomics has also been used to explore the difference between exfoliated breast cells (Klein et al., 2001). The normalization remains hazardous as the quantity of cellular material is often very low and frequently below the present level of detection by 2-D gel electrophoresis. Up to now, global proteomics is of little use to analyse exfoliated cells due to difficulties for proper labeling of cell populations frequently below 1,000. Such a labelling technique on 1,000 cells has been described by Sitek et al. (2008) by performing fluorescence dye saturation labeling in combination with high resolution two-dimensional gel electrophoresis, but the methodology is not yet applied to exfoliated cells. Reddy et al. (2006) have pooled the liver of six mice to obtain reliable resultson the circadian rhythmicity of hepatocyte proteome by two-dimensional difference gel electrophoresis. Again current animal models may help to circumvent scarcity of reference material as well as to allow efficient pooling of exfoliated epithelial cells from the same mucosa in order to obtain a sufficient amount of cellular material for 2-D gel electrophoresis.

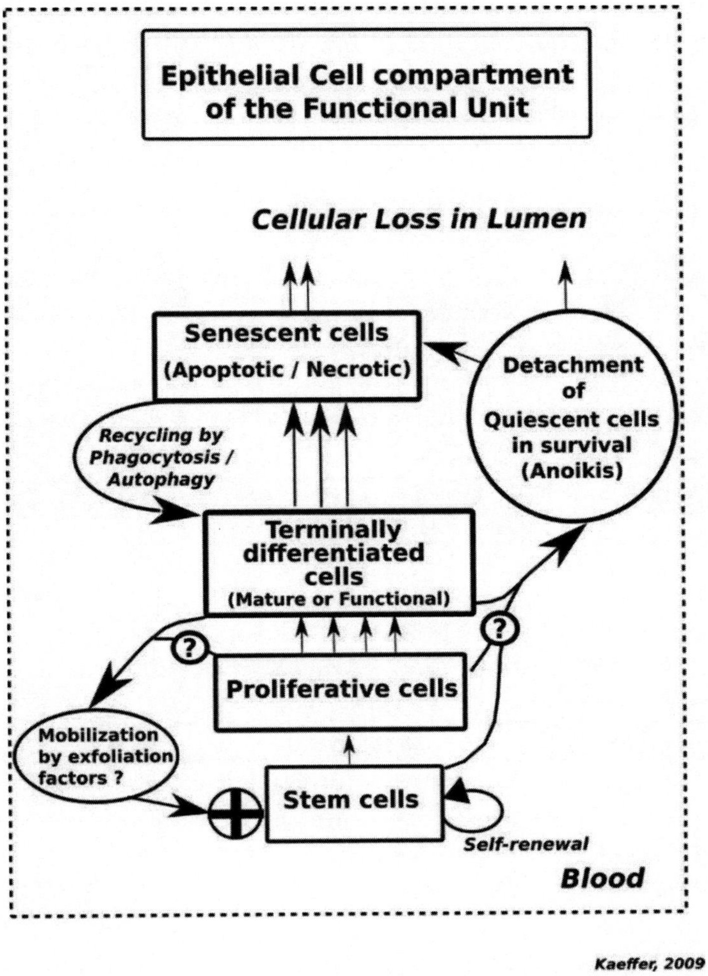

Figure 2. Main parameters of exfoliation of epithelial cells. Epithelial cell functional units are organized in four compartments (stem cells, proliferative, terminally differentiated, senescent). The isolation of exfoliated cells is dependent on the techniques (manual or induced by some stress), the epithelium itself, the health status and the species. A major difference in the autophagy / phagocytosis process of apoptotic and senescent cells has been found between rat or mouse exfoliated colonocytes which are mainly lost in the lumen and human colonocytes which are only rarely lost in the lumen but actively recycled by neighboring cells (Van Lieshout et al., 2004). Quiescent epithelial cells can be isolated from different body fluids (breast milk, urines, digestives fluids). They are believed to enter rapidly in anoikis after exfoliation. Anoikis is considered as an autophagic state promoting epithelial cell survival after a timely loss of contact with extracellular matrix and cell neighbors (Lock & Denab, 2008). Primary cultures can be derived from these cell populations indicating that some phenotypes retain clonogenic potentials. Molecular components of circadian clocks are detected in these cells like in any other somatic cells and could provide molecular data on the expression ratios between these proteins provided sampling procedures are standardized. Works are needed to check for any different rates of exfoliation according to the time of day, seasons as well as to differences in phenotypes. Circadian rhythmicity is known to be dependent on a lot of individual parameters like sex (Abizaid et al., 2004), adolescence status (Carskadon et al., 1998) as well as age of birth (full-term or preterm infants; Rivkees, 2003). Senescent cells observed at the top of human colonic crypts have been described as unreliable to study clock components (Pardini et al., 2005).

1.5. DNA and Associated Histones of the Regulatory Transcription Complex:

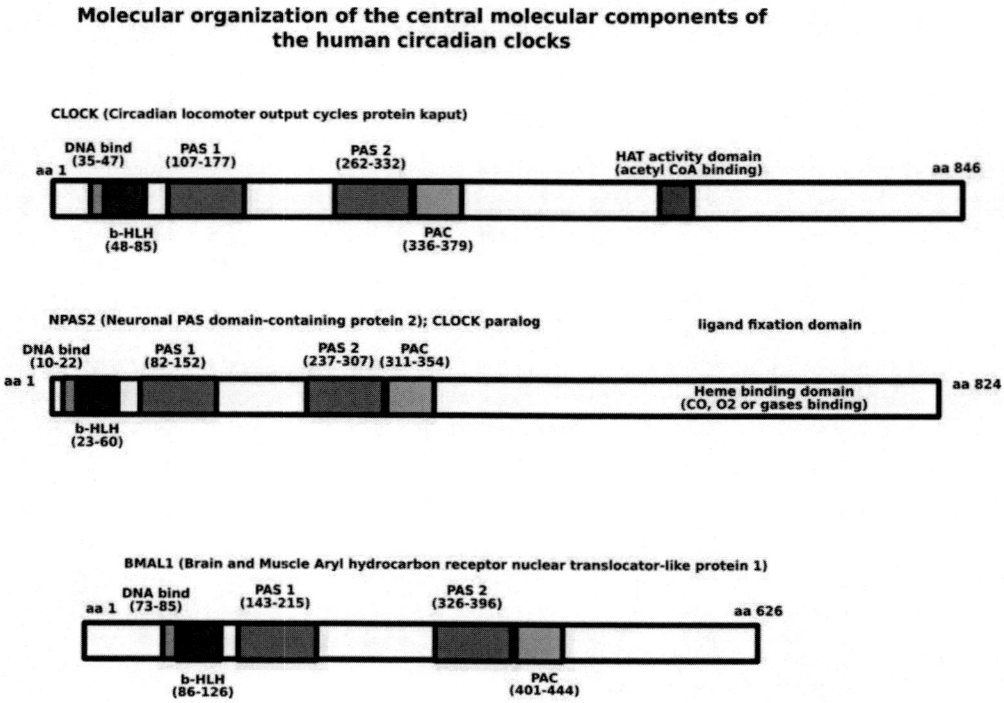

Figure 3. Molecular organization of the central components of the human circadian clocks. The clock system is believed to regulate around 10% of the transcriptome (Panda et al., 2002) and 20% of the proteome (Reddy et al., 2006). It can be conceptualize as a genetic system with 3 entry points: period genes, members of the immediate response familly which are believed to integrate signals linked to cellular stress, genes coding for the orphan receptors of retinoic acids which integrate signals from the retinol – vitamin A system, and genes of the central oscillator system (clock, npas2 and bmal1 as well as cryptochrome) which integrate the effects of glucose on the cell metabolism and gases tension. The link between circadian clock and the energy metabolism, mainly diet glucose catabolism may be the Redox status of NAD(H) and NADP(H) cofactors which directly influence on the capacity of binding on cell DNA of CLOCK:BMAL1 and NAPS2:BMAL1 complexes. Clock components are mainly interacting in large multiprotein complexes such that each protein's activity can regulate and be regulated by other members of the complex. CLOCK : BMAL1 and NPAS2 : BMAL1 constitute the master node of a multiprotein complex. Depending on the specific target gene, CLOCK and BMAL1 either remain bound to the promoter throughout the circadian cycle or vary rhythmically in their ability to bind DNA (Ripperger JA & Schibler U, 2006). In both cases, transcription inhibition is associated with the methylation of histone 3. Activation of transcription by CLOCK:BMAL1 is associated with histone acetylation (Etchegaray et al., 2003). Histone acetyl transferase activity of CLOCK is working with desacetylase activity of SIRT1 directly under the control of NAD+ binding (Nakahata et al., 2008; Nakahata et al., 2009). However NPAS2 can also heterodimerize with BMAL1 to largely compensate for or overlap with CLOCK in brain (Reick et al., 2001), vasculature (Curtis et al., 2004) or liver (Bertolucci et al., 2008) as well as to control of the gaz balance through its heme (Dioum et al., 2002) along with PERIOD2 (Kaasik et al., 2004; Figure 3). Mitogen-activated protein kinase interacts with and phosphorylates BMAL1 at multiple sites thereby decreasing the ability of the CLOCK:BMAL1 heterodimer to activate transcription from circadian promoters (Sanada et al., 2002); the equilibrium being post-translationally regulated by small ubiquitin modified proteins on BMAL1 through CLOCK-induced activity (Cardone et al., 2005). Transcription of BMAL1 is also controlled by nuclear receptor Rev-erb-alpha. This repression is dependent on recruitment of the nuclear receptor corepressor-histone deacetylase 3 (NcoR-HDAC3) complex directly to the promoter of Bmal1 gene (Raghuram et al., 2007; Yin et al., 2007; Rambaud et al., 2009). Rev-erb-alpha and beta may be acting as physiological sensors of intracellular heme.

Ways to monitor exfoliation of host DNA have been designed for genomic studies on total DNA of human breast epithelial cells (Thompson et al., 1998) or by Real-time PCR of host DNA on colonocytes of humans or rats (Van Lieshout et al., 2004). Histone acetylation has not been explored on exfoliated cells but it constitutes a future field of active investigation. Histone acetylation has been linked to the acquisition of epigenetic profiles during normal development or following perinatal denutrition and can be directly evaluated by taking advantage of the histone acetyltransferase activity harbored by CLOCK (Doi et al., 2006; Figure 3). Methodologies to check all these parameters will be crucial to broaden the usefulness of exfoliated epithelial cells in chronobiology as well as in physiology. Finally, with low potentials to measure chronobiological parameters but widely used in genotoxic studies, the formation of micronuclei in exfoliated cells, which emerge during mitosis of the basal layers of the epithelium from different organs (like oral and nasal cavity, bladder, cervix and esophagus), has been used as an end point to detect exposure to genotoxins or bioactive food components (Fenech et al., 1999; Majer et al., 2001; Holland et al., 2007).

In summary, the recovery of exfoliated cells is still difficult because of (1) their relatively low numbers, of (2) the need to relate these cells' phenotypes with the intact tissue structure of the donor and finally (3) on the degradation of biological information which depends on the bioactive compounds present in the biological fluids (some like digestive fluids can be heavily loaded with enzymes). The disposal of animal models of exfoliation will be of paramount importance to develop *in vivo* studies on autophagy during anoikis to evaluate the stability of biological molecular information in exfoliated cells. The next section will give a definition of exfoliation according to recent works in the field.

2. Definition of Exfoliation

Exfoliation is a loss of cellular material retaining the basic cytological features of typical cells (plasma membrane, cytoplasm and nucleus). Exfoliated epithelial cells can be obtained from a wide range of mucosae which line body passages and cavities communicating directly or indirectly with the exterior like mammary glands, oral, bronchial, urothelial or gastrointestinal epithelia. From an histological point of view, epithelia are organized in functional units containing different cellular compartments (stem, proliferative, mature or functional and senescent). These functional units are directly at the interface with the environment. Epithelia can be divided in simple cylindrical cell monolayers like colon or pseudo-stratified like urothelium (Karam 1999). At a given time point, a mucosal epithelium is supposed to loose different categories of cells by different mechanisms of exfoliation (Figure 2). The cell turn-over is driven by a delicate balance between cellular loss and proliferation. Proliferation is running on two cellular compartments, the proliferative cells capable of rapid mitosis to amplify tissue regeneration and the stem cells which are giving rise to all phenotypes through assymetric mitosis. The speed of mitosis in proliferative compartment is dependent on cellular loss at the top of the structure and on tightly regulated cell migration along the functional units (Potten and Loeffler, 1990; Kaeffer, 2002). Cell migration in the small and large bowels of mice shows a strong circadian rhythm, with cell velocity maximal at 9 a.m. and minimal at 5 p.m (Qiu et al., 1994). Other rhythms which could be controlled by circadian clocks have been observed in the intestine like cellular proliferation (Buchi et al., 1991, Marra et al., 1994) or apoptosis (Iwakiri et al., 2001). Cell proliferation is also believed to be under the control of clockwork not only in hepatocytes (Matsuo et al., 2000) but also along the rat's gut (Polidarov et al., 2009). Seasonal rhythms of proliferation have been described in adult rats by Haus et al., 1984 and Gomes et al, 2005. Circadian as well as seasonal rhythms in cell proliferation seem clearly relevant to the recovery of exfoliated quiescent cells retaining specific and functional biomarkers (Figure 2).

2.1. Exfoliation, a Possible Role in the Epithelium Homeostasis?

Exfoliation has been described as an active biochemical process linked to the epithelium homeostasis (Kitazawa et al., 2004, Kaeffer et al., 2007, Aoyama et al., 2008). It is believed that epithelial cells, loosing contact with companion cells (like fibroblasts) as well as extracellular matrix, enter anoikis (Frisch and Francis, 1994). The detachment of epithelial cells triggers both pro and antiapoptotic signals, such as nuclear factor kappa-B and inhibitor of apoptosis protein family members; these antiapoptotic mechanisms presumably delay the onset of apoptosis and allow cells to survive (Gilmore 2005, Yan et al., 2005, Liu et al., 2006). The balance between these signals and the duration of detachment determine further fate of these cells. Antiapoptotic signals presumably delay the onset of anoikis, allowing cells to survive provided that they can reestablish extracellular matrix contact in a timely manner (Lock & Debnath, 2008). Loss of extracellular matrix contact induces autophagy in normal epithelial cells, autophagy promotes the survival of detached cells during both anoikis and lumen formation in 3D epithelial cell culture (Debnath et al., 2002; Fung et al., 2008). Under these assumptions, exfoliation may be understood as a natural process to remove external cells from the luminal surface of an epithelium. Alternatively, exfoliation can also be one of the means of cell population size control. Anyway, exfoliation has a physiological role in the architecture of epithelium by allowing the formation of a lumen and we can surmise by providing sufficient flexibility to preserve the physical integrity of epithelia and allow its growth. In three-dimensional epithelial cell cultures, both autophagy and apoptosis are observed during lumen formation (Debnath et al., 2002; Fung et al., 2008). By loosing contact with the original mucosa, exfoliated epithelial cells have to activate autophagy as a survival mechanism to endure starvation. Starving cells are degrading cytoplasmic material to generate both nutrients and energy (Levine and Klionsky, 2004). Indeed, quiescent exfoliated epithelial cells without signs of apoptosis can be recovered under specific conditions in gastric fluid aspirates (Kaeffer et al., 2007) or by succion from breast glands (Thompson et al., 1998; Klein et al., 2001) or extensive rinsing at the end of routine colonoscopy (Davidson et al., 2003). Many exfoliated quiescent epithelial cells can be cultured suggesting that detachment-induced autophagy contributes to the viability of these cells. Human mammary epithelial cells die after 24-48 hours of detachment, certain epithelial cells, notably primary mouse mammary epithelial cells and rat intestinal epithelial cells, perish within a few hours following substratum detachment (Kaeffer, 2002; Fung et al., 2008).

Animal models have been developed to study the inducing effect of cycles of nutrient intake on the exfoliation of epithelial cells in the digestive lumen. On adult rats fastened for 24 hours and refed for one hour, the feeding intake induces exfoliation of quiescent parietal cells at the top of the gastric gland through an unknown exfoliation factor (Aoyama et al.,

2008). Under these conditions, stem cells located in the neck region of gastric glands are believed to be recruited actively to repopulate the surface of the adult rat stomach. By contrast, on adult laboratory mice, fatty acids (like palmitate) are inducers of intectine, a protein implicated in the exfoliation of apoptotic cells at the top of the villus of the small intestine within an hour after meal (Kitazawa et al., 2004). These models of nutritional manipulations to induce exfoliation on small intestine of mice (Kitazawa et al., 2004) or gastric mucosa of rats (Aoyama et al., 2008) indicate that the mechanism of exfoliation is highly context-dependent but they also open the possibility to develop *in vivo* studies of anoikis and autophagy in relation with the functioning of peripheral circadian clocks.

2.2. Exfoliation Obtained by Applying Mechanical Constraints

In its simple acceptation, exfoliation is a loss of cells in the environment due to external mechanical forces like brushing or friction. Such forces are deeply altering the epithelium architecture but yield rapidly high amounts of so-called normal epithelial cells. Manual exfoliation has been reported with brushing or scraping technique on oral epithelium of cheek or tongue (Holland et al., 2007), cervical (Steinau et al., 2005) or rectal swabbings (Loktoniov et al., 2007); airway epithelial cells in sputum and buccal mucosal cells obtained by rinsing the mouth or chewing-gum (betel chewers - Stitch et al., 1984), esophageal cells (Munoz et al., 1987), mammary by nipple aspirate, ductal lavage (Klein et al., 2001), breast milk (Thompson et al., 1998) or bladder urothelial cells present in urine samples (Felix, 1980; Robine et al., 1988). Manual exfoliation has also been proposed as a way to recover intact, normal epithelial cells on tissue biopsies made on colon resection (Mojica et al., 2005; Mojica et al., 2007) but a major drawback is that there is no control on mechanical constraints applied to the mucosa (an excessive force can damage the tissue resulting in obtaining biopsy-like material). The technique partially purifies the cell preparation by taking advantage of the cell's inherent biology. Epithelial cells remained in small groups or sheets, detached from any stromal elements that may have been scraped off (Mojica et al., 2005). This observation is also true for gastric epithelial cells (Kaeffer et al., 2007; Figure 1 A). As shown on figure 1 B, the recovery of mechanically exfoliated cells by gentle swabbing can be used to obtain typical epithelial cells immediately available to analysis and may constitute a reliable alternative to oral mucosa punches used by Bjarnason et al (2001) to study clock gene variations in time series on human volunteers.

In summary, exfoliation is a broad term which includes many different physiological situations. The proof that exfoliation of quiescent cells is following a circadian rhythm is still missing. Probably because, the set of physiological parameters leading to the induction of active exfoliation is difficult to handle and the interpretation of data obtained from manual exfoliation is also highly context-sensitive. The next section will introduce the regulation of peripheral circadian clocks and elaborate on the stability of chronobiological information in exfoliated epithelial cells.

3. Stability of Biological Information in Exfoliated Epithelial Cells

The biological information that we can extract from exfoliated epithelial cells depends on the techniques used to isolate cellular material or the manipulation of physiological parameters (Figure 2) and on the affinity of interactions of clock molecular components with stable molecules like DNA or the persistence of the physiological effects that they are inducing (Figures 3 and 4). Cells have to process diverse signals such as temperature, pH, and nutrient concentrations in order to maintain a normal physiology. *In vivo*, cells are believed to use clocks to organize and adapt cellular metabolisms and coordinate three-dimensional macromolecular organization (their phenotype) in a noisy molecular environment (molecular signals criss-crossing in and between cells and irrelevant chemical messages). This process has been formalized by Tan et al. 2007 in a genetic model useful to understand signal processing and the influence of cellular noise on the fidelity and speed of frequency-signal transmission. Exfoliated epithelial cells in anoikis can be seen as a way to obtain chronobiological information dating back from the time of cells leaving the top of the functional units. According to the mathematical model of Tan et al (2007), relevant biological information can be encoded in the amplitude or the frequency domains of a signal. Signals encoded in the amplitude domain are predominantly based on concentrations of signaling molecules, a parameter difficult to measure on exfoliated cells in anoikis without proper normalization related to the single cell level on highly purified cellular phenotypes. In addition, some highly labile biochemical modifications like phosphorylation are probably lost during the storage of biological fluids as well as in the isolation. Likewise, the physiological status of these cells can be irrelevant to the pathophysiology of the mucosa, due to the turn-over of the signal by the cellular machinery. However according to Gallego & Virshup (2007), the addition of phosphatase inhibitors into cellular extracts can preserve period proteins. Significant improvements in the quality of cellular extracts may be achieved by using extraction bufffers suitable to preserve clock gene products. The biological information encoded in the frequency domain of an oscillatory signal can be transmitted as concatenated signals with multiple biologically significant signals to gene behind the regulatory sequence within the promoter. On transgenic mammals, oscillations can be measured by recording rhythms of light emissions by cells in which the promoter of some clock gene is linked to a luciferase reporter. On human or non-transgenic mammals, oscillations per se cannot be measured in absence of spectrofluorimetric methods applicable on freshly recovered living cells but indirect evidence of gene-circuit activation may be recovered. Long-lasting or resilient information may be accessible either through (1) the machinery of transcription at the site of fixation on the DNA of the cells or (2) by the histone code as these epigenetics modifications are believed to be acquired with a stability related to the original tissue (Waterland and Michels, 2007) and to the time of day (Alenghat et al., 2008; Figure 4).

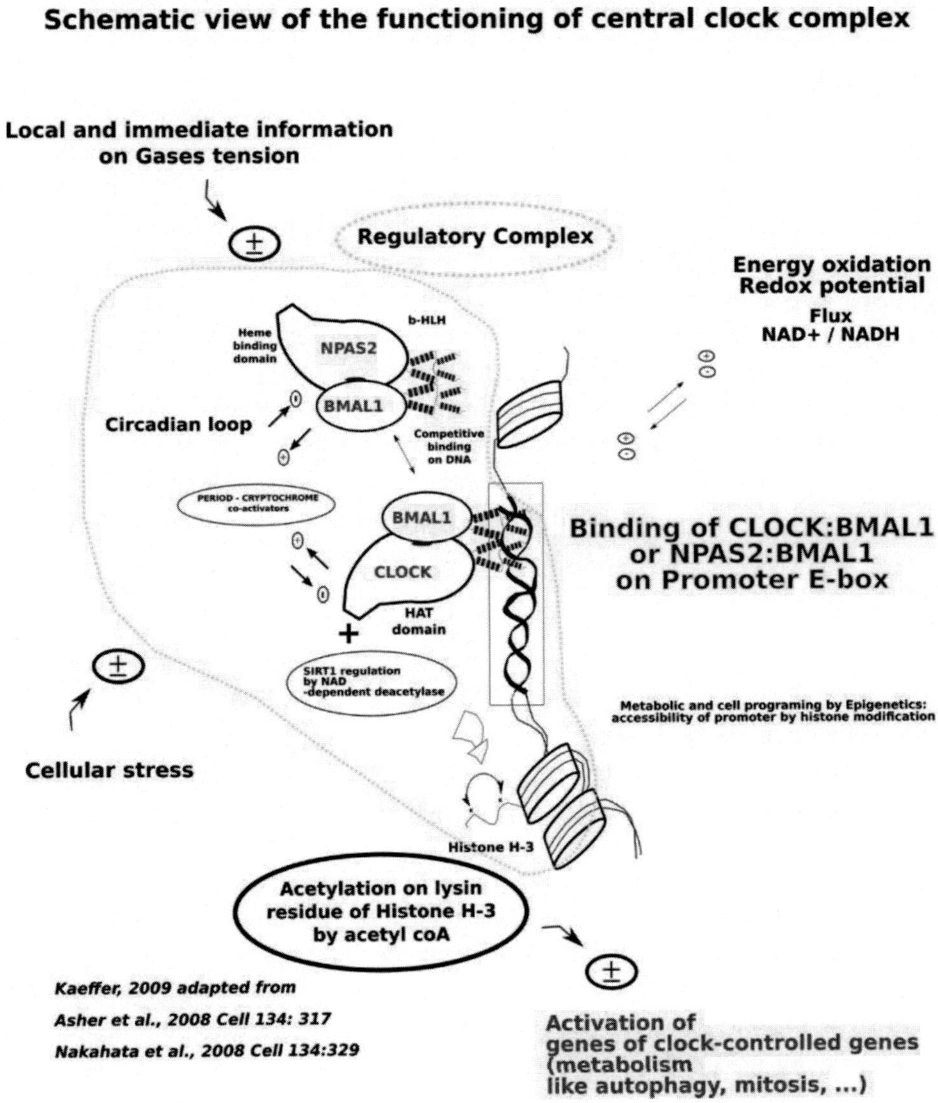

Figure 4. Stability of biological information in exfoliated epithelial cells with special emphasis on clock molecular components. Clocks are buffered against changes in temperature which typically speed up or slow down macromolecular association even if the mechanism of such a regulation is still unknown (Ruby et al., 1999; Brown et al., 2002; Tsuchiya et al., 2003). Limit cycle oscillations are stable to minor perturbations of regulator concentrations and they persist over time even when the organism is not subjected to periodic driving forces (e. g. light exposure, feeding rhythms). The capacity to evaluate clock functioning on exfoliated cells will depend on the stability of complex toward progressive and partly revertible anoikis, gases tension disruption as well as Redox potential, and temperature shifts. Clocks are stable molecular systems regarding the classic conditions (less than 1 hour) to isolate exfoliated cells. Data on clock genes expression by exfoliated cells has been reported (Kaeffer et al., 2007) however molecular information is still lacking like expression of npas2 transcripts or proteins. Gating and uncoupling of peripheral clocks with the central one are largely unexplored and need new methods to address the question. Exfoliated cells retain expression of several molecular clock components but are believed to loose rapidly metabolic coordination after isolation.

We may speculate that quiescent cells like epithelial gastric cells are retaining fully functional clocks. A crucial question for the use of exfoliated cells in chrononutrition is to know whether these cells retain consistent information with their time at exfoliation and/or their subsequent cell survival out of the organism. The induction of gastric cell exfoliation by nutrient cycle developed in rats by Aoyama et al (2008) and that we have adapted on lactating rat pups (Kaeffer et al., 2009) can be used in the future to address questions about the stability of information relevant to clock's functioning during anoikis. Exfoliated epithelial cells can be followed by microscopic examination from the initial step of loss of contact at the mouth of the gastric gland to the recovery of cells in the stomach lumen. In addition, the model is clearly relevant to clinical situation in which patients are equipped with nasogastric tubing. However, there may be tissue-specific differences in the molecular composition of the circadian clock, and clock components that have subtle effects on the central clock function may play a more prominent role in the regulation of peripheral clocks. Yagita et al., 2001 have used spontaneously immortalized mouse embryo fibroblasts to explore the main clock components (proteins and mRNA) suggesting that peripheral clocks in cultured cells may be similar in composition and regulation as central clock but all these components are not always present in cells depending on their tissular origin. The most stricking example is the apparent redundancy of clock with its homolog npas2, which are largely equivalent molecules with strict structural differences (Reick et al., 2001; Rutter et al., 2001; Figure 3). Exfoliated epithelial cells after loosing contact with extracellular matrix are surviving by activating autophagy. Three main physiological pathways have been described associated to the regulation of autophagy: AKT (energy-sensing), EGR-R (growth factor-sensing) and bcl-2 (stress-related programmed-cell death). According to Gan et al. (2006), mTOR downregulation is also observed following extracellular matrix detachment. There are no data on the relation between autophagy and clock regulation, it would be crucial for the use of exfoliated cells to explore the stability of clock compounds in the hours following exfoliation.

By contrast, explants isolated from rat transgenic for period1 gene, have shown that this gene displays a chaotic expression (Stokkan et al., 2001). The recording of luminescence emitted *in vitro* by the explants maintained in classic tissue culture conditions (37°C, 5% CO2) has clearly shown that the chaotic light emissions by the transgene system stands up to 12-14 hours. Thereafter the rhythms of light emissions by liver explants are organized according to an oscillatory model reminiscent of period1's *in vivo* oscillations (as of Stokkan et al. (2001), the phase of the peak has been recorded during the first subjective day in culture i. e. between 12 and 36 hours, in order to compare explants from rats maintained on *ad libitum* feeding with rats on restricted feeding). To avoid such chaotic evolution with the loss of chronobiological information, experimenters are using mechanical punches of mucosae which are directly snap frozen in liquid nitrogen. This strategy is a reliable but invasive solution to study clock gene expression in time series. Otherwise, tissue biopsies can be explanted in culture to derive primary cells or cell lines and record clocks functioning just like with exfoliated cells. However, cell lines have lost contact with body's network and their clock systems are probably quickly reorganized to tune up with their new *in vitro* environment (Nagoshi et al., 2004; Brown et al., 2005).

In summary, clock-related information is not yet extensively studied in exfoliated cells. Recent data on the molecular biology of clock components indicate that central and peripheral clocks differ in their coupling with the different categories of synchronizers as well as in their output on rodent models (Sumova et al., 2006) as well as on human data (Gallego and Virshup, 2007). In addition, molecules of the rev-erb-alpha or ror-a families have yet to be fully characterized to understand the molecular link between circadian clocks and endocrine metabolism. Molecular variants of rev-erb-alpha with similar structural properties and different half-lifes have been shown to help in the fine tuning of the circadian periodicity (Rambaud et al., 2009). Future works may take advantage of epigenetic tags to delineate the onset of acquisition of a stable epigenetic profile to explore clock gene regulation in peripheral organs.

PERSPECTIVES

The development of non-invasive methods is crucial to allow easy sampling of human populations in nutritional/clinical intervention studies. A specific demand is to develop studies in exfoliation in order to improve nursing care in preterm infants. Development of peripheral clocks depends on maturation of the organ housing the clock as well as on maturation of the molecular clockwork. The first appearance of molecular oscillations might be thus highly organ-specific and/or species-specific. A clear-cut example is the difference between oscillations of clock genes in the heart (Sakamoto et al., 2002) and the liver (Sladek et al., 2007). The possibility that subtle environmental influences during development cause persistent changes in epigenetic regulation may be a critical public health importance in the prevention of metabolic syndrome (Waterland & Michels, 2007). The recovery of exfoliated epithelial cells from pups or infants suffering from perinatal denutrition at the onset of the problem or later in adult life may help to know whether the histone acetyltransferase's activity of CLOCK can be used to explore the stability of epigenetic profile in exfoliated epithelial cells of mucosa.

Up to now, liver has been the organ of choice to study peripheral clock functioning but a better understanding of circadian rhythms implies to integrate also information from complex tissue architecture like lung or gut which are at the interface with the external environment. An analysis of exfoliated cells to determine proportions of cells in anoikis can be rapidly achieved by direct examination of nuclei integrity and relevant cell populations may be purified through antibody methods. Unique data on coupling / decoupling capacity of peripheral clocks may be obtained in long-term experiments by combining the recovery of low cell numbers with analytical cytology and microarray technologies. However there is a lack of biomarkers to study exfoliated epithelial cells and the role of clock genes, if any, in autophagy. Among many unsolved questions which can be listed, we can wonder what is the biological information retained, altered or lost during anoikis? Future works may focus on the mTOR signaling pathway which has been found down-regulated in detached epithelial cells (Gan et al., 2006) and, in adipocytes, linked to diurnal gene expression and metabolic regulation (Loboda et al., 2009).

REFERENCES

Abizaid A, Mezei G, Sotonyi P, Horvath TL. (2004). Sex differences in adult suprachiamastic nucleus neurons emerging late prenatally in rats. *European Journal of Neuroscience 19*: 2488-2490.

Ahmed FE, James SI, Lysle DT, Dobbs LJ Jr, Johnke RM, Flake G, Stockton P, Sinar DR, Naziri W, Evans MJ, Kovacs CJ, Allison RR. (2004). Improved methods for extracting RNA from exfoliated human colonocytes in stool and RT-PCR analysis. *Digestive Diseases and Sciences 49*: 1889-1898.

Albaugh, GP, Iyengar V, Lohani A, Malayeri M, Bala S, Nair PP. (1992). Isolation of exfoliated colonic epithelial cells, a novel non-invasive approach to the study of cellular markers. *Int. J. Cancer 52*: 347-350.

Albeck JG, MacBeath G, White FM, Sorger PK, Lauffenburger DA, Gaudet S. (2006). Collecting and organizing systematic sets of protein data. *Nature Review Molecular Cell Biology 7*: 803-812.

Alenghat T, Meyers K, Mullican SE, Leitner K, Adeniji-Adele A, Avila J, Bucan M, Ahima R, Kaestner KH, Lazar MA. (2008). Nuclear receptor corepressor and histone deacetylase 3 govern circadian metabolic physiology. *Nature 456*: 997-1000.

Aoyama F, Sawaguchi A, Ide S, Kitamura K, Suganuma T. (2008). Exfoliation of gastric pit-parietal cells into the gastric lumen associated with a stimulation of isolated rat gastric mucosa in vitro: a morphological study by the application of cryotechniques. *Histochem. Cell Biol. 129*: 785-793.

Bader GM, Papanicolaou GN. (1952). The application of cytology in the diagnosis of cancer of the rectum, sigmoid, and descending colon. *Cancer 5*: 307-14.

Bandaletova T, Bailey N, Bingham SA, Loktionov A. (2002). Isolation of exfoliated colonocytes from human stools as a new technique for colonic cytology. *APMIS 110*: 239-246.

Bertolucci C, Cavallari N, Colognesi I, Aguzzi J, Chen Z, Caruso P, Foà A, Tosini G, Bernardi F, Pinotti M. (2008). Evidence for an overlapping role of CLOCK and NPAS2 transcription factors in liver circadian oscillators. *Molecular and Cellular Biology 28*: 3070-3075.

Bjarnason GA, Jordan RC, Wood PA, Li Q, Lincoln DW, Sothern RB, Hrushesky W JM, Ben-David Y. (2001). Circadian expression of clock genes in human oral mucosa and skin. Association with specific cell-cycle phases. *Am. J. Pathol. 158*: 1793-1801.

Brown SA, Fleury-Olela F, Nagoshi E, Hauser C, Juge C, Meier CA, Chicheportiche R, Dayer J-M, Schibler U. (2005). The Period length of fibroblast circadian gene expression varies widely among human individuals. *PLOS Biology 3*: 1813-1818.

Brown SA, Zumbrunn G, Fleury-Olela F, Preitner N, Schibler U. (2002). Rhythms of mammalian body temperature can sustain peripheral circadian clocks. *Current Biology 12*: 1574-1583.

Buchi KN, Moore JG, Hrushesky WJ, Sothern RB, Rubin NH. (1991). Circadian rhythm of cellular proliferation in the human rectal mucosa. *Gastroenterology 101*:410-415.

Cardone L, Hirayama J, Giordano F, Tamaru T, Palvimo JJ, Sassone-Corsi P. (2005). Circadian clock control by SUMOylation of BMAL1. *Science 309*: 1390-1394.

Carskadon MA, Wolfson AR, Acebo C, Tzischinsky O, Seifer R. (1998). Adolescent sleep patterns, circadian timing, and sleepiness at a transition to early school days. *Sleep 21*: 871-881.

Curtis AM, Seo S-B, Westgate EJ, Rudic RD, Smyth EM, Chakravarti D, FitzGerald GA, McNamara P. (2004). Histone acetyltransferase-dependent chromatin remodeling and the vasculature clock. *The Journal of Biological Chemistry 279*: 7091-7097.

Davidson LA, Lupton JR, Miskovsky E, Fields AP, Chapkin RS. (2003). Quantification of human intestinal gene expression profiles using exfoliated colonocytes: a pilot study. *Biomarkers 8*: 51-61.

Davis CD. (2003). Use of exfoliated cells from target tissues to predict responses to bioactive food components. *Journal of Nutrition 133*: 1769-1772.

Debnath J, Mills KR, Collins NL, Reginato MJ, Muthuswamy SK, Brugge JS. (2002). The role of apoptosis in creating and maintaining luminal space within normal and oncogene-expressing mammary acini. *Cell 111*: 29-40.

Dioum EM, Rutter J, Tuckerman JR, Gonzalez G, Gilles-Gonzalez M-A, McKnight SL. (2002). NPAS2: a gas-responsive transcription factor. *Science 298*: 2385-2387.

Doi M, Hirayama J, Sassone-Corsi P. (2006). Circadian regulator CLOCK is a histone acetyltransferase. *Cell 125*: 497-508.

Etchegaray J-P, Lee C, Wade PA, Reppert SM. (2003). Rhythmic histone acetylation underlies transcription in the mammalian circadian clock. *Nature 421*: 177-182.

Felix JS, Littlefield JW (1980). Human newborn urine as a source of epithelial cells. *Birth Defects Orig Artic Ser. 16*: 231-237.

Fenech M, Holland N, Chang WP, Zeiger E, Bonassi S. (1999). The HUman project: an international colloborative study on the use of the micronucleus technique for measuring DNA damage in humans. *Mutat. Res. 428*: 271-283.

Frisch SM, Francis H. (1994). Disruption of epithelial cell-matrix interactions induces apoptosis. J Cell Biol *124*: 619-626.

Fung C, Lock R, Gao S, Salas E, Debnath J. (2008). Induction of autophagy during extracellular matrix detachment promotes cell survival. *Mol Biol Cell 19*: 797-806.

Gallego M, Virshup DM. (2007). Post-translational modifications regulate the ticking of the circadian clock. *Nat Rev Mol Cell Biol. 8*: 139-148.

Gan B, Yoo Y, Guan JL. (2006). Association of focal adhesion kinase with tuberous sclerosis complex 2 in the regulation of s6 kinase activation and cell growth. *J Biol Chem. 281*:37321-9

Gilmore AP. (2005). Anoikis. *Cell Death Differ. 12 (suppl 2)*, 1473-1477.

Gomes JR, Pereira AA, Barth L, Silva JS, Leite ML, Wille AC, Soares MAM. (2005). Circadian variation of the cell proliferation in the jejunal epithelium of rats at weaning phase. *Cell Prolif 38*: 147-152.

Haus E, Lakatua DJ, Sackett-Lundeen L, White M. (1984). Circannual variation of intestinal cell proliferation in BDF1 male mice on three lighting regimens. *Chronobiol. Int. 1*: 185-194.

Holland N, Harmatz P, Golden D, Hubbard A, Wu YY, Bae J, Chen C, Huen K, Heyman MB. (2007). Cytogenetic damages in blood lymphocytes and exfoliated epithelial cells of children with inflammatory bowel disease. *Pediatric Research 61*: 209-214.

Iwakiri R, Gotoh Y, Noda T, Sugihara H, Fujimoto K, Fuseler J, Aw TY. (2001). Programmed cell death in rat intestine: effect of feeding and fasting. *Scandinavian Journal of Gastroenterology 36*: 39-47.

Kaasik K, Lee CC. (2004) Reciprocal regulation of haem biosynthesis and the circadian clock in mammals. *Nature 430*: 467-471.

Kaeffer, B. (2002) Mammalian intestinal epithelium cells in primary culture: a mini review. *In vitro Cellular and Developmental Biology – Animal 38*: 123-134.

Kaeffer B, Des Robert C, Alexandre-Gouabau MC, Pagnièz A, Amarger V, Legrand A, Küster A, Piloquet H, Champ M, Le Huërou-Luron I, Rozé J-C. (2007). Recovery of exfoliated cells from the gastro-intestinal tract of premature infants: a new tool to obtain "non-invasive biopsies"? *Pediatric Research 62*: 564-569.

Kaeffer B, Rozé JC. NeoExfolCel Data base of exfoliated cells for gastric and colonic exfoliated cells or preterm infants. French agency for the data base protection (Base de données: Biomarqueurs de cellules exfoliées digestives de l'enfant prématuré. Sigle: NeoExfolCel (NEC), 2009, IDDN.FR.001.230013.000.R.P.2009.000.10300).

Kaeffer B, Qi H, Matos RJB, Legrand A, Chauty A, Darmaun D, Morel F, Gouthière L, Bolanos-Jimenez F, Champ M, Rozé J-C. (2009). Exfoliated epithelial cells, a source of information on clock genes expression by preterm infants to explore the onset of metabolic syndrome. *XI Congress of the European Biological Rhythms Society*, Strasbourg, August 22-28.

Kamra A, Kessie G, Chen JH, Kalavapudi S, Shores R, McElroy I, Gireesh T, Sudhakaran PR, Dutta SK, Nair PP. (2005). Exfoliated colonic epithelial cells: surrogate targets for evaluation of bioactive food components in cancer prevention. *Journal Nutrition 11*, 2719-2722.

Karam SM. (1999). Lineage commitment and maturation of epithelial cells in the gut. *Frontiers in Bioscience 4*, 286-298.

Kitanishi K, Igarashi J, Hayasaka K, Hikage N, Saiful I, Yamauchi S, Uchida T, Ishimori K, Shimizu T. (2008). Heme-binding characteristics of the isolated PAS-A domain of mouse Per2, a transcriptional regulatory factor associated with circadian rhythms. *Biochemistry 47*: 6157-6168.

Kitazawa H, Nishihara T, Nambu T, Nishzawa H, Iwaki M, Fukuhara A, Kitamura T, Matsuda M, Shimomura I. (2004). Intectin, a novel small intestine-specific glycosylphosphatidylinositol-anchored protein, accelerates apoptosis of intestinal epithelial cells. *The Journal of Biological Chemistry 279*: 42867-42874.

Klein P, Glaser E, Grogan L, Keane L, Lipkowitz S, Sobalie P, Brooks L, Jenkins J, Steinberg SM, DeMarini DM, Kirsch I. (2001). Biomarker assays in nipple aspirate fluid. *Breast J. 7*: 378-387.

Lenaerts K, Bouwman FG, Lamers WH, Renes J, Mariman EC. (2007). Comparative proteomic analysis of cell lines and scrapings of the human intestinal epithelium. *BMC Genomics 8*: 91 doi: 10.1186/1471-2164/8/91

Levine B and Klionsky DJ. (2004). Development by self-digestion: molecular mechanisms and biological functions of autophagy. *Dev Cell 6:* 463-477.

Liu Z, Li H, Wu X, Yoo BH, Yan SR, Stadnyk AW, Sasazuki T, Shirasawa S, LaCasse EC, Korneluk RG, Rosen KV. (2006). Detachment-induced upregulation of XIAP and cIAP2 delays anoikis of intestinal epithelial cells. *Oncogene 25*: 7680-7690.

Loboda A, Kraft WK, Fine B, Joseph J, Nebozhyn M, Zhang C, He Y, Yang X, Wright C, Morris M, Chalikonda I, Ferguson M, Emilsson V, Leonardson A, Lamb J, Dai H, Schadt E, Greenberg HE, Lum PY. Diurnal variation of the human adipose transcriptome and the link to metabolic disease. *BMC Med Genomics*. 2009 Feb 9;2:7.

Lock R, Debnath J. (2008). Extracellular matrix regulation of autophagy. *Current Opinion in Cell Biology 20*: 583-588.

Loktionov A. (2007). Cell exfoliation in the human colon: myth, reality and implications for colorectal cancer screening. *Int J Cancer 120*: 2281-2289

Loktionov A., Bandatelova T, Llewelyn A, Ferrett C, Lywood RCG, Lywood HGG. (2006). Colorectal cell sampling device. *World Intellectual Property Organisation*. International Publication Number WO 2006/003447 A1.

Majer BJ, Laky B, Knasmuller S, Kassie F. (2001). Use of the micronucleus assay with exfoliated epithelial cells as a biomarker for monitoring individuals at elevated risk of genetic damage and in chemoprevention trials. *Mutat Res 489*: 147-172.

Marra G, Anti M, Percesepe A, Armelao F, Ficarelli R, Coco C, Rinelli A, Vecchio FM, D'Arcangelo E. (1994). Circadian variations of epithelial cell proliferation in human rectal crypts. *Gastroenterology 106*: 982-987.

Matsuo T, Yamaguchi S, Mitsui S, Emi A, Shimoda F, Okamura H. (2003). Control mechanism of the circadian clock for timing of cell division *in vivo*. *Science 302*: 255-259.

Miller CL, Diglisic S, Leister F, Webster M, Yolken RH. (2004) Evaluating RNA status for RT-PCR in extracts of postmortem human brain tissue. *Biotechniques 36*: 628-633.

Mojica WD, Rapkiewicz AV, Liotta LA, Espina V. (2005). Manual exfoliation of fresh tissue obviates the need for frozen sections for molecular profiling. *Cancer Cytopathol 105*: 483-491.

Mojica WD, Stein L, Hawthorn L. (2007). An exfoliation and enrichment strategy results in improved transcriptional profiles when compared to matched formalin fixed samples. *BMC Clinical Pathology 7*:7 doi 10.1186/1472-6890/7/7

Moldenhauer G, Momburg F, Möller P, Schwartz R, Hämmerling GJ. (1987) Epithelium-specific surface glycoprotein of Mr 34,000 is a widely distributed human carcinoma marker. *Br. J. Cancer 56*: 714-721.

Moore-Ede MC. (1986). Physiology of the circadian timing system: predictive versus reactive homeostasis. *Am. J. Physiol. 250*: R737-52.

Munoz N, Hayashi M, Bang LJ, Wahrendorf J, Crespi M, Bosch FX. (1987) Effect of riboflavin, retinol and zinc on micronulei of buccal mucosa and of esophagus: a randomized double-blind intervention study in China. *J. Natl Cancer Inst. 79*: 687-691.

Nakahata Y, Kaluzova M, Grimaldi B, Sahar S, Hirayama J, Chen D, Guarente LP, Sassone-Corsi P. (2008). The NAD+-dependent deacetylase SIRT1 modulates CLOCK-mediated chromatin remodeling and circadian control. *Cell*. 2008 Jul 25;134(2):329-40.

Nakahata Y, Sahar S, Astarita G, Kaluzova M, Sassone-Corsi P. (2009). Circadian control of the NAD+ salvage pathway by CLOCK-SIRT1. *Science 324*(5927):654-7

Nagoshi E, Saini C, Bauer C, Laroche T, Naef F, Schibler U. (2004). Circadian gene expression in individual fibroblasts: cell-autonomous and self-sustained oscillators pass time to daughter cells. *Cell 119*: 693-705.

Osborn NK, Ahlquist DA. (2005). Stool screening for colorectal cancer: molecular approaches. *Gastroenterology 128*: 192-206.

Panda S, Antoch MP, Miller BH, Su AI, Schook AB, Straum M, Schultz PG, Kay SA, Takahashi JS, Hogenesh JB. (2002). Coordinated transcription of key pathways in the mouse by the circadian clock. *Cell 109*: 307-320.

Pardini L, Kaeffer B, Trubuil A, Bourreille A, Galmiche J-P. (2005). Human intestinal circadian clock: expression of clock genes in colonocytes lining the crypt. *Chronobiology International 6*: 951-961.

Phillips J, Eberwine JH. (1996). Antisense RNA amplification: a linear amplification method for analyzing the mRNA population from single living cells. *Methods 10*: 283-288.

Polidarova L; Sotak M; Sladek M; Pacha J; Sumova A. (2009). Temporal Gradient in the Clock Gene and Cell-Cycle Checkpoint Kinase *Wee1* Expression along the Gut. *Chronobiology International 26*: 607-620.

Portaluppi F, Touitou Y, Smolensky MH. (2008). Ethical and methodological standards for laboratory and medical biological rhythm research. *Chronobiology International 25*: 999-1016.

Potten CS, Loeffler M. (1990). Stem cells attributes, cycles spirals, pitfalls and uncertainties-lessons for and from the crypt. *Development 110*: 1001-1020.

Qiu JM, Roberts SA, Potten CS. (1994). Cell migration in the small and large bowel shows a strong circadian rhythm. *Epithelial Cell Biology 3*: 137-148.

Raghuram S, Stayrook KR, Huang P, Rogers PM, Nosie AK, McClure DB, Burris LL, Khorasanizadeh S, Burris TP, Rastinejad F. (2007). Identification of heme as the ligand for the orphan nuclear receptors REV-ERBalpha and REV-ERBbeta. *Nat Struct Mol Biol. 14*:1207-13.

Rambaud J, Triqueneaux G, Masse I, Staels B, Laudet V, Benoit G. (2009) Rev-erb-alpha2 mRNA endodes a stable protein with a potential role in circadian clock regulation. *Molecular Endocrinology 23*: 630-639.

Reddy AB, Karp NA, Maywood ES, Sage EA, Deery M, O'Neill JS, Wong GK, Chesham J, Odell M, Lilley KS, Kyriacou CP, Hasting MH. (2006). Circadian Orchestration of the Hepatic Proteome. *Current Biology 16*: 1107-1115.

Reick M, Garcia JA, Dudley C, McKnight SL. (2001). NPAS2: an analog of Clock operative in the mammalian forebrain. *Science 293*: 506-509.

Reinberg AE. (2003). *Chronobiologie médicale, chronothérapeutique.* Médecine-Sciences Flammarion, 298 pp.

Ripperger JA, Schibler U. (2006). Rhythmic CLOCK-BMAL1 binding to multiple E-box motifs drives circadian Dbp transcription and chromatin transitions. *Nature Genet. 38*: 369-374

Rivkees SA. (2003). Developing circadian rhythmicity in infants. *Pediatrics 112*: 373-381.

Robine N, Relier JP, Le Bars S. (1988). Urocytogram, an index of maturity in premature infants. *Biol. Neonate 54*: 93-99.

Ruby NF, Burns DE, Heller HC. (1999). Circadian rhythms in the suprachiasmatic nucleus are temperature-compensated and phase-shifted by heat pulses in vitro. *The Journal of Neuroscience 19*: 8630-8636.

Rutter J, Reick M, Wu LC, McKnight SL. (2001). Regulation of Clock and NPAS2 DNA binding by the Redox state of NAD cofactors. *Science 293*: 510-514.

Sakamoto K, Oishi K, Nagase T, Miyasaki K, Ishida N. (2002). Circadian expression of clock genes during ontogeny in the rat heart. *Neuroreport 13*: 1239-1242.

Sanada K, Okano T, Fukada Y. (2002). Mitogen-activated protein kinase phosphorylates and negatively regulates basic helix-loop-helix-PAS transcription factor BMAL1 *J. Biol. Chem. 277*: 267-271.

Sheibani K, Shin SS, Kezirian J, Weiss LM. (1991). Ber-Ep4 antibody as a discriminant in the differential diagnosis of malignant mesothelioma versus adenocarcinoma. *Am J Surg Pathol 15*: 779-784.

Silveira L Jr, Betiol Filho JA, Silveira FL, Zângaro RA, Pacheco MT. (2008). Laser-induced fluorescence at 488 nm excitation for detecting benign and malignant lesions in stomach mucosa. *J Fluorescence 18*: 35-40.

Sitek B, Sipos B, Klöppel G, Schmiegel W, Hahn SA, Meyer HE, Stühler K. (2008). Application of fluorescence dye saturation labeling for differential proteome analysis of 1,000 microdissected cells from pancreatic ductal adenocarcinoma precursor lesions. *Methods Mol Biol. 425*:1-14.

Sladek M, Jindrakova Z, Bendova Z, Sumova A. (2007). Postnatal ontogenesis of the circadian clock within the rat liver. *Am. J. Physiol. Regul. Integr. Comp. Physiol. 292*: R1224-9.

Steinau M, Lee DR, Rajeevan MS, Vernon SD, Ruffin MT, Unger ER. (2005). Gene expression profile of cervical tissue compared to exfoliated cells: impact on biomarker discovery. *BMC Genomics 1*, 64.

Stitch HF, Stitch W, Rosin MP, Vallejara MO. (1984). Use of the micronucleus test to monitor the effect of Vitamin A, beta-carotene and canthaxanthin on the buccal mucosa of betel nut/tobacco chewers. *Int. J. Cancer 34*: 745-750.

Stokkan K-A, Yamazaki S, Tei H, Sakaki Y, Menaker M. (2001). Entrainment of the circadian clock in the liver by feeding. *Science 291*: 490-493.

Sumova A, Bendova Z, Sladek M, El-Hennamy R, Laurinova K, Jindrakova Z, Illnerova H. (2006). Setting the biological time in central and peripheral clocks during ontogenesis. *FEBS Letters 580*: 2836-2842.

Tan C, Reza F, You L. (2007). Noise-limited frequency signal transmission in gene circuits. *Biophysical Journal 93*: 3753-3761.

Thompson PA, Kadlubar FF, Vena SM, Hill HL, McClure GH, McDaniel LP, Ambrosone CB. (1998). Exfoliated ductal epithelial cells in human breast milk: a source of target tissue DNA for molecular epidemiologic studies of breast cancer. *Cancer Epidemiology, Biomarkers & Prevention 7*: 37-42.

Tsuchiya Y, Akashi M, Nishida E. (2003). Temperature compensation and temperature resetting of circadian rhythms in mammalian cultured fibroblasts. *Genes to Cells 8*: 713-720.

Van Lieshout EM, Van Doesburg W, Van der Meer R. (2004). Real-time PCR of host DNA in feces to study differential exfoliation of colonocytes between rats and humans. *Scand J Gastroenterol 39*: 852-857.

Waterland RA, Michels KB. (2007). Epigenetic epidemiology of the developmental origins hypothesis. *Annu. Rev. Nutr. 27*: 363-388.

Yagita K, Tamanini F, van der Horst G, Okamura H. (2001). Molecular mecanisms of the biological clock in cultured fibroblasts. *Science 292*: 278-281.

Yan SR, Joseph RR, Rosen K, Reginato MJ, Jackson A, Allaire N, Brugge JS, Jobin C, Stadnyk AW. (2005). Activation of NF-kappa B following detachment delays apoptosis in intestinal epithelial cells. *Oncogene 24*: 6482-6491.

Yang GC, Hoda SA. (1997). Combined use of the "Scratch and Smear" sampling technique and ultrafast Papanicolaou stain for intraoperative cytology. *Acad Cytol 41*: 1513-1518.

Yin L, Wu N, Curtin JC, Qatanani M, Szwergold NR, Reid RA, Waitt GM, Parks DJ, Pearce KH, Wisely GB, Lazar MA. (2007). Rev-erbalpha, a heme sensor that coordinates metabolic and circadian pathways. Science. *318*: 1786-9.

Zanello SB, Jackson DM, Holick MF. (2000). Expression of the circadian clock genes clock and period1 in human skin. *J. Invest. Dermatology 115*: 757-760.

In: Daughter Cells
Editors: A. Hitomi et al., pp. 243-266

ISBN: 978-1-60876-790-8

Chapter VII

CHANCE FOR PLANTS: WIDER POSSIBILITIES AND CAPABILITIES

Károly Bóka **and Norbert Orbán***
Eötvös Loránd University, Budapest, Hungary

ABSTRACT

Plants can show flexible capacity of coping with the impairing factors of their abiotic and biotic environment and their high developmental plasticity, and easily forced cellular totipotency is a general feature. Plants have a notable potential to maintain separate cell groups in their apical meristems to assure growing. In addition to these proembryonic cell division centers, plants posses various manners (e.g., dormant meristemoids and residual stem cells) to acquire cell division capability in their bodies and, as a result, the daughter cells achieve an altered differentiation status there. The initiated growth and regeneration might also be based on newly formed dividing cells entering again into the cell cycle, and the different cell fate of unequal daughter cells is usually fundamental in the realization of altered development. In the latter case, a precondition of the subsequent events is dedifferentiation of mature, specialized cells and remodeling of their genetic program.

Our results demonstrate that cells may attain disparate dedifferentiation statuses depending on their ancestry, and the initiated redifferentiation processes might be arrested even before entering cell divisions or continued up to complete root formation. Transfer cells of leaf veins share function in short-distance transport under normal conditions and stem cell formation under forced organogenesis. The *Rubia tinctorum* in vitro model system allowed us to detect the main symptoms of the rapid morphological changes during dedifferentiation. Modification of cell wall structure, enhanced autophagic activity, asymmetric plastid fission and unequal cell division seem to be basic steps of meristemoid

* Correspondence concerning this chapter should be addressed to: Károly Bóka, Eötvös Loránd University, Department of Plant Anatomy, Budapest, Pázmány Péter sétány 1/C, H-1117 Hungary. Tel: +3613812165; Fax: +3613812166; E-mail address: bokak@ludens.elte.hu.

(stem cell) formation. The theoretical background and the importance of the cytological characteristics mentioned above are also discussed.

INTRODUCTION

Life Strategy of Plants and Its Impact on Their Organization

Species of the biosphere, this global ecological system integrating all living beings and their interactions, have acted during the past 3.5 billion years to fill accessible niches. Their chance to survive depended merely on the adaptation of the given species—its structural and physiological competency to occupy an ecological space. The various living creatures required different strategies to survive, and even today they must change their features to maintain their best accomplishments.

Plants invented a sessile life strategy, which allowed them to explore the local sources at a high efficiency, although, for survival, they had to cope with numerous harmful factors of the environment. So, plants evolved different manners to persist among unfavourable conditions. In addition, life on the land demanded a more differentiated body and a higher degree of specialization [1]. Growth and the long-lasting possibility to renew their organs permitted plants to exist among adverse circumstances (injury, impairing factors, etc.). The extended interval of cell divisions in specialized regions of their body, differentiation of new cells and easily forced totipotency ensure wider capabilities for plants. Daughter cells of different origins specialized to fulfil the new demands of changing environment. Continuous or restarted cell divisions ensured better adaptation for plants by developing new parts, additional organs with novel functions or changing plant activity [2].

Cell Divisions, Growth and Development

Cell divisions fundamentally determine the growth and progress of a plant body. Shoot and root apices assure longitudinal accession while vascular cambium (VC) activity is responsible for lateral expansion of plant organs, mainly the shoots and roots. These apical and cambial cell divisions, together with differentiation events (e.g., cell elongation), also decide on growth pattern and body proportions of plants. Cell division centers (meristems) can act permanently, but active and dormancy periods also can alternate depending on the lifestyle of the given species. In developing organs, intercalary cell lines are involved in local growth and development events, like petiol formation or stomatal complex establishment [3].

Apart from the processes mentioned above, irregularly occurring cell divisions as well may influence the function and structure of plant organs. Injuries, stress, extreme conditions and human activity might force regaining of the cell division capacity in certain tissues of mature parts. The base of these events is that cells or cell groups are able to return to the cell cycle and manifest their totipotency easily [4,5].

Cells having the capability to produce specialized tissues might be termed stem cells according to their role in renewing plant organs or in whole plant regeneration [4-6]. Time, place, intensity and ratio of permanent, temporal and occasional divisions of stem cells are pivotal in plant body configuration and effective functioning under normal and extreme conditions.

Meristems, Meristemoids and Stem Cells

Embryophytes possess a special, early developmental stage in their ontogeny, termed *embryo*, which origins from the zygote and occurs as a transient phase leading to the fully matured sporophyte. The embryo's complexity and state of development might be strongly diverse in various taxons, but the embryo generally shows strong axial polarity and has progenitors of basic plant organs. The most intricat structural and physiological level is evolved at the spermatophytes [7]. In angiosperms, well-defined subsequent ontogenetic phases occur (proembryo, globular stage, heart stage, torpedo stage, mature embryo) during embryogenesis [8], although in some parasitic species the embryo is reduced to a small disc of meristematic cells [9]. In general, cells of the seed-bearing plant embryo maintain their mitotic activity, arrested only temporally during dormancy.

The plant zygotic embryo contains mainly undifferentiated cells. A few cell types are determined; however, most of the embryonic cells divide during formation of the young sporophyte. Apical meristems (shoot apex and root tip) are maintained through the sporophyte development and their number is multiplied by lateral shoot and lateral root formation [1]. Permanent mitotical activity of apical meristems lengthens plant body and permits primary structure construction.

Root tips continuously grow the root system while shoot apex acts more intricately. Vegetative shoot apex augments stem and forms new leaf primordia. Size and shape of shoot apex dome are not equal during the meristematic activity and primordium occurrence is periodical [10]. Along with plant body growth and development, shoot apex might switch its function. Changing its form and size, it can perform as a reproductive shoot apex producing a modified shoot, flower or inflorescence. According to the altered task, generation of primordia of flower parts has modified time and spatial pattern [11]. All of the abovementioned events are closely connected with cell divisions and the maintained meristematic identity of apical cell groups. In the long existence of the postembryonal meristematic activity, the unequal cell division has the basic role where daughter cells differ strongly in their appearance, genetical activity and post-cytokinetic fate [2].

Similarly, VC cells ensuring secondary growth (mainly in stems and roots) produce new cells of parenchyma, phloem and xylem tissues. Depending on the position of the daughter cells, their differentiation is determined but in all cases definitely distinct from the cells remaining in the cambial cell layer, which retain their cell division activity and undifferentiated state [2,3].

Apical and lateral meristems, producing high number of cells, guarantee primary and secondary growth. In young organs cells do not differentiate at the same time [2]. Groups of cells with limited cell number or even separated cells might hold their undifferentiated state for a starched period in a differentiating tissue environment and can give rise to intercalary growth or formation of specialized structures (stomatal complex, secreting units). In developing organs, these meristemoids, referred by some authors to stem cells [5], extend differentiation time giving a possible extra plasticity important for adaptation.

Apart from the regularly active meristems, plants have dormant meristematic cell groups also. These inactive meristems (e.g., axillary buds) may exist for much of the time and, in accordance with plant necessities (e.g., loss of the operating shoot meristem), might be activated and form new organs [12]. As in the regular meristems, cell division ability and stem cell status is preserved in a small cell group of previously dormant meristem while daughter cells divide frequently and differentiate into specified tissues. Normally, the working shoot meristem keeps in dormancy these axillary buds (apical dominancy) and governs them hormonally [1]. It is suggested that auxin, released from the active shoot apex, arrests cell divisions in the axillary bud meristems in spite of the fact that they are also able to produce a base level of auxin [13].

Neogenesis of Meristematic Cells

Plants cannot usually avoid unfavourable factors of their biotic and abiotic environments and have to cope with the harmful effects. In special cases, plants might lack active or dormant meristems (e.g., in organs with limited growth) and specialized cells of their stable tissues might be activated. Certain cells can return to the cell cycle and divide mitotically. They are able to produce undifferentiated cells, which form meristematic cell groups [14]. These initiating cells with small volume, possessing a dense cytoplasm retain their meristematic properties but their daughter cells differentiate to form organs or even whole plants (organogenesis or regeneration, respectively). Alterations in hormonal control after injury or stress can provoke differentiated cells to regain their cell division capacity and to manifest their totipotency [15]. Their daughter cells are able to specialize into differentiated cells of various tissues of an organ or a whole organism. So, in plants, certain somatic cells may act as residua, stem cell forming units. In plant biotechnology, plant factories are often established on somatic embryogenesis, the mass production of plant regenerants grown up from somatic cell suspensions through the forced embryogenetic developmental route [16,17].

The processes mentioned above are all connected with the retained or regained cell division capacity and the preserved and for long-time maintained meristem/stem cell identity. However, the background of this feature is not fully elucidated yet in plants. It seems to be correlated with gene expression pattern of meristematic cells and their interaction with their daughter cells developing into stable tissues in which cell cycle is arrested.

Understanding Meristem Operation

Although the molecular mechanism of the plant stem cell status remains elusive, the growing body of evidence allows outline the basic process of plant meristem operation. Stem cell maintenance and action in animals offers a positive paradigm for plant physiologists but cytological and evolutionary specificities of plants allow us to make only cautious comparison of the two groups [4,6]. The main principle is that stem cell state means a special functional and structural existence of undifferentiated cells in a microenvironment, in a specific niche in the organism, where the stem cells communicate with their neighbouring cells to establish their individuality and there is a strict signalling control to maintain their own status quo. The surrounding differentiating cells ensure physical and physiological entity of the niche [6,18,19]. Probably, apart from the known signalling mechanisms (feed-back, feed-forward, cross-talk, etc.), the quantity of signalling factors and the duration and frequency of their action is also important [18,20,21].

Both intrinsic and external factors control the generation of daughter cell types. Specific molecular elements of cell cycle control may be associated with the factors governing stem cell renewal, longevity and differentiation. In plants, shoot apex and root tip meristems, as active stem cell reservoirs, are the best studied. In the apical meristems, a large number of undifferentiated progeny is produced and these cells surrounding plant stem cell niches, located in the apical meristems, provide signals to maintain the stem cells in an undifferentiated state, the "stemness", a term which was created by some authors to characterize a specific cell state rather than to assign a defined entity [6].

Unequal cell division resulted by stem cell activity is a common feature, during which two daughter cell types are produced [6,22-25]. Unequal cell division of apical stem cells of plants is a precondition of the renewal of stem cell population and production of cells with transit-amplifying role [26]. Beside the ontogenic importance, appearance of the unequal cell division was also a perquisite for vascular plant evolution because it enabled invention of specialized cells of tissues parallel with the longstanding apical stem cell activity. Cytological requirements of these processes were the occurrence of preprophase band, phragmoplast, symplastic communication (plasmodesmata), cell polarity, and ability to change cell plate position [1].

The above described features and requirements are important events of the apical meristem function but do not administer information about the control of meristem purpose. During the last decades studies of various mutants pointing to certain genes appear to specify apical meristems and differences in shoot and root apical meristem organization. Analysis of meristematic activity-defective mutants identified some essential genes responsible for normal shoot and root apex functioning. *Arabidopsis thaliana* (Brassicaceae) proved to be the most useful model plant because of the high number of natural and induced mutants and its exceptional values in laboratory work; however, nowadays a growing number of hints are published on other objects (e.g., maize, rice, tomato, potato, etc.) to recognize the role of genes in the specification and organization of meristems [7]. Now it is suggested that large and diverse set of genes is important in the constitution of meristems, though the growing amount of data is connected to the apical meristems and we know less about other types of meristems.

The emerging sketch of the genetic control of meristem activity shows the orchestrated activity of genes of which some proved to encode transcriptional regulators (e.g., in *shoot meristemless* (*stm*) mutant in *A. thaliana*) [27]. It is supposed that homeobox genes containing DNA-binding homeodomain play fundamental role in controlling meristematic status. Activity of homeobox genes was detected in different species like KNOTTED 1 (KN1) from *Zea mays*, KN1-type gene from *Oryza sativa*, STM in *A. thaliana*, NO APICAL

MERISTEM (NAM) from *Petunia* and SHOOTLESS (SHL) from *O. sativa* [27-31]. Beside the homeobox genes, role of numerous genes was confirmed in correct meristem maintenance and daughter cell differentiation. The best-characterized signalling interaction between genes is documented in the case of CLAVATA (CLV) and WUSHEL (WUS) genes. CLV gene family promotes the transition of stem cells into specialized developmental stage (cells with transit-amplifying role) while WUS correctly specifies the shoot apical meristem. Activity of CLV and WUS genes is suggested to set the dynamic equilibrium between a rising pluripotent stem cell population and a daughter cell population displaced towards the periphery guaranteeing the integrity of the shoot apical meristem for a long lasting activity [7].

The stem cell population of the root apical meristem is composed of cells of different developmental origin, from the apical and basal cells of proembryo [7]. Initials arising from two clonal groups are separately existing and acting autonomously, though in an integrated manner. Coordinating interactions are poorly understood but cell-to-cell communication seems to play an elementary role. Laser ablation experiments underlined the root tip integrity and also pointed to the importance of auxin signalling inasmuch as auxin peak is occurring after ablation of initial cells [32]. It is supposed that as in the shoot apex, auxin has also a central role in the organization of root apical meristem. Investigation of root tip meristem mutants sheds light on stem cell origin and their functions usually connected with the differences in cell division pattern and daughter cell positioning [7].

Much less is known about the signalling events in the VC, meristemoids, cork cambium and *in vitro* forced meristematic cell groups from which organogenesis or whole plant regeneration may start. Morphogenetic information gained from the secondary meristems and artificially induced stem cell groups foreshadow high similarity to the apical meristem organization principle and signalling processes; however, knowledge on gene activities in these objects is strongly limited [2]. VC is likely a single cell layer bordered inside by xylem and outside by phloem. VC stem cells divide asymmetrically and one of the daughter cells retain a stem cell, while the second one becomes a mother cell with a transit-amplifying role and during the subsequent divisions and differentiation produce the cellular components of xylem or phloem depending on the direction and the cellular environment. Intercellular signalling is predicted to be required for the maintenance of the stem cell population in parallel with the production of conductive tissue elements [2,33].

Cell Plasticity and Transdifferentiation

Similarity in signalling events is also supported by the conformity of morphogenetic processes during somatic embryogenesis. Natural or *in vitro* somatic embryos repeat the sequence of zygotic embryonic developmental stages. Transdifferentiation (dedifferentiation and subsequent redifferentiation) demands reorganization of the gene expression pattern and necessitates the remodelling of chromatin and control of chromatin [34,35]. This obvious example of cell plasticity implies a link between cell fate, cellular memory and cell division. Cell cycle re-entry events underscore importance of Polycomb group (PcG) proteins, which are necessary to maintain the totipotent stem cell state, the “stemness” [4]. Some factors or their action-pattern induce proliferation linking environmental signals to cell cycle progression [18]. Epigenetic processes create a cell memory and the epigenetically decorated chromatin is the object of chromatin remodelling processes, so, different cells of the various tissues of an organism have different ability to transdifferentiate [4,18].

RESULTS OF OUR EXPERIMENTS: STEM CELL INDUCTION IN AN *IN VITRO* MODEL

General Notes

Some outcomes of our investigations were published in a prior paper [36] in which a detailed description of experimental conditions, used methods and plant object are included. In the present chapter, in addition to the explanation of necessary basic information on the experiments, we focus on the exhaustive portrayal of the complex morphological progress only outlined previously.

Conditions, Decisive Factors and Sampling

We used an in vitro system to induce callus formation to model the mitotic activity stimulation in a differentiated organ without persistent or dormant meristems. In aseptic culture, on artificial substrate (Murashige-Scoog (MS) medium with 30g/L sucrose, 1mg/L indol acetic acid (IAA), 0.2mg/L naphtil acetic acid (NAA), 0.2mg/L kinetin, solidified with 0.6% agar) expanded leaves of *Rubia tinctorum* were used as explants. Cultures were grown in the laboratory at dim natural light to avoid the high light stress. Callus induction in the surfaces-sterilized detached leaves was followed for 14 days with every day sampling. Later stage of callus development and the subsequent organogenesis were checked up to six weeks. High sugar content of the medium apart of its morphogenetic effect served also as a carbon source for the explants to keep off starvation effect. Samples were investigated with stereomicroscope (SM), light microscope (LM) equipped with differential interference contrast (DIC) unit, transmission electron microscope (TEM), spectrophotometer and high performance liquid chromatography (HPLC) to detect morphological and physiological changes during dedifferentiation and redifferentiation.

Stereomicroscopic Observations

Lamina of the surface sterilized leaves was green while midrib region and main veins were pale green-coloured. The intensive green colour of lamina faded but up to the end of observation leaf blade remained green (Figure 1A). Midribs and primary lateral veins of the explants turned to yellow (Figure 1B) and after 5-7 days callus cells ripped the leaf epidermis at the yellow veins (Figure 1C). Root organogenesis appeared at the places of callus induction after 2-4 weeks (Figure 1D). Large leaves were dissected into smaller cuts and the wounded edges of lamina callus formation was usual. However, callus growth on wounded explants was described in the former literature; we omitted it in our investigations.

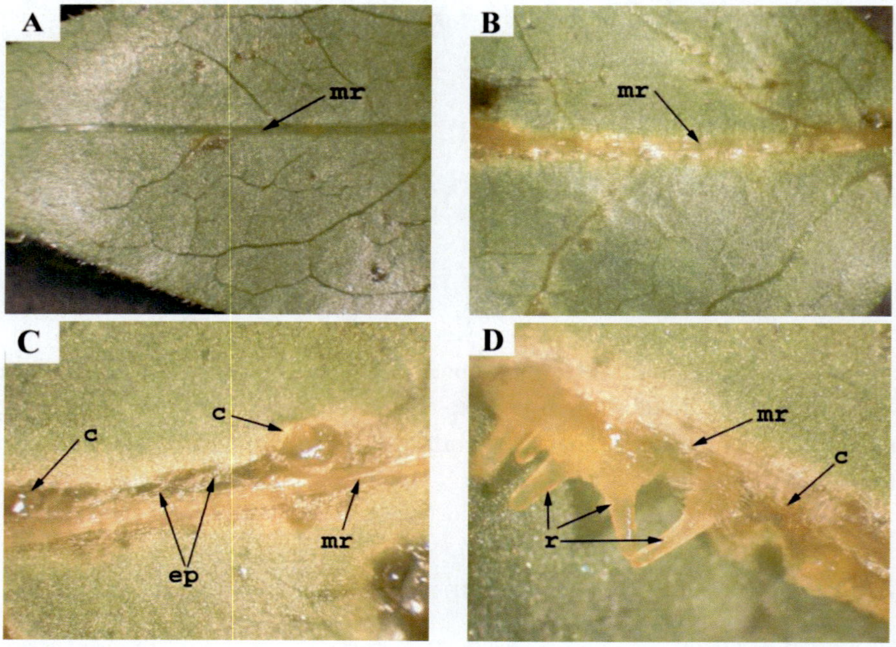

Figure 1. Micrographs of *Rubia tinctorum* control leaf and leaves after callus induction (aci). A: control leaf; B: leaf 3 days aci; C: leaf 7 days aci; D: leaf 4 weeks aci. Mangification: A, B: 20x; C: 40x; D: 30x. Legends: mr – midrib; c – callus; ep – remnants of epidermis; r – root.

Alterations in Pigment Content

Amount of the photosynthetic pigments was lowered with the time of callus generation, up to 20–25% of the chlorophyll content of control leaves. Decrease of chlorophyll a+b level was quick: at the beginning (0–3 days after callus induction [ACI]) reduction was sharp but from 7 days ACI the chlorophyll level was almost steady. A light greenish colour of explants was retained to the end of observation, more than 6 weeks.

Roots and rhizome of *R. tinctorum* contained high amount and various compounds of anthraquinones in vacuoles; however, leaves are practically free of them. In a few days, at the region of yellow-coloured veins accumulation of anthraquinones had started and the yellow callus cells contained heightened concentration on anthraquinones. Sum anthraquinone content increased in the leaves up to 14 days ACI and showed a two-step dynamism. At 7–8 days ACI, a local minimum of pigment content occurred demonstrating a short metabolic change. Dynamism of the various anthraquinone compounds differed, so, anthraquinone composition pattern considerably varied in the samples harvested at different times. Changes of anthraquinone content and composition seemed to run parallel with redifferentiation process.

LM Observations

For cell divisions, we screened the three main constituents of leaves: epidermis, mesophyll, and vascular system. The wounded lamina edge was excluded.

We could not detect cell divisions or any other structural alterations through the whole period of experiments in the epidermis.

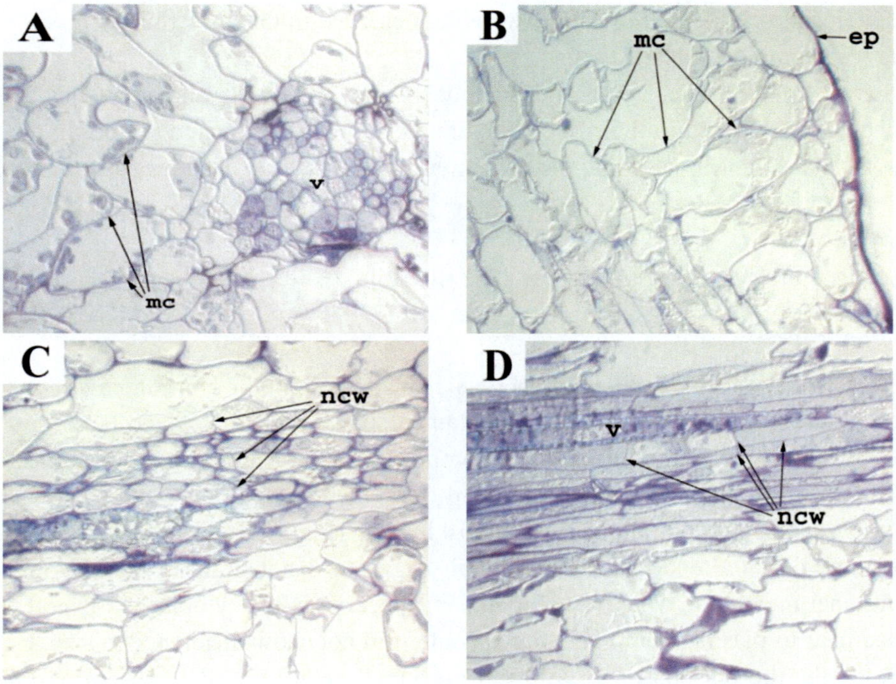

Figure 2. Micrographs of *Rubia tinctorum* leaves after callus induction (aci). A: mesophyll cells 3 days aci; B: mesophyll cells 7 days aci; C: cell divisions 3 days aci; D: cell divisions 7 days aci. Mangification: A–D: 300x. Legends: ep – epidermis; mc – mesophyll cell; ncw – new cell wall; v – vein.

Mesophyll cells showed mitotic inactivity. In spite of this fact, mesophyll cells were structurally changed. Vacuolar volume was slightly decreased and the amount of cytoplasm seemed to be higher with numerous organelles (Figure 2A-B). The originally flat lamina folded and curled to some extent; however, we were not able visualize cell divisions in the palisade or spongy tissues.

Deformation of flat lamina seemed to be the consequence of torsions at the veins. Thickenings of the vascular bundles and new cell formation were also observed. Presence of the first new cell walls indicating the mitotic activity was obvious after three days ACI (Figure 2C). Along the conducting cells in the vascular bundles, new and thin cell walls were visible occasionally. The compact cytoplasm of daughter cells contained flocks of small vesicles but larger vacuolar compartments were lacking. In the 5–7 day-old samples, cell divisions were frequent. Veins contained several cell layers near the vascular elements and progress in callus formation was evident in the longitudinal section of veins (Figure 2D).

Ultrastructural Features

In the fortnight ACI period, samples studied by LM were also observed with TEM. As it was described at the LM level, epidermis, mesophyll and vascular bundle region was monitored. The degree of ultrastructural changes was strikingly different in the three main domain of leaf. Epidermis cells were not changed structurally, there was not altered ratio between cytoplasm and vacuolar space and we were not able to find differences in organelle density or fine structure. Cell divisions were not observed in the epidermis as it was expected from the results of LM investigations.

Ultrastructural changes were conspicuous in the mesophyll and the veins, numerous differences were detected in their fine structure during the investigated period. In the transfer cells of vascular bundles, we could observe impressive structural changes and cell divisions.

Chronology of Events in the Mesophyll Cells (Figure 3A-F)

Mesophyll cells had large central vacuole characteristic in the control leaves. Cytoplasm-vacuole volume ratio was low. Vacuole contained sparse precipitates and the tonoplast was undamaged. Mesophyll cells had several chloroplasts in their cytoplasm arranged near the cell wall. Chloroplasts contained a well-developed inner membrane system and a low number of electron-dense plastoglobuli. Starch was almost lacking (Figure 3A).

After one day, the endoplasmic reticulum began to disintegrate in the mesophyll cells, prevalent inner membrane twists (myelin bodies) appeared in the cytoplasm. Several vesicules developed near to plasma membrane, the vacuoles did not show difference in size. Compared to control cells, the plastids of 1 day ACI mesophyll cells seemed to be normal; however, considerable quantity of starch appeared in them (1–3 medium sized grains). Plastoglobuli of these chloroplasts became larger and their number increased (Figure 3B–C).

In the 3-day-old samples, grana of chloroplasts in the mesophyll cells were slightly higher, starch content decreased. A number of small vacuoles appeared in the cytoplasm near to nucleus, which probably derived from vesicules of smooth endoplasmic reticulum. The cytoplasm became lightly tenuous, and the organelles begun to aggregate.

Five days ACI, plastids of mesophyll cells contained starch occasionally, while their grana were enlarged. Plastoglobuli were numerous and increased in diameter. Mitochondria were visible in high number in the cytoplasm and many vesicules and small vacuoles were discernible. Apart from the central vacuole and cytoplasmic vacuolar structures some myelin bodies were present indicating continued autophagic activity (Figure 3D).

On the 7th–9th day, the number of plastoglobuli significantly increased, and they formed clusters. Starch content in plastids of mesophyll cells disappeared. Grana were noticeable but deteriorating. The aspect of the cytoplasm was brighter, less organelles and ribosomes were observable. Flocks of small cytoplasmic vacuoles became slightly larger; there were still autophagic areas in the cytoplasm.

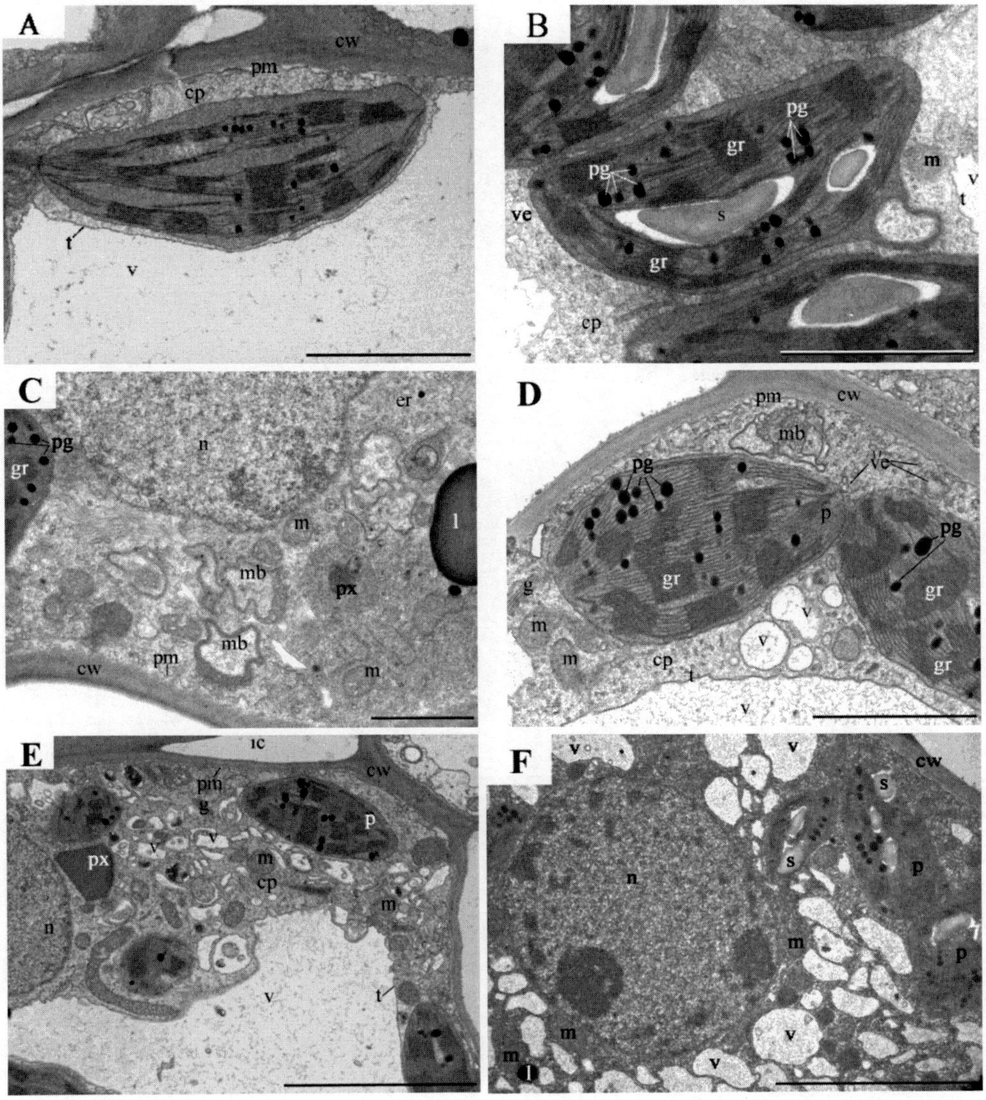

Figure 3. Electron micrographs of control mesophyll cells and after callus induction (aci). A: control mesophyll cells, bar=2μm. B and C: mesophyll cells 1 day aci, bar=2μm. D: 5 days aci, bar=2μm. E: 11 days aci, bar=5μm. F: 13 days aci, bar=5μm. Legends: cp – cytoplasm; cw – cell wall; er – endoplasmic reticulum; g – Golgi; gr – granum; ic – intercellular space; l – lipid droplet; m – mithochondrium; mb – myelin body; n – nucleus; p – chloroplast; pg – plastoglobuli; pm – plasma membrane; px – peroxisome; s – starch; t – tonoplast; v – vacuole; ve – vesicle.

In the vacuoles of mesophyll cells next to vascular bundles fine precipitate was detectable after 11- to 13-day ACI. The small cytoplasmic vacuoles arranged often around nucleus or fused together and with the extant central vacuole. Chloroplasts had well-developed thylakoids, high grana. Starch did not appear in plastids of the 11-day ACI samples but it was visible in small quantities in the 13-day ACI leaves. Number of plastoglobuli decreased, their cluster formation left off (Figure 3E-F).

Sequence of Events in Transfer Cells (Figure 4A-F)

In the vascular bundles typical transfer cells were visible. Transfer cells showed characteristic cell-wall protrusions deep into the cell lumen. These local thickenings showed more or less the same density and structure as the rest of the cell wall. The central part of these cells contained one or few vacuoles, plastids were small or medium sized and only few grana were observable inside them. Grana consisted of only a few (2–3) thylakoids (Figure 4A).

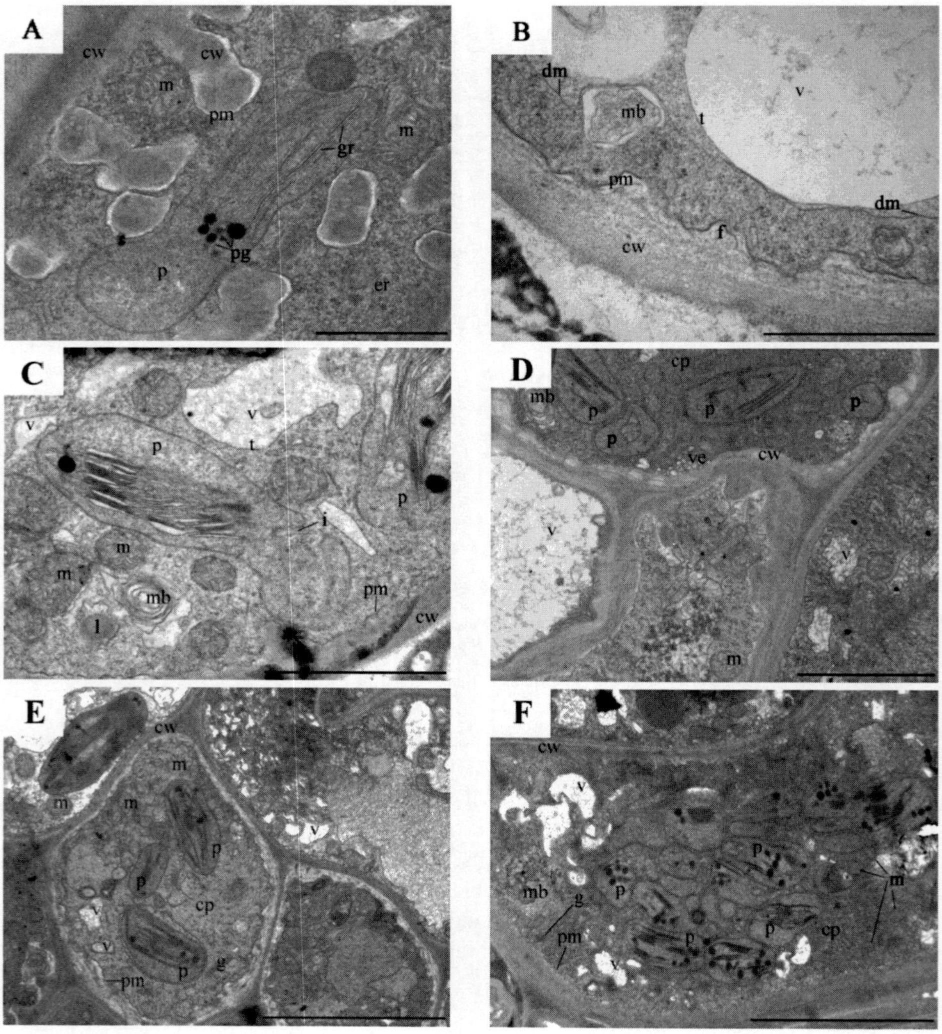

Figure 4. Electron micrographs of control tranfer cells and after callus induction (aci).
A: control transfer cells, bar=1μm. B: transfer cells 1 day aci, bar=1μm. C: transfer cells 3 days aci, bar=2μm. D: transfer cells 5 days aci, bar=2μm. E: transfer cells 11 days aci, bar=5μm. F: transfer cells 13 days aci, bar=5μm. Legends: cp – cytoplasm; cw – cell wall; dm – isolation (delimiting) membrane; er – endoplasmic reticulum; f – fibrillar region of cell wall; g – Golgi; gr – granum; i – invagination; l – lipid droplet; m – mithochondrium; mb – myelin body; ml – middle lamella; p – chloroplast; pg – plastoglobuli; pm – plasma membrane; t – tonoplast; v – vacuole; ve – vesicle.

Ultrastructure of the transfer cells had already changed after one day. Some of these cells showed symptoms of lytic activity in the cell wall. As a consequence of this process, the middle lamella became loosen. The inner cell wall layer bearing protrusions and the thickened regions seemed to be electron transparent with fibrillar central region. As a morphological symptom of autophagy, delimiting membrane isolated vacuole(s) with the central part of cytoplasm and myelic structures appeared. The isolation (delimiting) membrane in some points seemed to be a double membrane structure (Figure 4B).

In the 3-day-old sample, the delimited central part of the plasma of certain transfer cells looked to disappear autophagically while in others only smaller areas showed similar feature. Plastid divisions seemed to be unequal in several cases. The cell wall next to the plasmalemma turned into a light but more compact layer without protrusions (Figure 4C). However, some cells remained structurally almost unchanged.

In the 5-day ACI explants, the cytoplasm of transfer cells was dense, and size of plastids varied strongly, probably as a result of further unequal divisions of these organelles. The small plastids had weakly developed thylakoids (Figure 4D).

At 7- to 9-day ACI, cytoplasm of transfer cells was dense and compact, in some cells almost without vacuoles. The inner cell wall layer became denser but it could be distinguished. Mitochondria were numerous but hardly visible because of the low contrast between them and cytoplasm.

Cytoplasm of transfer cells was dense near the nucleus in the 11- to 13-day ACI leaves and contained several organelles. The high amount of smooth endoplasmic reticulum, number of Golgi and numerous mitochondria and plastids various in size were characteristic. Nuclei were enlarged, often with dense caryolympha. In particular regions of the dedifferentiated transfer cells groups of small vacuoles were visible. At these areas high number of mitochondria were detectable (Figure 4E-F).

Detailed Description of the Characteristic Cytological Alterations of Completely Dedifferentiated Transfer Cells

Cell Wall Reshuffle

One of the most characteristic cytological features of transfer cells is the presence of well-developed cell wall ingrowths [37]. According to Pate and Gunning [38], they are classified to the A-type transfer cells. In veins of *R. tinctorum*, transfer cells are present along with the conductive cells. In control leaf explants, transfer cells have all cell wall features previously described, the outer electron-translucent and central electron-dense parts of ingrowths are recognizable (Figure 4A). This distinctive cell wall form might be dynamically reorganized, after 1 day ACI we could see serious changes in transfer cell wall structure. In several transfer cells (but not in all of them) ingrowths are reduced. Remnants of the protrusions are present but they show advanced degradation in size and complexity. Probably, at least part of them, the myelin bodies may arise from the plasma membrane (PM) reorganization and deterioration of cell wall protrusions (Figure 4B). Parallel with PM surface reduction and degradation of ingrowths, middle layer of dedifferentiating transfer cells is also modified, its "compact" appearance become more mellowed (Figure 5). Middle lamella metamorphosis may result in reduction of cell-cell contacts important to keep on cell identity. Quick transformation of the cell wall (reduction of ingrowths is fundamental to decrease the cell surface and reorganization of middle lamella and probably plasmodesmata in it) allows to change cell-to-cell communication and to liquidate transfer cell niche, opening through it the way for dedifferentiation process. Similar ingrowths, as there are in control samples, are not detectable up to the end of our ultrastructural observations. In some cells, the inner cell wall layer forms a light coat but it never holds protrusions (Figure 4D-E). Middle lamella structure seems to be normal in samples from 5 days ACI.

Autophagy and Vacuolar System

Cytoplasm organization of transfer cells in the control samples is characteristic. In the peripheral part of cytoplasm organelles are densely packed. Among the cell wall ingrowths RER, SER cisterns and mitochondria are frequent, undifferentiated plastids can also be seen there. A large central vacuole or a few smaller but centrally arranged vacuoles existed (Figure 4A). After 1 day ACI cytoplasm reorganization is impressively advanced. SER become fragmented, small segments of its cisterns are observable in the cytoplasm, while mitochondria, plastids and other organelles are relocated. Two main autophagic processes can be distinguished. Rather peripherally, myelin bodies arise. PM surface reduction requires engulfment of redundant membrane areas and cross-section of degrading ingrowths can have similar appearance. Another autophagic process is detectable in the central part of dedifferentiating transfer cells. An isolation membrane delimits the vacuole(s) positioned in the central part of cell with some cytoplasm portion around it/them (Figure 4B). This isolation membrane looks to consist two thin membranes bordering an almost organelle-free but vacuolated central cytoplasm region (Figure 6). In parallel with the autophagic processes, the cytoplasm production is ahead; so, after 3–5 days ACI, dedifferentiating transfer cells have small vacuoles and a rather compact cytoplasm (Figure 4C-E). After 5 days, autophagic events become rare and small vacuoles/vesicles show a higher frequency. Flocks of vesicles/vacuoles are near the nucleus in a peculiar region of plasma (Figure 4F).

Plastids of Transfer Cells

Plastids of transfer cells are undifferentiated in the control samples. They are small in size and their inner membrane system is undeveloped. Only a few thylakiods are present, grana are low (2–3 thylakoids). Shape of plastids is elongated but often irregular (Figure 4A). During 1–7 days ACI, plastid size and thylakoid organization is changed in the transfer cells. Various plastid size is observable, dividing plastids are recurrent. The two plastid-progenies are often different in size (Figure 4C-D). Plastids dividing in this way are numerous. The high number of plastids and their variability in size and inner structure suggest plastidial unequal fission by which in organization and size proplastid-like plastids might be produced and delivered to the meristematic cells. In samples 10–14 days ACI, plastids are more differentiated, their inner membrane system is developed, though fare from chloroplast complexity (Figure 4F)

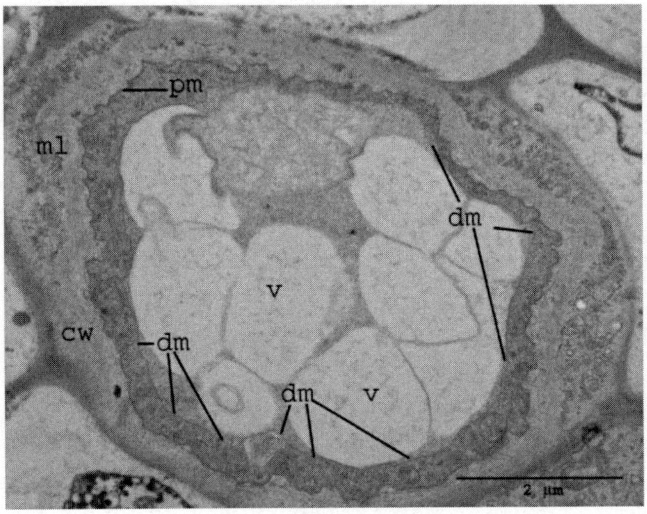

Figure 5. Micrograph of transfer cell 1 day aci. Central cytoplasm region is delimited by the isolation membrane. bar=2μm. Legends: cw – cell wall; dm – isolation (delimiting) membrane; ml – middle lamella; pm – plasma membrane; v – vacuole.

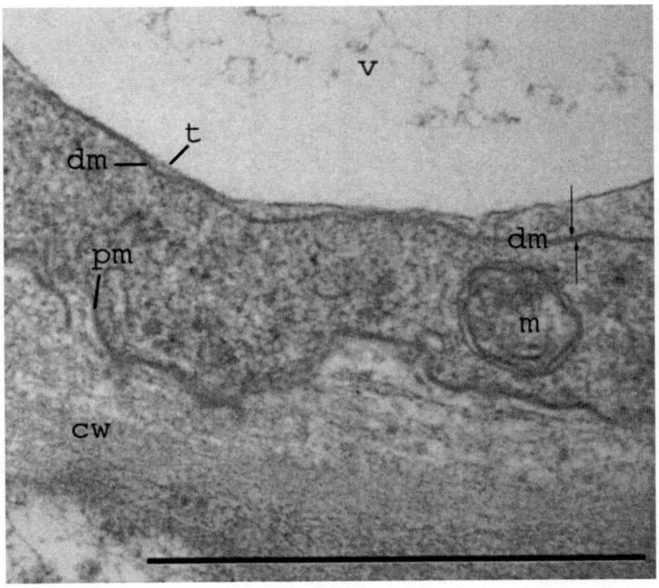

Figure 6. Micrograph of a transfer cell 1 day aci with delimited central cytoplasm region at a higher magnification. bar=1μm. Legends: cw – cell wall; dm – isolation (delimiting) membrane; m – mithochondrium; pm – plasma membrane; t – tonoplast; v – vacuole.

Unequal Cell Division

LM investigations of vein areas shows that after 3 days ACI cell divisions occur in the explants (Figure 2C-D). The previously elongated cells divided unequally with transverse cell walls forming a longer and a sorter cell. Apart from the fact that the daughter cells are different in size, as it was demonstrated by TEM, they differ also in their ultrastructure. Throughout unequal cell divisions small, non-vacuolated cells can arise with compact cytoplasm and undifferentiated plastids, and might form stem cells from which meristems may develop. The longer cells have larger vacuoles and fewer organelles.

Root Formation by Organogenesis

A representative evidence of meristem formation and functionality is the induced organogenesis. In case of *R. tinctorum,* intensive root formation is detectable after a short callus production period. Roots originate from small clamps of yellow callus cells. After unequal cell divisions, intensive cell division activity is observed and differentiation of the progenitors takes place. Callus cells elongate and a well-defined central vacuole develops. Cells accumulate anthraquinones. However, some of the cells form meristematic cell groups and according to the intrinsic hormonal equilibrium and exogenous hormone administration the offspring cells may form SAM or RAM. In *R. tinctorum*, root tip formation is dominating and roots appear on the calli (Figure 1D).

CONCLUSION

Meristems in Nature and in *In Vitro* Systems

For the life strategy of plants, a prerequisite element of organization and progression is the presence of meristematic cell groups and their extended though sometimes periodical or seemingly hectic activity. Meristematic cells in SAM and RAM, meristemoids, VC, ontogenetically occurring or wound forced cork cambium and calli in regeneration places or in vitro cultures act as stem cells. They show totipotency (e.g., SAM), which is limited in the developing but dividing yet daughter cells to unipotency, the capability to form specialized cell type of a tissue at the end of a differentiation path determined by intrinsic factors and external items [1]. Genetic background of the cell together with the cellular habitat and the environmental compounds determine on the fate of a daughter cell [2,4].

Mutations influencing meristem organization and activity offer an excellent tool to elucidate the role of genetic context. Investigation of impaired or inhibited meristematic cell division, stem cell identity preservation or daughter cell differentiation reveal mechanisms, fundamental principles of plant body organization [7], the doctrine of persistent potential for renewal of plant body as a counterpart of the specialization and differentiation urged by land-life-being of plants [1]. Their evolutionary strategy allows plants to show high plasticity in *in vitro* systems, among conditions created by human. Plants can be regenerated even from a single somatic cell or a microspore and a long list of plant objects from plant species exist showing organogenetic capability, and most of them are widely used by biotechnologists nowadays [15-17,39-44].

Results of mutants and comparative studies suggest that probably all meristems are organized analogously. Permanently and temporally acting apical meristems continuing the embryonic developmental progress and the VC stem cells with their regained but long-lasting cell division capability cope with the maintenance of stem cell niche formed by stem cells and their progenies together [2,6,33]. However, it is not obvious if residual meristems (e.g., dormant buds) are organized in an identical way. In dormancy, cell divisions are arrested and cells show low metabolic activity in these meristematic cell groups but a proper hormonal signal (e.g., loss of the functioning apical meristem) may activate some of them in short [12]. The natural loss or the artificial removal of the active apical meristem provokes intermission of latent meristem existence and hibernated cells become active and latent meristems sprout off. Among normal conditions, apical dominancy is not complete and certain axillary buds can be activated (e.g., lateral shoot growth) according to the pattern characteristic for the given species and the effects of environmental factors modifying this intrinsic program. Intriguingly, crown gall *ipt* gene expression in transgenic *Nicotiana tabacum* plants abolishes apical dominance pointing to the importance of cytokinin content, increased up to 137 times in the transgenic shoots [45].

Phellogen of secondary dermal tissue (periderm) can act for long time in plants and produce at least two daughter cell types, progenitors of the phellem and the phellogen itself. Wound initiated cork cambium forced and formed at natural conditions has a limited lifetime. In a first step, injuries stimulate (often cortical and/or vascular) cells to return to the cell cycle and produce a pack of relatively undifferentiated natural callus cells. Cork cambium after wounds, as a secondary meristem, covers the damaged area with callus and, in the second step, subsequently produces the highly specialized secondary dermal protecting tissue layer, the phellem, with strongly suberinized cell walls [3].

Stem Cell Pools, Transfer Cells of *Rubia*

While callus induction at the wounded surfaces is better studied, we focused on the callus production events observed at the veins. In our work, we artificially initiated a process similar to the first phase previously described. *Rubia* leaf explants showed vigorous callus formation at the injured edges of lamina of cuttings and also at the midrib and main veins of all explants. The yellow callus was contrasted in its appearance and differentiation stage with the rest of green lamina. At the cut edges, hormonal changes were evolved by injury, acting mainly on mesophyll cells and the growth regulator content of the medium, while midrib callus formation was based on the fate of transfer cells influenced by the intrinsic hormonal equilibrium of the detached leaf and the exogenous regulators together. In spite of their different cellular ancestors and hormonal activity, calli from both explant regions evolved similarly and RAM formation and root development proceeded identically.

New plantlets might arise on the "mother" plants' reproductive parts (e.g., *Poa bulbosa*, *Poa alpina*, *Dentaria bulbosa*, *Rhisophora* species) but in some taxons, accelerated regeneration capability on leaves is common (e.g., *Crassulaceae*, *Sansaveria*) and its preference to venation leading to viviparity is also known (*Asplenium viviparum*, *Ceratopteris thalictroides*, *Nymphaea micrantha*, etc.). The exceptional regenerative ability of leaf segments (*Gladiolus*, *Sansaveria*, *Lachenalia*, etc.) and especially their midrib or other veins is a well-known fact in horticulture. *Begonia*, *Peperomia* and *Santalum* regenerants may be produced from leaves on wet sand and new plantlet initiation of *Saintpaulia* also starts at the petiole vascular bundles [14,15,41-44]. The preexisting regenerative capacity is enhanced by perpendicular cuts on the vein, often applied also in other *in vitro* micropropagation systems. Veins of intact leaves might possess exceptional regenerative features. In addition, vascular bundles show higher responsibility to stresses and enhanced H_2O_2 production capability [46,47]. Some authors define vascular bundles as "cryptic meristems" because of their atypical morphogenetic capacity [6]. Histological investigations of these works did not enlighten fully the reason for this valuable feature, but proved that cell divisions always precede plant regeneration and auxin-cytokinin balance is fundamental to trigger transdifferentiation [41-45]. In the case of *R. tinctorum*, we revealed that transfer cells of vascular bundles dedifferentiate and return to the cell cycle. They behave as residual stem cells; however, their primary role in the leaf was the short distance transport of solvents [37]. Although transfer cells have a specialized structure from which the cell wall ingrowths and enlarged PM surface are obvious, they show an unexpected dynamism. Placental, epidermal and vascular transfer cells equally are able to elaborate or demolish their cell wall protrusions quickly, almost in some hours, according to the needs and conditions [48-55]. Degradation can be influenced by heavy metal pollution [56], mutation [57], or alteration of available nutrient level [58], while increased cell wall ingrowth formation is observed at Fe deficiency [59] or salinity [60-61] through the enhanced assembly of primary cell wall-like material of ingrowths by the wall synthesizing machinery [62].

The powerful plasticity of transfer cells in the *Rubia* model probably is based on their innate morphogenetic dynamism. This attribute allows an easy transition of transfer cells of leaf explants to stem cell status and establishment of new meristems from the primarily produced progenitor cells. As the roots produced through organogenesis show normal morphology and growth, we can suppose that root tip meristem organization is identical to embryogenous RAMs of plants. In that case, signalling events governing root meristem formation and maintenance also should be alike, and root tip complexity ought to be achieved from a single cell or a cell group with stem cell-like properties.

In the *Rubia* model system, we can see that transfer cells dedifferentiated into stem cells and their daughter cells could differentiate into callus cells or even to develop into normal roots. These morphogenetic events point to the re-acquired totipotency of transfer cells and their transdifferentiation (redifferentiation of daughter cells) of new cells to obtain organogenic capacity. During the whole process, some embryonic cells develop into conductive tissue (and transfer cells among them) and the complex outcome of the intrinsic and exogenous factors allow the transfer cells to form new meristems and the organogenesis from them. The new conditions alter cell fate and a novel morphogenic program is approved; the previous transfer cells form callus and root tissues. At the same time, meristematic identity is retained while RAM organization and its activity are steadily maintained.

Several authors described the manifestation of regained cell division activity or—as termed nowadays—stem cell potential, the "stemness". We can accept that in all the cases in which enforcement of cells in a complex organ or tissue contribute to recovery of the totipotency, the cells, which were able to achieve it under the actual conditions, were the ones best equipped for mobilization, genetic remodelling and new specialization. Their skill for re-entering the cell cycle depends on "cell memory", the epigenetically modified plan of action, and hormonal events.

The first one is well documented in our *Rubia* system because the epidermal, mesophyll and transfer cells behaved differently under the seemingly same conditions reaching distinct stages of dedifferentiation. Different cells of *Rubia* leaf explants could attain a range of dedifferentiation levels and according to it their fate was absolutely dissimilar. In epidermis, we could not detect any changes; the mesophyll cells returned to a certain incomplete level of dedifferentiation, although it was not enough to return to the cell cycle. From the conductive tissue cells, transfer cells showed the ability to regain and manifest totipotency and re-enter the cell cycle in order to start a new morphogenic process. Obvious evidence for re-specialization was the root formation (structural) and anthraquinone production (metabolic) detected in leaf explants of *Rubia*.

As a hormonal factor, the presence of auxin and cytokinin seemed to be pivotal. In fact, hormone concentrations might differ at least in the vascular bundles of explants and different organogenetic capacity was described in the various explants and their parts depending on their age, structure and even position [14-17,39-44]. Vivipary was acquired in *Nicotiana* leaves that were genetic mosaics for the expression of the *ipt* gene involved in cytokinin synthesis (*ipt* codes an isopenthenyltransferase enzyme). Adventious bud formation could be triggered by the enhanced endogenous cytokinin content [45]. Interestingly, the transformed tobacco plant leaves showed leaf bulges similar to ones observed on *Rubia* leaf explants after 3–5 days ACI, so we could expect cytokinin effects at the beginning of the transfer cell transdifferentation progress. In *Chenopodium rubrum* cells, the addition of cytokinin plus auxin resulted in a temporal increase of cyclin transcript levels, demonstrating a strong correlation between cell cycle machinery and hormonal control [63].

Dedifferentiation events are closely related with augmented autophagic activity in the cells. The housekeeping function of basal autophagy allows a cell to retain its structural integrity [64], while stress effects or change of cell fate enhance autophagic activity [65,66]. The described conserved autophagy machinery [67] and the possibility of isolation membrane formation [68] confirm our results that bulk degradation processes are involved in the dedifferentiation mechanism of transfer cells.

Autophagy is one of the effectors, a process through which rejuvenation and adaptation of plant tissues to adverse conditions is achievable [68]. However, a connection between autophagy and vacuolar system remodelling remains elusive; data show that vacuole genesis, autophagosome formation and endomembrane transport are linked [69-71]. During transfer cell dedifferentiation of *Rubia* leaf explants, early cytological events involve at least two types of autophagic processes, i.e., myelin body formation and isolation membrane delimited cytoplasm elimination, resulting in dedifferentiated cells with compact cytoplasm. Changes in plastid structure also characterise the progression of rejuvenation. Unequal plastid divisions caused by lowered MinD homologue amount [72] and/or alterations in signalling [73-75] result in a plastid population strongly varying in size and differentiation level. Segregation of proplastid-like organelles into the daughter cells with reduced volume and compact cytoplasm may be a relevant event during dedifferentiation, as a prominent indication of plastid binary fission machinery and cell division regulation relationship [76-77].

In sum, the final result of morphogenic events depends on the genetic plasticity of species, the age and origin of explant (its morphogenic capacity) and the new, orchestrated control factors of the environment. The very first outset of the morphogenic way through organogenesis from transfer cells (*Rubia*), regeneration (*Begonia*) or somatic embryogenesis (*Gentiana*, *Gladiolus*) is possibly alike. However, the different daughter cells, the morphogenic start of explant, specific endogenous controls and the various dedifferentiation levels will have an influence; so, the divergent cell fates manifested in the explants is a result of the factors mentioned above, which determine their common outcome.

References

[1] Graham LE, Cook ME, Busse JS. The origin of plants: body plan changes contributing to a major evolutionary radiation. *PNAS* 2000; 97(9): 4535-40.

[2] Bao Y, Dharmawardhana P, Arias R, Allen MB, Ma C, Strauss SH. WUS and STM-based reporter genes for studying meristem development in poplar. *Plant Cell Rep* 2009; 28: 947-62.

[3] Esau K. *Plant Anatomy*. John Wiley & Sons, Inc., Chapman & Hall, Ltd., New York. 1953.

[4] Costa S, Shaw P: 'Open minded' cells: how cells can change fate. *Trends Cell Biol* 2007; 17(3): 101-6.

[5] Geisler M, Nadeau J, Sack FD. Oriented asymmetric divisions that generate the stomatal spacing pattern in Arabidopsis are disrupted by the *too many mouths* mutation. *Plant Cell* 2000; 12: 2075-86.

[6] Singh MB, Bhalla PL. Plant stem cells carve their own niche. *Trends Plant Sci* 2006; 11(5): 241-6.

[7] Raghavan V. The coming of age of plant embryology. *Curr Sci* 2001; 80(2): 244-51.

[8] Dodeman VL, Ducreux G, Kreis M. Zygotic embryogenesis versus somatic embryogenesis. *J Exp Bot* 1997; 48(313): 1493-509.

[9] Deeks SJ, Shamoun SF, Punja ZK. Tissue culture of parasitic flowering plants: methods and applications in agriculture and forestry. *In Vitro Cell Dev Biol - Plant* 1999; 35: 369-81.

[10] Medford J, Behringer FJ, Callos JD, Feldmann KA. Normal and abnormal development in the *Arabidopsis* vegetative shoot apex. *Plant Cell* 2006; 4(6): 631-43.

[11] Kwiatkowska D. Flower primordium formation at the Arabidopsis shoot apex: quantitative analysis of surface geometry and growth. *J Exp Bot* 2006; 57(3): 571-80.

[12] Beveridge CA. Axillary bud outgrowth: sending a message. *Curr Opin Plant Biol* 2006; 9(1): 35-40.

[13] Halperin W. Organogenesis at the shoot apex. *Annu Rev Plant Physiol* 1978; 29: 239-62.

[14] Attfield EM, Evans PK. Developmental pattern of root and shoot organogenesis in cultured leaf explants of *Nicotiana tabacum* cv. Xanthi nc. *J Exp Bot* 1991; 42(1): 51-7.
[15] Sudarshana MS, Niranjan MH, Girisha ST. *In vitro* flowering, somatic embryogenesis and regeneration in *Boerhaavia diffusa Linn.* - A medicinal plant. *Glob J Biotechn Biochem* 2008; 3(2): 83-6.
[16] Iantcheva A, Vlahova M, Atanassov A: Somatic embryogenesis from leaf explants 1 – Somatic embryogenesis from leaf explants of *Medicago truncatula* cv. Jemalong genotype 2HA. [cited 2009 Jun 27] [10 screens] Available from: URL: http://www.noble.org/MedicagoHandbook/pdf/SomaticEmbryogenesisfromLeafExplants.pdf
[17] Iantcheva A, Vlahova M, Atanassov A: Somatic embryogenesis from leaf explants 2 – Somatic embryogenesis from leaf and petiole explants of *Medicago truncatula* cv. Jemalong genotype 2HA. [cited 2009 Jun 27] [11 screens] Available from: URL: http://www.noble.org/MedicagoHandbook/pdf/SomaticEmbryogenesisfromLeafExplants.pdf
[18] Basak O, Taylor V. Stem cells of the adult mammalian brain and their niche. *Cell Mol Life Sci* 2009; 66: 1057-72.
[19] Harrisingh MC, Lloyd AC. Ras/Raf/ERK signalling and NF1. *Cell Cycle* 2004; 3(10): e69-e72.
[20] Trewavas A. Plant cell signal transduction: The emerging phenotype. *Plant Cell* 2002 14: s3-s4.
[21] Bóka K, Orbán N. New aspect of H_2O_2 signaling. *Plant Signal Behav* 2007; 2(6): 498-500.
[22] Łotocka B. Vascular system within developing root nodules of *Lupinus luteus* L. Part 1. Juvenile stage. *Acta Biol Crac Ser Bot* 2008; 50(1): 79-88.
[23] Appezzato-da-Glória B, Machado SR. Ultrastructural analysis of *in vitro* direct and indirect organogenesis. *Revista Brasil Bot* 2004; 27(3): 429-37.
[24] Mikuła A, Tykarska T, Zielińska M, Kuraś M, Rybczyński JJ. Ultrastructural changes in zygotic embryos of *Gentiana punctata* L. during callus formation and somatic embryogenesis. *Acta Biol Crac Ser Bot* 2004; 46: 109-20.
[25] Charzynska M, Murgia M, Cresti M. Microspore of *Secale cereale* as a transfer cell type. *Protoplasma* 1990; 158: 26-32.
[26] Dinneny JR, Benfey PN. Plant stem cell niches: standing the test of time. *Cell* 2008; 132(4): 553-7.
[27] Jackson D, Veit B, Hake S. Expression of maize KNOTTED1 related homeobox genes in the shoot apical meristem predicts patterns of morphogenesis in the vegetative shoot. *Development* 1994; 120: 405-13.
[28] Sentoku N, Sato Y, Kurata N, Ito Y, Kitano H, Matsuoka M. Regional expression of the rice KN1-type homeobox gene family during embryo, shoot and flower development. *Plant Cell* 1999; 11(9):1651-63.
[29] Barton MK, Poething RS. Formation of the shoot apical meristem in *Arabidopsis thaliana*: an analysis of development in the wild and shoot meristemless mutant. *Development* 1993; 119(3): 823-31.
[30] Souer E, van Houwelingen A, Kloos D, Mol J, Koes R. The NO APICAL MERISTEM gene of Petunia is required for pattern formation in embryos and flowers and is expressed at meristem and primordial boundaries. *Cell* 1996; 85(2): 159-70.

[31] Satoh N, Hong S-K, Nishimura A, Matsuoka M, Kitano H, Nagato Y. Initiation of shoot apical meristem in rice: characterization of four SHOOTLESS genes. *Development* 1999; 126(16): 3629-36.

[32] Souter M, Lindsey K. Polarity and signalling in plant embryogenesis. *J Exp Bot* 2000; 51(347): 971-83.

[33] Sablowski R. The dynamic plant stem cell niches. *Curr Opin Plant Biol* 2007; 10(6): 639-44.

[34] Gross E, Casagrande LIT, Caetano FH. Ultrastructural study of ectomycorrizas on *Pinus caribaea* Morelet. var. *hondurensis* Barr. and Golf. seedlings. *Acta Bot Bras* 2004; 18(1): 1-7.

[35] Shukla A, Chaurasia P, Bhaumik SR. Histone methylation and ubiquitination with their cross-talk and roles in gene expression and stability. *Cell Mol Life Sci* 2009; 66: 1419-33.

[36] Orbán N, Boldizsár I, Bóka K. Structural and chemical study of callus formation from leaves of *Rubia tinctorum*. *Biol Plant* 2007; 51(3): 421-9.

[37] Gunning BES, Pate JS. "Transfer cells" – Plant cells with wall ingrowths, specialized in relation to short distance transport of solutes – Their occurrence, structure, and development. *Protoplasma* 1969; 68: 107-33.

[38] Pate JS, Gunning BES. Vascular transfer cells in Angiosperm leaves – A taxonomic and morphological survey. *Protoplasma* 1969; 68: 135-56.

[39] Maraschin SF, de Priester W, Spaink HP, Wang M. Androgenic switch: an example of plant emryogenesis from the male gametophyte perspective. *J Exp Bot* 2005; 56(417): 1711-26.

[40] Zoriniants S, Tashpulatov AS, Heberle-Bors E, Touraev A. The role of stress in the induction of haploid microspore embryogenesis. In: Palmer CE, Keller WA, Kasha KJ. eds. *Haploids in crop improvement II*. Berlin: Springer-Verlag; 2005. p: 35-53 (Biotechnology in Agriculture and Forestry; vol 56).

[41] Harris GP, Hart EMH. Regeneration from leaf squares of *Peperomia sandersii* A. DC: a relationship between rooting and budding. *Ann Bot* 1964; 28: 509-26.

[42] Bowes BG, Curtis EW. Conservation of the British National *Begonia* Collection by micropropagation. *New Phytol* 1991; 119 (1): 169-81.

[43] Mujib A. In vitro regeneration of sandal (*Santalum album* L.) from leaves. *Turk J Bot* 2005; 29: 63-7

[44] Shimada Y, Mori G, Oda M, Ishida G. Effects of BA and leaf piece orientation on adventitious bud formation in leaf cutting of *Begonia* tuberhybrida group. *J Japan Soc Hort Sci* 2007; 76(2): 157-62.

[45] Estruch JJ, Prinsen E, Van Onckelen H, Schell J, Spena A. Viviparous leaves produced by somatic activation of an inactive cytokinin-synthesizing gene. *Science* 1991; 254: 1364-7.

[46] Maeda H, Song W, Sage TL, DellaPenna D. Tocopherols play a crucial role in low-temperature adaptation and phloem loading in *Arabidopsis*. *Plant Cell* 2006; 18: 2710-32

[47] Mullineaux PM, Karpinski S, Baker NR. Spatial dependence for hydrogen peroxide-directed signalling in light-stressed plants. *Plant Physiol* 2006; 141: 346-50.

[48] Gunning BES, Pate JS. Cells with wall ingrowths (transfer cells) in the placenta of ferns. *Planta* 1969; 87: 271-4.

[49] Lal M, Narang A. Ultrastructural and histochemical studies of transfer cells in the callus and apogamous sporophytes of *Physcomitrium coorgense* Broth. *New Phytol* 1985; 100: 225-31.

[50] Lal M, Chauhan E. Transfer cells in the sporophyte – gametophyte junction of *Physcomitrium cyathicarpum* Mitt. *Protoplasma 1*981; 107: 79-83.

[51] Lee Y-I, Yeung EC, Lee N, Chung M-C. Embryo development in the lady's slipper orchid, *Paphiopedilum delenatii*, with emphasis on the ultrastructure of the suspensor. *Ann Bot* 2006; 98: 1311-9.

[52] Schikora A, Schmidt W. Formation of transfer cells and H^+-ATPase expression in tomato roots under P and Fe deficiency. *Planta* 2002; 215: 304-11.

[53] Tiwari SC. An intermediate-voltage electron microscopic study of freeze-substituted generative cell in pear (*Pyrus communis* L.): features with relevance to cell-cell communication between the two cells of germinating pollen. *Sex Plant Reprod* 1994; 7: 177-86.

[54] Vassilyev AE. Dynamics of ultrastructural characters of *Drosophyllum lusitanicum* Link (Droseraceae) digestive glands during maturation and after stimulation. *Taiwania* 2005; 50(3): 167-82.

[55] McCurdy DW, Patrick JW, Offler CE. Wall ingrowth formation in transfer cells: novel examples of localized wall deposition in plant cells. *Curr Opin Plant Biol* 2008; 11: 653-61.

[56] Vecchia FD, La Rocca N, Moro I, De Faveri S, Andreoli C, Rascio N. Morphogenetic, ultrastructural and physiological damages suffered by submerged leaves of *Elodea canadensis* exposed to cadmium. *Plant Sci* 2005; 168: 329-38.

[57] Borisjuk L, Wang TL, Rolletschek H, Wobus U, Weber H. A pea seed mutant affected in the differentiation of the embryonic epidermis is impaired in embryo growth and seed maturation. *Development* 2002; 129: 1595-607.

[58] Greenwood JS, Helm M, Gietl C. Ricinosomes and endosperm transfer cell structure in programmed cell death of the nucellus during *Ricinus* seed development. *PNAS* 2005; 102(6): 2238-43.

[59] Schmidt W, Bartels M. Formation of root epidermal transfer cells in *Plantago*. *Plant Physiol* 1996; 110: 217-25.

[60] Kramer D: The possible role of transfer cells in the adaptation of plants to salinity. *Physiol Plant* 1983; 58: 549-55.

[61] Winter E. Salt tolerance of *Trifolium alexandrinum* L. III. Effects of salt on ultrastructure of phloem and xylem transfer cells in petioles and leaves. *Aust J Plant Physiol* 1982; 9: 239-50.

[62] Vaughn KC, Talbot MJ, Offler CE, McCurdy DW. Wall ingrowths in epidermal transfer cells of *Vicia faba* cotyledons are modified primary walls marked by localized accumulations of arabinogalactan proteins. *Plant Cell Physiol* 2007; 48(1): 159-68.

[63] Fountain MD, Valdes O, Fettig S, Beck E. Expression of cell cycle control factors in non-dividing and ageing photoautotrophic plant cells. *Physiol Plant* 2003; 119(1): 30-9.

[64] Bassham DC. Plant autophagy – more than a starvation response. *Curr Opin Plant Biol* 2007; 10: 587-93.
[65] Bassham DC, Laporte M, Marty F, Moriyasu Y, Ohsumi Y, Olsen LJ, Yoshimoto K. Autophagy in development and stress responses of plants. *Autophagy* 2006; 2(1): 2-11.
[66] Contento AL, Xiong Y, Bassham DC. Visualization of autophagy in *Arabidopsis* using the fluorescent dye monodansylcadaverine and a GFP-AtATG8e fusion protein. *Plant J* 2005; 42: 598-608.
[67] Su W, Ma H, Liu C, Wu J, Yang J. Identification and characterization of two rice autophagy associated genes, OsAtg8 and OsAtg4. *Mol Biol Rep* 2006; 33: 273-8.
[68] Mizushima N. The pleiotropic role of autophagy: from protein metabolism to bactericide. *Cell Death Differ* 2005; 12: 1535-41.
[69] Ishihara N, Hamasaki M, Yokota S, Suzuki K, Kamada Y, Kihara A, Yoshimori T, Noda T, Ohsumi Y. Autophagosome requires specific early Sec proteins for its formation and NSF/SNARE for vacuolar fusion. *Mol Biol Cell* 2001; 12: 3690-702.
[70] Marty F. Cytochemical studies on GERL, provacuoles, and vacuoles in root meristematic cells of *Euphorbia. PNAS* 1978; 75(2): 852-6.
[71] Yano K, Hattori M, Moriyasu Y. A novel type of autophagy occurs together with vacuole genesis in miniprotoplasts prepared from tobacco culture cells. *Autophagy* 2007; 3(3): 215-21.
[72] Colletti KS, Tattersall EA, Pyke KA, Froelich JE, Stokes KD, Osteryoung KW. A homologue of the bacterial cell division site-determining factor MinD mediates placement of the chloroplast division apparatus. *Curr Biol* 2000; 10(9):507-16.
[73] Fujiwara MT, Nakamura A, Itoh R, Shimada Y, Yoshida S, Moller SG. Chloroplast division site placement requires dimerization of the ARC11/AtMinD1 protein in *Arabidopsis. J Cell Sci* 2004; 117: 2399-410.
[74] Aldridge C, Moller SG. The plastid division protein AtMinD1 is a Ca^{2+}-ATPase stimulated by AtMinE1. *J Biol Chem* 2005; 280(36): 31673-8.
[75] Maple J, Moller SG. An emerging picture of plastid division in higher plants. *Planta* 2005; 223: 1-4.
[76] Fujiwara MT, Hashimoto H, Kazama Y, Abe T, Yoshida S, Sato N, Itoh RD. The assembly of the FtsZ ring at the mid-chloroplast division site depends on a balance between the activities of AtMinE1 and ARC11/AtMinD1. *Plant Cell Physiol* 2008; 49(3): 345-61.
[77] Maple J, Moller SG. Plastid division: evolution, mechanism and complexity. *Ann Bot* 2007; 99: 565-79.

In: Daughter Cells
Editors: A. Hitomi et al., pp. 283-292
ISBN: 978-1-60876-790-8

Chapter VIII

DAUGHTER CELLS SELF-RENEW THROUGH ASYMMETRIC DIVISION

V. V. Terskikh**, *V. A. Vasiliev and E. A. Vorotelyak
Koltzov Institute of Developmental Biology
Russian Academy of Sciences, Russia

ABSTRACT

To escape senescence and dying, cells endure continual proliferation, as only dividing cells can self-renew. In the course of evolution, asymmetric division was used by multicellular organisms to solve the problem of replicative aging of stem cells. There are data showing that aged tissue-specific stem cells retain their intrinsic proliferative potential and that age-related changes in the systemic environment and niche in which stem cells reside can modulate the molecular signaling pathways critical to the function of stem cells. In unicellular and multicellular organisms, asymmetric division enables segregation of damaged molecules (biological garbage) into one daughter cell. This appears to be an evolutionarily conserved mechanism preventing transmission of damaged proteins to the longest-lived daughter cell. Thus, dividing stem cells have a simple physiological mechanism to free themselves of toxic proteins destined for degradation during asymmetric cell division. We suggest that partition of damaged proteins and organelles and segregation of template DNA may function together to produce long-lived stem cells.

INTRODUCTION

The hallmark of stem cells is their ability to self-renew while generating daughter cells that are committed to differentiation. The self-renewing ability of stem cells is tightly related to their capacity to undergo asymmetric divisions. Therefore, self-renewing asymmetric

* Correspondence concerning this chapter should be addressed to V. V. Terskikh, terskikh@bk.ru.

division is a central question in stem cell biology. Though the concept of asymmetric cell division was formulated more than a hundred years ago, only recently data were obtained that allow understanding of asymmetric cell division at a molecular level. Exciting progress has been made during the past years in our understanding of how cells establish polarity and how cell-fate determinants are segregated during asymmetric cell division. Inactivation of any of several molecular mechanisms that control cell polarity and asymmetry of the partition of cell fate determinants during stem cell division may result in hyperproliferation of the stem cell compartment and could contribute to cancer (Bilder, 2004; Gonzalez, C. 2007).

It is reasonable to suggest that polarity and asymmetric division might have arisen originally as a solution to the problem of self-renewing progeny, but in multicellular organisms the functions of asymmetric division have diversified, particularly to produce tissue-specific cells or cells committed to apoptosis. The biological impact of asymmetric stem cell division on the development and maintenance of tissue homeostasis is of great importance. In epidermis, a thoroughly studied model system, there is a gradient of differentiation from basally located stem/progenitor cells to the most specialized cells, and different types of differentiating keratinocytes are organized along a basal-apical axis. Some basal epidermal cells divide asymmetrically, generating a committed suprabasal cells and proliferative basal cells (Lechler, Fuchs, 2005).

Asymmetric Division and Stem Cells

Neuroblasts of the fruit fly *Drosophila melanogaster* comprise perhaps one of the best-characterized model cell systems with asymmetric division (Lin, Schagat, 1997). Neuroblasts undergo asymmetric divisions, generating two daughter cells of different size and fate. The larger daughter cell retains neuroblast identity and can continue to divide asymmetrically, whereas the smaller daughter, termed the ganglion mother cell, is committed to differentiate and divides terminally to produce two neurons or glial cells. Neuroblasts resemble stem cells in that they undergo self-renewing divisions and are multipotent, capable of generating several different types of neurons and glia. However, they make a limited number of daughter cells and eventually undergo apoptosis. The size of neuroblast clones produced during the embryonic phase of neurogenesis varies from two cells to more than 40 cells (Maurange and Gould, 2005).

Self-renewing asymmetric division of a stem cell is controlled by both extrinsic cues and intrinsic mechanisms. Stem-cell populations are established in niches—specific anatomic locations that regulate their behavior. The role of the stem cell niche is regulating the balance between stem cells and differentiating progeny. Niche cells produce factors that facilitate stem cells' maintenance over time and saves them from depletion, while protecting the organism from unrestrained stem-cell proliferation. It constitutes a basic unit of tissue physiology, integrating signals that mediate the balanced response of stem cells to the needs of organisms. The well-illustrated role of the niche in regulating asymmetric stem cell division comes from the study of germline stem cells in the *Drosophila* ovary and testis (Lin, 2002; Chen, McKearin, 2003). Both female and male germ line stem cell systems contain somatic niche cells that secrete ligands producing necessary signaling to maintain stem cells.

Asymmetric distribution of proteins and other molecules within cells leads to cell polarization. Par proteins play a central role in producing cell polarity. They organize themselves either by inclusion or exclusion into complexes that localize to specific domains in the cell, which is critical for the establishment of other polarized aspects of the cells (Macara, 2004). In *Drosophila*, particularly, the differential segregation of the Numb protein to one of the two daughter cells during asymmetric division is required for the daughters to adopt distinct fates during the development of the nervous system, muscle, and malpighian tubules (Uemura et al., 1989; Rhyu et al., 1994; Knoblich et al., 1995; Spana et al., 1995; Carmena et al., 1998; Wan et al., 2000). In neuroblasts, for example, Numb becomes localized to the cytosolic face of the basal plasma membrane during mitosis, and the mitotic spindle orients along the basal–apical axis, so that only the basal daughter cell inherits Numb (Rhyu et al., 1994). Numb is thought to influence cell fate, at least in part, by inhibiting Notch signaling (Frise et al., 1996; Guo et al., 1996; Spana and Doe, 1996).

Generation of epithelial cell polarity begins with the establishment of physical and molecular asymmetry at the cell surface. The current model of segregation of cell fate determinants in asymmetric mitosis is that apical and basolateral protein complexes are mutually antagonistic, which leads to the distribution of these proteins in a polarized fashion (Bilder et al., 2003). The asymmetric mode of neuroblast division has been intensively studied and many cell-polarity proteins and other components regulating it are now well characterized (Jan and Jan (2001; Chia and Yang, 2002; Roegiers and Jan, 2004; Chia et al., 2008). The apically localized proteins comprise two complexes. The evolutionary conserved Par protein complex includes Bazooka/Par3, atypical protein kinase C (aPKC), and is linked by the adaptor protein Inscuteable to Par6. This protein complex is the first to localize to the neuroblast cell cortex and is involved in excluding the basally localized proteins from the apical cortex. This protein complex regulates the activity of the tumor suppressor lethal giant larvae (Lgl), which is also essential for correct targeting of the basal protein complexes. Par6 can associate with Lgl, after that aPKC is believed to inactivate Lgl by phosphorylation. Miranda is then recruited to the basal cell cortex by the active nonphosphorylated Lgl. Miranda associates with and facilitates the asymmetric localization of the translational repressor Brain tumor (Brat), the homeodomain transcription factor Prospero, and the double-stranded RNA-binding protein Staufen, which itself can bind *prospero* transcripts. The second basal complex contains the Notch antagonist Numb and its binding Partner of Numb (Pon). Upon segregation into the GMC, Miranda is degraded, allowing Prospero to translocate into the nucleus to activate genes involved in differentiation and repress genes involved in proliferation.

The second apical protein complex contains proteins involved with heterotrimeric G protein signaling, including **Gαi**, Partner of Inscuteable (Pins), and Locomotion defects (Loco). This complex is thought to mediate a receptor-independent heterotrimeric G protein signaling mechanism involving the regulation of **Gαi** through interactions with the cytoplasmic guanine nucleotide exchange factor (Ric-8) and guanine nucleotide dissociation inhibitors (Loco and Pins). The **Gαi**–Pins–Loco complex mediates mitotic spindle formation and alignment to ensure that the cleavage plane is orthogonal to the apical-basal polarity axis. The geometry of the neuroblast mitotic spindle is asymmetrical; the spindle length is longer in the apical side, and the entire spindle is displaced toward the basal cortex. The centrosomes are also nonequivalent, with the larger mother centrosome emanating more extensive astral microtubules and being preferentially retained within the neuroblast through subsequent divisions. Pins can associate with the centrosome- and apical cortex–associated nuclear mitotic apparatus protein-related protein mushroom body defective (Mud), which is essential for proper spindle alignment, as well as Discs large (Dlg) and the astral microtubule plus end protein Khc-73 to induce cortical polarity. The actin/myosin cytoskeleton also plays an important role in the assembly of these apical/basal protein complexes. Actin filaments but not microtubules appear to play an essential role in cortical tethering of the proteins, and the *Drosophila* myosins II (Zipper) and VI (Jaguar) exist in mutually exclusive complexes with Miranda and are essential for correct asymmetric localization of the cell fate determinants.

Position of the spindle during asymmetric division dictates segregation of polarized cell-fate determinants. Asymmetry of the mitotic spindle is implicated in the interpretation of cortical polarity by the spindle and in spindle orientation (Kusch et al., 2003). Yamashita et al. (2007) demonstrated that asymmetric outcome of stem cell division can be specified by an oriented spindle and local self-renewal signals from stem cell niche. In the *Drosophila* male germ line the mother centrosome remains anchored near the niche in the cell that retains stem cell identity while the daughter centrosome migrates to the opposite side of the cell before spindle formation.

Aging of Stem Cells

Aging, defined as a decline in reproduction rate or in regenerative potential with age, is a fundamental characteristic of all living organisms and may be the result of age-related changes in stem cells. Most adult stem cells are maintained throughout the entire life of an organism. For example hematopoietic stem cells (HSCs) sustain lifelong production of mature blood cells. In fact, it has been documented that HSCs can even outlive their original donor upon repeated serial transplantation in lethally irradiated recipients (Harrison, 1979), the most widely used model to study the process of HSCs exhaustion. However, it has been documented that serial transplantation results in a permanent loss of self-renewal capacity of HSCs. Most of these studies have suggested that the limit to serial transplantation is caused by exhaustion of stem cells, but this finding has been challenged by experiments in which it was argued that decline was caused by the transplantation procedure (Iscove, Nawa 1997).

In many experiments, HSCs from old and young donors have repopulated recipients equally well, implying that changes with age are not intrinsic. D. E. Harrison (1983) studied the long-term erythropoietic repopulating ability of stem cells of different age and the results obtained suggest that stem cells do not age. Cho et al., (2008) found that clonal analysis of repopulating HSCs demonstrated that lymphoid-biased HSCs were lost and long-lived myeloid-biased HSCs accumulated in the aged mice. And myeloid-biased HSCs from young and aged sources behaved similarly in all aspects tested. This indicates that aging does not change individual HSCs. Rather, aging changes the clonal composition of the HSC compartment.

Studies in systems such as the *Drosophila* gonad have indicated that mechanisms must exist for replacement of stem cells lost due to natural turnover (Wallewnfang et al., 2006;

Margolis, Spradling, 1995). One obvious strategy that has been observed in the *Drosophila* ovary is symmetric division of resident stem cells to repopulate the niche (Margolis, Spradling, 1995). Studies in both the *Drosophila* and mouse male germ line have demonstrated aging of the stem cell niche, including an apparent decline in self-renewal factors normally expressed by key niche support cells Therefore, immortal germ line stem cells appear to reside within a mortal, somatic niche that is susceptible to aging (Boyle et al.; Ryu et al., 2006; Zhang et al., 2006). Another possible mechanism is obtaining stem cell status by partially differentiated progenitor cells as a result of programmed de-differentiation.

Human embryonic blastocyst-derived stem cell lines do not age and may be cultivated in vitro practically indefinite under special conditions (Thomson et al., 1998; Reubinoff et al., 2000). Studies in murine skin epidermis demonstrated that over an average lifetime there was no measurable loss in the physiologic functional capacity of epidermal stem cells. It seems probable that epidermal stem cells themselves are intrinsically aging resistant and that local environmental or systemic factors modulate skin aging (Stern, Bickenbach, 2007; Giangreco et al., 2008).

Conboy et al. (2005) suggested that decline of regenerative potential of aged muscle may be a result of decline in skeletal muscle stem cell (satellite cell) activity due to a loss of Notch signaling. Conboy and colleagues used heterochronic parabiosis between old and young mice and demonstrated that decline of regenerative potential of aged satellite cells can be reversed through the shared circulatory system. The exposure of satellite cells from old mice to young serum enhanced the expression of the Notch ligand Delta, increased Notch activation, and enhanced proliferation of satellite cells *in vitro*. These data suggest that aged tissue-specific stem cells retained their intrinsic proliferative potential and that age-related changes in the systemic environment and niche in which stem cells reside can modulate the molecular signaling pathways critical to the function of stem cells.

In stem cells of renewing tisssues multiple mechanisms operate that allow to self-renew and to escape senescence. For example *Bmi1*, a *Polycomb* group repressor, is essential for the self-renewal of adult murine HSCs and neuronal stem cells, partially via repression of genes involved in senescence (Park et al., 2004). Asymmeric division may be another mechanism used to prevent senescence of stem cells.

Aging in Bacteria and Yeast

Significant progress has been made in understanding of self-renewing in bacteria and budding yeast. Unicellular organisms subjected to aging provide a simple system to study how daughter cells reset age and cell strains remain immortal. In these organisms mother cells have a limited capacity to produce daughter cells, and the decline in this capacity with each generation is referred to as replicative aging which is a fundamental characteristic of all living organisms down to bacteria. It was proposed that age asymmetry produced by asymmetric division depends on partition of active and undamaged cellular components to the daughter cell.

The life cycle of asymmetrically dividing bacterium *Caulobacter crescentus* begins with a nonreproductive swarmer cell that after a period of free swimming differentiates into a sessile and reproductive stalked cell. The stalked cell starts to produce progeny. In this bacterium, division gives rise to two cells that can readily be distinguished by their morphology. The reproductive output of stalked cells decreased at an accelerating rate with increasing age, indicating that senescence was observed only in cells with old poles (Ackermann et al., 2003).

In the outwardly symmetrically dividing *Escherichia coli*, the cell inheriting functionally distinct old pole after division grows more slowly and divides less frequently, therefore exhibiting aging (Stewart et al., 2005). In *Escherichia coli*, Lindner et al. (2008) using fluorescently tagged chaperone (IbpA) involved in aggregate processing showed that aggregated proteins accumulated upon cell division in cells with older poles. These cells partially loose reproductive ability relative to the new-pole progeny, devoid of inclusion. This suggests an asymmetric strategy whereby dividing cells inheriting the aggregates exhibit diminished growth and reproductive potential. Thus, in bacteria, aggregated proteins accumulated upon cell division in cells with older poles represent one of the prominent markers of aging factor leading to bacterial senescence and death (Linder et al., 2008; Maisonneuve et al., 2008).

The yeast *Saccharomyces cerevisiae* reproduces by asymmetric cell division, or budding. A mother cell gives rise to a small daughter cell each division. Life span can thus be determined by following the fate of a mother cell through multiple rounds of division until cessation of proliferation. In each cell division, the daughter cell is usually smaller and younger than the mother cell, as defined by the number of divisions it can potentially complete before it dies. Although individual yeast cells have a limited life span, this age asymmetry between mother and daughter ensures that the yeast strain remains immortal. Daughters from old mothers have shortened life span potential and appear to be prematurely old themselves (Kennedy et al., 1994). But this shortening of life span is reversible: by the second or third asymmetric division of daughters from old mothers, their life span potential is fully restored.

Replicative aging is a complex process. One plausible mechanism for aging is accumulation of mutations in the mitochondrial genome (Krishnan et al., 2007). Mutations on the mitochondrial genome are thought to arise due to the close proximity of mtDNA to the oxidative phosphorylation system located on the inner mitochondrial membrane, making the mitochondrial genomes vulnerable to damage through the leakage of reactive oxygen species during the oxidative phosphorylation process. Lai et al. (2002) found that segregation of active mitochondria to daughter cells is important for maintenance of age asymmetry and raise the possibility that mitochondrial dysfunction may be a normal cause of aging. The finding that dysfunctional mitochondria accumulated in yeasts as they aged support this notion. These authors isolated temperature-sensitive mutant in *ATP2* that had limited growth capacity. In this mutant, senescence phenotype of clones resulted from a loss of age asymmetry at 36°, such that daughters were born old. It was characterized by a loss of mitochondrial membrane potential followed by the lack of proper segregation of active mitochondria to daughter cells. The results indicated that segregation of active mitochondria to daughter cells was important for maintenance of age asymmetry and raise the possibility that mitochondrial dysfunction may be a normal cause of aging.

Sinclair and Guarente (1997) proposed that one pathway leading to replicative aging in *Saccharomyces cerevisiae* was caused by progressive accumulation of extrachromosomal rDNA circles in yeast mother cells. According to this model, rDNA circles are produced stochastically by intrachromosomal homologous recombination at the rDNA locus and are inherited asymmetrically by mother cells, which leads to their replicative senescence. Sinclair and Guarente have estimated that the extrachromosomal rDNA content of old cells exceeded the content of the linear genome. In *Saccharomyces cerevisiae* Falcón and Aris (2003) found a correlation between plasmid accumulation and life span reduction. These findings support the view that accumulation of DNA episomes, including episomes such as extrachromosomal rDNA circles, cause cell senescence in yeast. It was shown (Shcheprova et al., 2008) that during yeast budding, a diffusion barrier is formed in the nuclear envelope that limits the translocation of DNA circles into the bud. In accordance with the role of diffusion barrier in the asymmetric segregation of DNA circles, the barrier mutant *bud6*Δ fails to properly reset age in buds.

Cellular ageing is characterized by the accumulation of oxidatively modified proteins which may be due to increased protein damage or decreased elimination of oxidized proteins. Oxidatively modified proteins are continuously produced in cells by reactive oxygen and nitrogen species generated as a consequence of aerobic metabolism. Carbonylation is an irreversible oxidative damage which increases with the yeast replicative age (Stadtman, 1992).

Aguilaniu et al. (2003) visualized carbonylated proteins in single cells of the budding yeast and revealed that they accumulated with replicative age and that ability of mother cells to retain oxidatively damaged proteins during cell division diminished with replicative age. Aguilaniu and colleagues observed uneven distribution of oxidized proteins between mother and daughter cells and showed that carbonylated proteins were not inherited by daughter cells during cell division and that old mother cells exhibited markers of oxidative stress. And a yeast strain lacking the *sir2* gene was defective in the segregation of oxidized protreins during cytokenesis.

Aging Cells and Oxidatively-Damaged Proteins

Oxidatively modified proteins that accumulate in aging cells can form large aggregates because of covalent cross-linking. These oxidized proteins are often toxic, and threaten cell viability. Protein aggregation is an inevitable consequence of cellular function. Probably all cells possess the molecular machinery to form protein aggregations but not all misfolded proteins are able to form aggregates in every kind of cells. So, aggregates do not represent a general response to protein misfolding (Beaudoin et al., 2008).

It was demonstrated by several authors that in *E. coli* the two poles of each cylindrical cell are intrinsically different, because one is created de novo in each cell cycle, while the other is retained from the mother cell. This form of polarity turns out to be of crucial importance: the cell that retains the mother pole through several generations ages and has a reduced growth rate. It was suggested that the bacterium uses polarity to ensure that deleterious material accumulate preferentially in the daughter that possesses the older pole. Over several generations, some cells will inevitably inherit an increasingly large number of damaged proteins associated with old poles. These cells would be less fit and have a reduced growth rate. Now there is strong evidence that oxidized proteins or protein aggregates are causative factors in replicative aging (Lindner et al., 2008; Nystrom, 2007; Maisonneuve et al., 2008)

Experiments with dividing human embryonic stem cells and other mammalian cultured cell lines demonstrated that in many supposedly equal mitoses the segregation of proteins destined for degradation (Smad1 phosphorylated by MAPK and GSK3, phospho-β-catenin, and total polyubiquitinylated proteins) was asymmetric (Fuentealba et al., 2008). In these cells no protein aggregation was shown, but proteins specifically targeted for proteasomal degradation were inherited preferentially by one mitotic daughter during somatic cell division. The authors concluded that many mitotic divisions are unequal, unlike what was previously thought.

Cells avoid accumulating potentially toxic aggregated proteins by mechanisms including the suppression of aggregate formation by molecular chaperones and the degradation of misfolded proteins by proteasomes. The total fraction of newly synthesized misfolded proteins can be estimated as a high proportion of all newly synthesized proteins. It is evident that normal cells must cope with a continuous flux of misfolded proteins that must be degraded (Fink, 1999; Schubert et al., 2000). In unstressed cells, protein aggregates do not accumulate, despite their continued production, by chaperoning nascent or unfolded proteins and by selectively degrading misfolded polypeptides before they can aggregate. Protein aggregates are regularly formed during thermal or oxidative stress. In animal cells, aggregated proteins are specifically delivered from the periphery of the cell to the centrosomes or the microtubule organizing center by dynein-dependent retrograde transport on microtubules and form pericentriolar cytoplasmic inclusions that contain misfolded ubiquitinated proteins encapsulated by intermediate filaments. This microtubule-dependent inclusion body is called an aggresome.

Once formed, protein aggregates tend to be insoluble and metabolically stable under physiological conditions. The fact that protein aggregates do not accumulate in unstressed cells, despite their continued production, is due in part to the existence of cellular "quality control" machinery which suppresses the formation of aggregates by ensuring the fidelity of transcription and translation, by chaperoning nascent or unfolded proteins, and by selectively degrading improperly folded polypeptides before they can aggregate. The prominence of protein aggregates in intracellular and extracellular lesions is associated with cell death in many degenerative diseases (e.g., amyloid diseases, Alzheimer's disease, Parkinson's disease, Huntington's disease and alcoholic liver disease (Tran and Miller, 1999).

To compare survival of growth-arrested confluent and proliferating human fibroblasts and astrocytes Stroikin et al. (2005) inhibited autophagic sequestration with 3-methyladenine or suppressed autophagic degradation by the protease inhibitor leupeptin. Their results support the role of accumulating aggresomes in triggering senescence in non-dividing postmitotic cells. In contrast, proliferating cells apparently escaped senescence by continuous segregation of aggresomes to only one of the two daughter cells.

To investigate the behavior of aggresomes in cycling cells, Rujano et al. (2006) expressed fluorescently-tagged fragments of the *huntingtin* gene, containing polyglutamine repeats, in cell culture lines. Expression of the polyglutamine protein fragment induced the formation of aggresomes. When followed through cell division, the aggresomes tended to asymmetrically segregate to one of the daughter cells. When a Huntingtin protein fragment was expressed in the neuroblast lineage, protein aggregates formed and associated with the centrosome that was inherited by the neuroblast (Rujano et al., 2006). Data now available indicate that aggregated proteins are asymmetrically partitioned in dividing cells of both prokaryotes and eukaryotes and that this segregation could provide a protective mechanism in all organisms. Thus, dividing cell populations have a simple physiological mechanism to remove proteins destined for degradation.

The partition of aggregated damaged proteins in stem cells is poorly undestood now. Rujano and colleagues (2006) investigated the inheritance of aggregated proteins in epithelial cells of the small intestine from two patients with the human neurodegenerative disease spinocerebellar ataxia type 3. It was found in both patients that aggregated mutant ataxin-3 protein was present in a large fraction of the shorter-lived differentiated epithelial cells, dividing transit amplifying cells, as well as the adjacent nondividing differentiated Paneth cells whereas these inclusions were absent in the long-lived stem cells of the intestinal crypts. These data suggest that stem cells may clean themselves of damaged undegraded proteins.

Cells like self-renewable tissue stem cells or germ-line cells that exist during the entire lifespan may use storage of aberrant proteins in aggresomes to preserve fitness and functionality. Because aggresomes are formed at the centrosome, which is an organelle crucial for proper spindle formation and cell division, it is conceivable that aggresome formation and irreversibly accumulated proteins interfere with proper cell division. Aggresomes still allow normal cell divisions in which irreversibly damaged, stored proteins are inherited by only one daughter cell. This asymmetric inheritance appears to be evolutionarily conserved mechanism preventing transmission of damaged proteins to the longest lived daughter cell. Perhaps proliferating cells that accumulate too much undegraded proteins undergo apoptosis. Thus, dividing cell populations may have a simple physiological mechanism to free themselves of proteins destined for degradation at cell division. If this were the case, nondividing cells such as neurons would be at a disadvantage and more susceptible to dysfunction.

In the course of evolution asymmetric division was used by multicellular organisms for self-renewing stem cells. In cell populations sustained by the function of stem cells, during asymmetric division stem cells free themselves of proteins destined for degradation that may be segregated to the specialized cells or cells doomed to apoptosis (Singhvi, Garrige, 2009).

Immortal DNA Strand Hypothesis and Self-Renewing Daughter Cells

In 1975, the "**immortal strand**" hypothesis was proposed by John Cairns as a strategy for adult stem cells to protect their genome from DNA replication errors. This hypothesis holds (Cairns, 1975, 2006) that when the stem cell divides asymmetrically, it selectively retains those sister chromatids containing the older template DNA strands in the daughter destined to be the renewed stem cell, thus passing the younger strands (with any mutations acquired during replication), to the tissue-committed cell. The cosegregation of immortal template DNA strands and their link with the asymmetry apparatus has important implications for stem cell biology and cancer. Recent data on nonrandom partition of sister chromatids during mitosis in epithelial stem cells in the small intestine (Potten et al., 2002), in epithelial cells in the mammary gland (Smith, 2004), in cultured neuronal cells (Karpowicz et al., 2005), in satellite cells in postnatal muscles (Shinin et al., 2006; Conboy et al., 2007) animated interest in the hypothesis. A mechanism was proposed to explain how sister chromatids could acquire epigenetic differences that make them segregate asymmetrically during mitosis (Lew et al., 2008).

It was demonstrated that in regenerating muscles, under conditions when stem cells are dividing asymmetrically, they may cosegregate newly synthesized DNA strands, which is consistent with the assumptions of the immortal strand hypothesis. Earlier, it was suggested that a satellite cell pool comprises a heterogeneous population of behaviorally and phenotypically distinct cells (Zammit, Beauchamp, 2001). Now, evidence is accumulating that indicates that the satellite cell pool is composed of true satellite stem cells ($Pax7^+/Myf5^-$) and satellite myogenic cells ($Pax7^+/Myf5^+$). In vivo, about 10% of satellite cells only express Pax7 and have never expressed Myf5 (Rudnicki et al., 2009).

By labeling newly synthesized DNA in sequential divisions of stem/progenitor cells during muscle regeneration in mice, it was observed that great number of dividing cells (79%) non-randomly sorted their chromatids (Conboy et al., 2007). These data suggest that template strand cosegregation may not be limited to stem cell self-renewal, but may actually occur more generally during muscle regeneration. It was found also that template strand cosegregation was strongly associated with asymmetric cell divisions, yielding daughters with divergent fates. Daughter cells that inherited the older templates retained the more immature phenotype, whereas daughters inheriting the newer templates acquired a more differentiated phenotype. These data provide compelling evidence that template strand cosegregation may be linked with template age and could be associated with cell fate determination and suggest that template strand age is monitored during stem cell lineage progression (Conboy et al., 2007).

It was shown in the lab of Shahragim Tajbakhsh that some of freshly isolated satellite cells in mice displayed a selective template DNA strand and Numb protein segregation during mitosis in vivo as well as in culture (Shinin et al., 2006). The asymmetry of satellite cell division was monitored by Numb segregation, although the frequency of Numb asymmetry was greater than that for template DNA cosegregation. Cells with symmetric segregation of Numb and asymmetric segregation of DNA strands were also frequently observed. But, in the case in which both events displayed asymmetry during division, approximately 90% of cells cosegregated template DNA and Numb asymmetrically to the same daughter cell, whereas Numb and non-template DNA did not cosegregate (80%).

Available data, although often incomplete, suggest that most stem cells have the ability to switch between asymmetric and symmetric modes of division (Morrison, Kimble, 2006). The pattern of satellite cell staining by BrdU suggests that asymmetric DNA segregation may be intervened with symmetric divisions when template strands are marked and can be monitored after asymmetry is resumed. Alternatively, the newly-labelled non-template DNA strands may adopt immortal template DNA status (Shinin et al., 2006). It was also demonstrated (Shinin et al., 2006) that Numb cosegregated with the more intensely stained "mother" centrosome, as identified by preferential ε-tubulin and γ–tubulin staining.

Thus, in satellite cells, during the majority of asymmetric divisions, template DNA strands cosegregate with the cell fate determining Numb protein. Although the role of Numb in asymmetric cell division in vertebrates remains unclear, it is unlikely that Numb asymmetry may be directly linked with the segregation of template DNA strands. As pointed out above, oxidated proteins destined for degradation are associated with mother centrosomes (Rujano et al., 2006). Data by Shinin and colleagues suggest that a significant number of satellite cells during asymmetric division segregate template DNA to cells with mother centrosomes—that is, to the cells that receive damaged proteins. But it seems probable that in a small number of asymmetric divisions template DNA may cosegregate to long-lived self-renewing stem cells (Conboy et al., 2007). It is possible to suggest that the two processes, DNA cosegregation and partition of damaged proteins and organelles, function together in such a way that, in the minority of self-renewing satellite cells that represent putative stem cells, template DNA segregates to the younger self-renewing daughter cells that are free from damaged proteins. But this remains an open question.

Alternatively, when a stem cell that receives the template DNA begins to age, it may be substituted by committed daughter that would dedifferentiate and acquire a stem cell identity. A similar situation was found in the yeast *Saccharomyces cerevisiae* when an old mother cell produced a daughter that was prematurely old and had a shortened life span potential, but this shortening was fully restored by the second or third asymmetric division (Kennedy et al., 1994). And, for example, T. Kai and A. Spradling demonstrated that four- and eight-cell *Drosophila* germline cystocytes generated either in second instar larval ovaries or in adults over-producing the BMP4-like stem cell signal Decapentaplegic efficiently converted into single stem-like cells. These de-differentiated cells can develop into functional germline stem cells and support normal fertility (Kai, Spradling, 2004).

We hypothesize that the principle of asymmetric disposal of damaged cellular proteins during adult stem cell division may represent an evolutionarily conserved mechanism to preserve the long-lived adult stem cells. This mechanism may be linked with cosegregation of template DNA that would reduce accumulation of errors and diminish aging of stem cells.

CONCLUSION

Based on observations of some tissues, it may be concluded that stem cells themselves are intrinsically aging resistant and that local environmental or systemic factors modulate the aging of tissues where stem cells reside. In unicellular and multicellular organisms, asymmetric division was shown to give rise to two cells of different age. Now a large body of data is presented that demonstrates that bacteria and yeasts accumulate deleterious material—particularly, oxidatively damaged proteins—preferentially in the daughter that possesses the older pole. These cells would have a reduced growth rate. Now there is strong evidence that accumulation of damaged molecules is a causative factor in replicative aging. In the course of evolution of multicellular organisms, asymmetric division was fitted to the self-renewal of stem cells. During asymmetric division, stem cells free themselves of proteins destined for degradation. It was shown in a murine model that a significant number of satellite cells during asymmetric division segregate template DNA. Although the relation between asymmetric division of stem cells and segregation of template DNA strands remains unclear, it is possible to suggest that the two processes, DNA cosegregation and partition of damaged proteins and organelles, may function together to sustain self-renewing stem cells.

Acknowledgements

I am cordially thankful to my colleagues in the Department of Embryology and Reproduction Development of Russian Academy of Science, Valentina E. Vasilyeva and Galina E. Titova for their valuable consultation and contribution to the work at its different stages, and to Y. Osadtchy for his active help in dealing with figures and to the translators O. Butuzova and I. Putro. Also, I am most grateful to the Komarov Botanical Institution and Embryology Department of St-Petersburg State University for their support of this work. I would like to express my high appreciation to Dr. Frank Columbus and Mrs. Nadya S. Gotsiridze-Columbus, as well as to Ms. Debbie Saskin of Nova Science Publishers for their friendly attitude and effective collaboration in this work.

References

Aguilaniu H, Gustafsson L, Rigoulet M, Nyström T. Asymmetric inheritance of oxidatively damaged proteins during cytokinesis. *Science.* 2003; 299: 1751–1753.

Beaudoin S., Goggin K., Bissonnette C., Grenier C., Roucou X. Aggresomes do not represent a general cellular response to protein misfolding in mammalian cells. *BMC Cell Biol.* 2008; 9:59 doi: 10. 1186/1471-2121-9-59.

Bilder D. Epithelial polarity and proliferation control: links from the Drosophila neoplastic tumor suppressors. *Genes Dev.* 2004; 18:1909-1925.

Bilder, D., Schober, M. and Perrimon, N. Integrated activity of PDZ protein complexes regulates epithelial polarity. *Nat. Cell.Biol.* 2003; 5, 53-58.

Boyle, M., Wong C., Rocha M. and Jones D. L. Decline in self-renewal factors contributes to aging of the stem cell niche. *Cell Stem Cell.* 2007; 1: 470-478.

Cairns J. Mutation selection and the natural history of cancer. *Nature*. 1975; 255: 197-200.

Cairns J. Somatic stem cells and the kinetics of mutagenesis and carcinogenesis. *Proc. Natl. Acad. Sci. USA*. 2002; 99: 10567-10570.

Chen D., McKearin D. Dpp signaling silences *bam* transcription directly to establish asymmetric divisions of germline stem cells. *Curr. Biol.* 2003; 13: 1786-1791

Chia W., Somers W.G. and Wang H. *Drosophila* neuroblast asymmetric divisions: cell cycle regulators, asymmetric protein localization, and tumorigenesis. *J. Cell Biol.* 2008; 180: 267-272.

Chia, W. and Yang, X. Asymmetric division of Drosophila neural progenitors. *Curr. Opin. Genet. Dev.* 2002; 12: 459–464.

Conboy I.M., Conboy M.J., Wagers AJ, Girma E.R., Weissman I.L., Rando T.A. Rejuvenation of aged progenitor cells by exposure to a young systemic environment. *Nature.* 2005; 433: 760-764.

Conboy M.J., Karasov A.O., Rando T.A. High incidence of non-random template strand segregation and asymmetric fate determination in dividing stem cells and their progeny. *PLoS Biol.* 2007; 5(5): e102.

Falcón A.A. and Aris J.P. Plasmid accumulation reduces life span in Saccharomyces cerevisiae *J. Biol. Chem.* 2003; 278: 41607–41617.

Fink A. L. Chaperone-mediated protein folding. *Physiol. Rev.* 1999. V. 79: 425–449.

Giangreco A., Qin M., Pintar J. E., Watt F. M. Epidermal stem cells are retained *in vivo* throughout skin aging. *Aging Cell.* 2008; 7: 250-259.

Gonzalez, C. Spindle orientation, asymmetric division and tumour suppression in *Drosophila* stem cells. *Nat. Rev. Genet.* 2007; 8: 462 – 472.

Harrison D. E., Mouse erythropoietic stem cell lines function normally 100 months: loss related to number of transplantation. *Mech Ageing Dev.* 1979; 9: 427-433.

Harrison DE. Long-term erythropoietic repopulating ability of old, young, and fetal stem cells. *J. Exp Med.* 1983; 157: 1496-1504.

Iscove N.N., Nawa K. Hematopoietic stem cells expand during serial transplantation in vivo without apparent exhaustion. *Current Biol.* 1997; 7: 805-808.

Jan, Y.N. and Jan, L.Y. Asymmetric cell division in the Drosophila nervous system. *Nat. Rev. Neurosci.* 2001; 2: 772–779.

Janzen V., Randolf Forkert R., Fleming H. E., Saito Y., Waring M. T., Dombkowski D.M., Cheng T., DePinho R. A., Sharpless N. E., Scadden D.T. Stem-cell ageing modified by the cyclin-dependent kinase inhibitor $p16^{INK4a}$. *Nature.* 2006; 443: 421-426.

Jones D. L. Aging and the germ line: where mortality and immortality meet. *Stem Cell Rev.* 2007; 3:192–200.

Kai T and Spradling A. Differentiating germ cells can revert into functional stem cells in *Drosophila melanogaster* ovaries. *Nature.* 2004; 428:564-569.

Karpowicz P., Morshead C., Kam A., Jervis E., Ramuns J., Cheng V., van der Kooy D. Support for the immortal strand hypothesis: neural stem cells partition DNA asymmetrically in vitro. *J. Cell Biol.* 2005; 170: 721-732.

Kennedy, B.K., Austriaco, N.R., and Guarente, L. Daughter cells of *Saccharomyces cerevisiae* from old mothers display a reduced life span. *J. Cell Biol.* 1994; 127: 1985–1993.

Kopito R.R. Aggresomes, inclusion bodies and protein aggregation. *Trends Cell Biol.* 2000; 10: 524-530.

Krishnan K. J., Greaves L.C., Reeve A. K., Turnbull D. The ageing mitochondrial genome. *Nucleic Acids Research.* 2007; 35: 399–7405.

Kuang S., Kuroda K., Le Grand F., Rudnicki M.A. Asymmetric self-renewal and commitment of satellite stem cells in muscle. *Cell.* 2007; 129: 859-861.

Kusch J., Liakopoulos D. and Barral Y. Spindle asymmetry: a compass for the cell. *Trends Cell Biol.* 2003; 13: 562-568.

Lai C-Y., Jaruga E., Borghouts C., Jazwinski S. M. A Mutation in the *ATP2* gene abrogates the age asymmetry between mother and daughter cells of the yeast *Saccharomyces cerevisiae. Genetics.* 2002; 162: 73-87.

Lechler T., Fuchs E. Asymmetric cell divisions promote stratification and differentiation of mammalian skin. *Nature.* 2005; 437: 275-280.

Lew D.J., Burke D. J., Dutta A. The immortal strand hypothesis: How could it work? *Cell* 2008; 133: 21-23.

Lin H. The stem-cell niche theory: lessons from flies. *Nature Rev. Genet.* 2002; 3: 931-940.

Lin H., Schagat T. Neuroblasts: a model for the asymmetric division of stem cells. *Trends in Genet.* 1997; 13: 33-39.

Lindner A. B., Madden R., Demarez A., Stewart E. J., Taddei F. Asymmetric segregation of protein aggregates is associated with cellular aging and rejuvenation. *Proc. Natl. Acad. Sci. USA.* 2008. V. 105. P. 3076-3081.

Macara I. G. Par proteins: Partners in polarization. *Current Biol.* 2004; V. 14, R160–R162

Macara I. G. and Mili S. Polarity and differential inheritance—Universal attributes of life? *Cell.* 2008; 135: 802-812.

Maisonneuve E., Ezraty B., Dukan S. Protein aggregates: an aging factor involved in cell death. *J. Bacteriol.* 2008; 190: 6070-6075.

Margolis, J., & Spradling, A. (1995). Identification and behavior of epithelial stem cells in the Drosophila ovary. *Development.* 121; 3797–3807.

Maurange C., Gould A.P. Brainy but not too brainy: starting and stopping neuroblast divisions in *Drosophila. Trends in Neurosci.* 2005; 28: 30-36.

Morrison S. J., Kimble J. Asymmetric and symmetric stem-cell divisions in development and cancer. *Nature.* 2006; 441: 1068-1014.

Morrison S.J., Weissman I.L. The long-term repopulating subset of hematopoietic stem cells is deterministic and isolatable by phenotype. *Immunity.* 1994; 1: 661-673.

Nystrom, T. A bacterial kind of aging. *PLoS Genet.* 2007; *3*, e224.

Park I.-K., Morrison S.J., Clarke M. F. *Bmi1*, stem cells, and senescence regulation. *J. Clin. Invest.* 2004; 113: 175-179.

Potten C.S., Owen G., Booth D. Intestinal stem cells protect their genome by selective segregation of template DNA strands. *J. Cell Sci.* 2002; 115:2381-2388.

Reubinoff B.E., Pera M.F., Fong C.-Y., Trouson A., Bongso A. Embryonic stem cell lines from human blastocysts: somatic differentiation in vitro. *Nature Biotechnol.* 2000; 18: 399-404.

Roegiers F. and Jan Y. N. Asymmetric cell division. *Current Opinion in Cell Biol.* 2004; 16: 195-205.

Rujano M.A., Bosveld F., Salomons F.A., Dijk F., van Waarde M.A. W. H., van der Want J. J.L., de Vos R.A.I., Brunt E. R., Sibon O.C.M., Kampinga H.H. Polarised asymmetric inheritance of accumulated protein damage in higher eukaryotes. *Plos Biol.* 2006. V. 4 (12): e417.

Ryu, B. Y., Orwig, K. E., Oatley, J. M., Avarbock, M. R., Brinster, R. L. (2006). Effects of aging and niche microenvironment on spermatogonial stem cell self-renewal. *Stem Cells.* 2006; 24: 1505–1511.

Schubert U. et al. (2000). Rapid degradation of a large fraction of newly synthesized proteins by proteasomes. *Nature.* 2000; 404: 770–774.

Shcheprova Z., Sandro Baldi S., Frei S.B., Gonnet G., Barral Y. A mechanism for asymmetric segregation of age during yeast budding. *Nature.* 2008; 454: 728-734.

Singhvi A., Garriga G. Asymmetric divisions, aggresomes and apoptosis. *Trends Cell Biol.* 2009; 19: pp. 1-7.

Smith G. H. Label-retaining epithelial cells in mouse mammary gland divide asymmetrically and retain template DNA strands. *Development.* 2004; 132: 681- 687.

Stadtman E. R. Protein oxidation and aging. *Science*, 1992; 257: 1220-1224.

Stewart EJ, Madden R, Paul G, Taddei F Aging and death in an organism that reproduces by morphologically symmetric division. *PLoS Biol.* 2005; 3:e45.

Stroikin Y., Dalen H., Brunk U.T. and Terman A. Testing the ''garbage'' accumulation theory of ageing: mitotic activity protects cells from death induced by inhibition of autophagy. *Biogerontology.* 2005; 6: 39–47

Thomson J. A., Itskovitz-Eldor J., Shapiro S. S., Waknitz M. A., Swiergiel J.J., Marshall V. S.,Jones J. M. Embryonic stem cell lines derived from human blastocysts. *Science.* 1998; 282: 1145-1147.

Tran, P.B. and Miller, R.J. Aggregates in neurodegenerative disease: crowds and power? *Trends Neurosci.* 1999; 22: 194–197.

Wallenfang, M., Nayak, R., & DiNardo, S. Dynamics of the male germline stem cell population during aging of Drosophila melanogaster. *Aging Cell.* 2006; 5: 297–304.

Yamashita Y. M., Mahowald A.P., Perlin J.R., Fuller M.T. Asymmetric inheritance of mother versus daughter centrosome in stem cell division. *Science.* 2007; 315: 518-521.

Zammit P., Beauchamp J. The skeletal muscle satellite cell: stem cell or son of stem cell? *Differentiation* 2001; 68: 193–204.

Zhang, X., Ebata, K. T., Robaira, B., Nagano, M. C. *Biology of Reproduction.* 2006; 74: 119–124.

In: Daughter Cells
Editors: A. Hitomi et al., pp. 283-292

ISBN: 978-1-60876-790-8

Chapter IX

DAUGHTER CELLS AND STEM CELLS: PROPERTIES AND CHARACTERISTICS

Chiara Bianchini[1], Andrea Ciorba[1], Roberta Piva[2], Stefano Pelucchi[1] and Antonio Pastore[1]
[1]University Hospital of Ferrara, Italy;
[2]University of Ferrara, Italy.

ABSTRACT

Stem cells (SC) research is an important part of biotechnology studies that could lead to the development of new therapeutic strategies in different fields of medicine.

It is known that stem cells exist in early embryos as well as in some individual adult tissues. A lot of efforts have been made to understand biology of stem cells, particularly the features concerning their proliferation, self-renewal and differentiation abilities (the three major characteristics of SC) and subsequently to find out the role of the genes and proteins involved in these processes.

Research on stem cells is advancing knowledge on how healthy cells replace damaged cells in adult organisms. This promising area of science is also leading scientists to investigate the possibility of cell-based therapies to treat disease, which is often referred to as regenerative or reparative medicine.

Stem-cell biology is in a phase of dynamic expansion and is forming connections with a broad range of basic and applied disciplines. Stem cell research has offered the promise of effective cell-based therapies in treating many debilitating diseases such as diabetes, cardiovascular and neurodegenerative diseases, and cancer. In the last few years, the therapeutic potential of stem cells has inspired the imagination, intense interest, and targeted investment of scientists, clinicians, and the general public toward this area of biology.

Thus, in this chapter we would like to summarize recent advances in stem cell biology, as well as to discuss how clinical medicine could take advantage from this fascinating field of biology.

Introduction and Basic Definitions

Medical and scientific interest in stem cells is based on a desire to find a source of new, healthy tissue to treat diseased or injured human organs. It is known that some organs, such as the skin and the liver, are able to regenerate themselves when damaged, but it is not yet understood why and how some tissues have this capability and others do not.

Stem cells are unspecialized cells that can self-renew indefinitely and that can also differentiate into different mature cells with specialized functions. In humans, stem cells have been identified in the inner cell mass of the early embryo; in some foetal tissues, such as umbilical cord and placenta; and in several adult organs. Stem cells that are able to differentiate into cell types beyond those of the tissues in which they normally reside are said to exhibit plasticity. When a stem cell is found to give rise to multiple tissue types associated with different organs, the stem cell is defined as multipotent.

Embryonic stem cells (ESCs) are derived from an early-stage embryo. The outer layer of the blastocyst is called the trophoblast, and the cluster of cells inside the sphere is called the inner cell mass. The cells of the inner cell mass are multipotent stem cells that give rise to all cell types of the major tissue layers of the embryo (ectoderm, mesoderm, and endoderm). To be useful for producing medical therapies, cultured ESCs will need to be differentiated into appropriate tissues for transplantation. Researchers are just beginning to learn how to achieve and control this differentiation.

Adult stem cells are undifferentiated cells that are present / resident in a differentiated tissue. They can renew themselves in the body, making identical copies of themselves for the lifetime of the organism, or become specialized to yield the cell types of the tissue of origin. Sources of adult stem cells are bone marrow, blood, the eye, skeletal muscle, dental pulp, liver, skin, the lining of the gastrointestinal tract etc (Blau HM, 2001). Studies suggest that at least some adult stem cells are multipotent. For example, it has been reported that stem cells from the bone marrow, a mesodermal tissue, can give rise to the three major types of brain cells, which are ectodermal derivatives (Mezey et al., 2000) and that stem cells from the brain can differentiate into blood cells and muscle tissue (Bjornson et al., 1999).

Finding ways to culture adult stems cells outside the body is a high priority of stem cell research.

Embryonic Stem Cells

Human Embryonic stem cells (ESCs) were successfully grown in the laboratory for the first time in 1998 (Thomson et al., 1998), while ESCs derived from mouse blastocysts have been studied for almost four decades and provide a critical baseline of knowledge about the biology and cultivation of these cells (Torres, 1998; Wobus and Boheler, 1999). Embryonic stem cells (ESCs) are found in the inner cell mass of the human blastocyst, an early stage of the developing embryo lasting from the 4th to 7th day after fertilization. ESCs extracted from the inner cell mass during the blastocyst stage can be cultured in the laboratory and under the right conditions and can proliferate indefinitely. ESCs growing in this undifferentiated state retain the potential to differentiate into cells of all three embryonic tissue layers (self-renewal and multipotency). Types of cells derived from in vitro cultured, mouse, ESCs are: adipocites, various brain and nervous system cells, insulin-producing cells of the pancreas, bone cells, hematopoietic cells, yolk sac, endothelial cells, primitive endodermal cells, and smooth and striated muscle cells, including cardiomyocytes—heart muscle cells (Odorico et al., 2001).

Further evidence of the multipotent capability of human ESCs is based on studies in *in vivo* settings. Human ESCs injected into mice form a type of benign tumor called teratoma that is made up of tissues from all three embryonic layers. The tissues that arise in the tumor are often advanced, organized, and complex, and include teeth, gut, hair follicles, skin, epithelium, muscle, bone, cartilage, lung tissue, and neural cells (Thomson et al., 1998). The experiments showed the capability of ESCs to produce a variety of tissues, but the results also highlight the complexity of the biological "program" of tissue development that can unfold in different biological environments. These results also emphasize the abnormal, potentially neoplastic potential of ESCs.

Studies of the genes turned on and off as cells begin to differentiate, which are already under way with ESCs, will permit a better understanding of the genetic controls important in tissue differentiation (Duncan et al., 1998). At this purpose, in vitro studies of ESCs also provide an opportunity to explore the role of biochemicals produced in the normal cellular environment that induce stem cells to differentiate, to migrate to a site needing repair, and to assimilate into tissues (Schuldiner et al., 2000).

It is important to underline that research involving human ESCs is at the center of the ethical debate about stem cell use and potential in regenerative medicine.

Adult Stem Cells

The history of research on adult stem cells began years ago. In the 1950s, researchers discovered that the bone marrow contains at least two kinds of stem cells. One population, called hematopoietic stem cells, forms all the types of blood cells in the body. A second population, called mesenchymal stem cells, was discovered few years later (1970s). Mesenchymal stem cells represent a small proportion of the stromal cells in the bone marrow, and can generate bone, cartilage, fat, and fibrous connective tissue.

So far, adult stem cells in mammals have been identified in several organs and tissues. Typically there is a very small number of stem cells in each tissue -stem cell niche-, and these cells have a limited capacity for proliferation, thus making it difficult to generate large quantities of these cells in a laboratory.

Adult stem cells can divide and can give rise to mature cell types that have characteristic profiles and specialized structures and functions of different tissues (i.e. mesenchymal stem cells can give rise to a variety of cell types: osteocytes, chondrocytes, adipocytes, but also neuron like cells or cord blood types) (Lagasse et al., 2000; Taniguchi et al., 1996; Thiese et al., 2001).

The idea of employing adult stem cells in certain therapeutic applications is appealing as adult stem cells are naturally balanced to generate a particular tissue, which might consist of several cell types, so they should be able to give rise to all the components of that tissue when transplanted into a patient.

A relevant limitation to immediate development of therapies based on adult stem cells is the inability to maintain these cells in culture for very long before they differentiate into their mature progeny.

Researchers are beginning to understand how an adult stem cell gives rise to another precursor cell, using biochemical signals: this information could be use to manipulate this process. The ability to "**reprogram**" a cell may be exploited someday to therapeutic ends; however, the reversal of the normal pathway of differentiation may have biological consequences not yet detected. Rigorous experimentation will be needed to evaluate the implications of this basic research finding for regenerative medicine.

Different Types of Adult Stem Cells

Hematopoietic Stem Cells (HSC)

As stated above, HSC are probably the stem cells most familiar to clinicians given the large amount of research devoted to their characterization as well as their successful application to current patient care (Poulsom R., et al, 2002, Lo KC et al 2003). Bone marrow is an ideal tissue for studying stem cells because of its relative easy accessibility and because proliferative responses of bone marrow–derived stem cells can be readily investigated. Furthermore, there are a number of well-defined mouse models and cell surface markers that allow effective study of hematopoiesis in healthy and injured mice. Because of these characteristics and the experience of bone marrow transplantation in the treatment of hematological cancers, bone marrow–derived stem cells have also become a major tool in regenerative medicine.

HSC can be divided into 2 classes of cells, namely long-term and short-term types. Long-term cells self-renew for life and give rise to short-term cells, which in turn self-renew for about 8 weeks. Short-term cells also produce more differentiated or multipotent progenitors, which then differentiate into the several blood cell types (Morrison SJ et al 1997) This population of HSC from bone marrow is heterogeneous, although it can be enriched by various protocols relying on the positive identification of various cell surface proteins as well as the removal of cells expressing markers of maturity. As first identified in the mouse model, human HSC are Thy-1 positive, CD34 positive, Lin negative and CD38 negative (Uchida N et al, 1998). The transplantation of histocompatible HSC, using an enrichment protocol, can completely reconstitute the entire hematopoietic cell line in a patient following whole body radiation or chemotherapy (Lo KC et al 2003).

Mesenchymal Stem Cells

The description of mesenchymal stem cells (MSC) firstly appeared during the 1970s from the research of Friedenstein et al. (Friedenstein AJ et al 1976). Since the 1990s, increasing experimental data have demonstrated that MSC are defined as multipotent adult stem/progenitor cells that can differentiate into connective skeletal tissue, bone, cartilage, marrow stroma, and adipocytes (Caplan AI 1991, Prockop DJ, 1997, Pittenger MF et al 1999). In addition, some controversial data suggest that MSC may give rise to other cells from mesodermal origin such as sarcomeric muscle (skeletal and cardiac) (Wakitani S et al, 1995, Makino S et al 1999, Planat-Bernard V et al 2004), endothelial cells (Oswald J et al,

2004), and even cells of nonmesodermal origin, such as hepatocytes (Chagraoui J et al 2003), neural cells (Woodbury D et al, 2000), and epithelial cells (Spees JL et al 2003, Ma Y et al 2005, Sensebè L et al, 2009).

Hepatic stem cells

The liver is well known for its regenerative ability. The whole organ can re-grow from as little as a third of its original tissue (Vessay CJ et al 2001,). This restorative response is attributed to the tremendous ability of resting hepatocytes to undergo cell division, that is up to 12 to 16 divisions per cell. In addition, the existence of endogenous and exogenous liver stem cells capable of forming hepatocytes and bile ducts has been demonstrated in animal models (Peterson BE et al 1998, Lo KC et al 2003). Protocols are emerging to purify and culture hepatic stem cells. With a supply of cultured hepatocytes, hepatocyte transplantation has been suggested as a bridge for patients awaiting liver transplantation (Strom SC et al 1997). Cultured hepatocyte transplantation has also been attempted with some success for various metabolic conditions, such as ornithine carbamoylase deficiency50 and Crigler-Najjar syndrome, type I (Fox IJ et al 1998).

Muscle Stem Cells

Postnatal skeletal muscle undergoes repair and regeneration through the function of so-called satellite cells. Located in the periphery between the muscle fiber sarcolemma and the surrounding basement membrane, these satellite cells account for only 5% of the nuclei present in muscle fiber (Goldring K et al 2002, Feldman JL 1992). Although they are quiescent most of the time, satellite cells are considered the major source of myogenic growth and regeneration (Feldman JL 1992). Recent studies also suggest that cells of other tissue origins, such as bone marrow, can contribute to muscle regeneration (Wakitani S et al 1995). With a renewable source of muscle stem cells the goals of myoblast transplantation for primary muscle diseases, such as muscular dystrophy, may be within reach (Rando TA et al 1997, Lo KC et al 2003, Ferrari G et al 1998).

Adipose Derived Stem Cells

The initial methods to isolate cells from adipose tissue were pioneered by Rodbell (Rodbell M, 1966) and Rodbell and Jones (Rodbell M, Jones AB, 1966) in the 1960s. Whereas multipotent stem cells are abundant within murine white adipose tissue, their numbers and differentiation potential are reduced in brown adipose tissue (Prunet-Marcassus B et al 2006). In humans, differences in stem cell content have been noted between subcutaneous white adipose tissue depots, with the greatest numbers recovered from the arm as compared with the thigh, abdomen, and breast (Gimble JM et al 2007). Furthermore, it is well established that differences exist with respect to preadipocyte and endothelial cell numbers between subcutaneous and omental white adipose depots in human subjects (Van Harmelen V et al, 2004). It remains to be determined as to which human adipose tissue depot should be harvested for optimal stem cell recovery. Nonetheless, it has been reported that adipose derived stem cells could differentiate in adipocytes, cardiomyocytes, chondrocytes, osteoblasts, endothelial cells, myocytes, neuronal like cells (Zuk PA 2001, Gimble JM et al 2007).

Neural Stem Cells

Stem cells have also been identified in the adult brain, an organ long thought to comprise non-renewable neurons. It has been reported that the highest concentration of dividing cells is localized to the subventricular zone and subgranular zone of the dentate gyrus of the hippocampus (Cameron HA et al 1994). Neuronal stem cells can be isolated from fetal and adult brains, cultured and induced to differentiate into the three major neuronal cell types, namely astrocytes, oligodendrocytes and neurons, as identified by antibody markers (Gage FH, 2000). Cultured stem cells can then be transplanted back into developing fetal brains as well as into adult brains. These cells have demonstrated the ability to migrate throughout the nervous system as well as the ability to differentiate into tissues ranging from hippocampus to olfactory bulb.(Gaiano N et al 1998, Gage FH et al 1995, Suhonen JO et al, 1996). These neuronal stem cells are characterized with various cell surface markers (i.e. they are CD133 positive, CD34 negative and CD45 negative) (Uchida N et al 2000, Lo KC et al 2003, Greenfield JP et al 2008).

Stem Cell Research for Regenerative Medicine

Both adult and embryonic stem cells can contribute to the development of regenerative medicine. Human embryonic and adult stem cells each have advantages and disadvantages regarding potential use for cell-based regenerative therapies.

Several Authors believe that tissues derived from embryonic and adult stem cells may differ in the likelihood of being rejected after transplantation. Adult stem cells, as well as tissues derived from them, are currently believed less likely to initiate rejection after transplantation. This is because a patient's own cells could be expanded in culture, designed into assuming a specific cell type (differentiation), and then reintroduced into the patient.

Embryonic stem cells (ESCs) have the advantage of multipotency and have shown themselves to be readily culturable in the laboratory. Although the degree of plasticity of adult stem cells is still unknown and there are difficulties in purifying and culturing them, the only proven stem cell-based medical therapies that are currently available rely on adult-derived stem cells from bone marrow and skin, and adult stem cells from other tissues might someday provide therapies that stimulate the body's own regenerative potential (Clark DL et al 2000).

Improvement of the knowledge in this field is strictly associated with: 1. the discovery and characterization of new cellular and molecular markers that distinguish specific types of stem cells and specific stages of stem cell differentiation; 2. the study of profiles of gene expression in stem cells and their progeny; 3. the analysis of the relationship between several molecular signalling implicated in supporting or inducing a specific cellular phenotype; 4. the set up of new procedures for isolating stem cells from the body, and new techniques to propagate them reliably; 5. the development of a suitable combination of adequate cell progenitors, biomaterials and biological response modifiers for tissue engineering; 6. the understanding of tissue functions following stem cell transplantation.

As knowledge of stem cells grows, investigators will be able to ask meaningful questions about therapeutic approaches, including whether to implant cells in an undifferentiated state or a differentiated state, and which of the various sources of stem cells are best suited to address a specific clinical need. It may become apparent that combined therapies (transplanting multiple stem cell types, or using gene therapy in combination with stem cells transplants and biomaterials) will be needed, depending on such factors as the stage of a particular disease or the age of a patient. For now, all of these questions must wait for the establishment of a more firm scientific fundament (Colter DC et al 2000, Ema H et al 2000).

Because observing the behaviour of tissue in vivo is difficult and the results can be confounded in many ways, sources of human stem cells that can be cultured in vitro are perhaps the most critical need of investigators. They will permit many more questions to be posed and answered. Many more experiments can be completed with cultured cells in the same amount of time and with the same degree of effort as in living organisms. Moreover, data from in vitro studies allow more insightful and better-defined experiments to be developed in living organisms. Access to ESCs is likely to ultimately determine the rate at which scientists make progress in this field. In fact, the successful cultivation of postnatal and adult sources of stem cells for regenerative medicine is likely to advance more rapidly if the study of ESCs proceeds and cells from different sources can be compared. ESCs exhibit many properties whose improved understanding could assist researchers in modifying adult stem cells to achieve better growth in culture and greater capacity for controlled differentiation (Li H et al 1998, Lumelsky N et al 2001).

Stem cell research is still in its infancy, and there are substantial gaps in knowledge that pose obstacles to the realization of new therapies from either adult- or embryo- derived stem cells.

CONCLUSIONS

In the face of extraordinary advances in the prevention, diagnosis, and treatment of human diseases, devastating illnesses such as heart disease, diabetes, cancer, and diseases of the nervous system, such as Parkinson's disease and Alzheimer's disease, continue to deprive people of health, independence, and well-being. Research in biology has led to the discovery of stem cells, embryonic and adult; future development of this field will gain immense new knowledge in the biology of human development that will likely hold remarkable potential for therapies and cures.

REFERENCES

Bjornson, C.R., Rietze, R.L., Reynolds, B.A., Magli, M.C., and Vecovi, A.L. (1999). Turning brain into blood: a hematopoietic fate adopted by adult neural stem cells in vivo. *Science 283*:534-7.

Blau, H.M., Brazelton, T.R., and Weimann, J.M. (2001). The evolving concept of a stem cell: entity or function? *Cell 1*05:829-41.

Cameron, H. A. and Gould, E.: Adult neurogenesis is regulated by adrenal steroids in the dentate gyrus. *Neuroscience, 61*: 203, 1994.

Caplan AI. Mesenchymal stem cells. *J Orthop Res* 1991; 9: 641.

Chagraoui J, Lepage-Noll A, Anjo AA, et al. Fetal liver stroma consists of cells in epithelial-to-mesenchymal transition. *Blood* 2003; 101: 2973.

Clarke, D.L., Johansson, C.B., Wilbertz, J., Veress, B., Nilsson, E., Karlstrom, H., Lendahl, U., and Frisen, J. (2000). Generalized potential of adult neural stem cells. *Science 288*:1660-3.

Colter, D.C., Class, R., DiGirolamo, C.M., and Prockop, D.J. (2000). Rapid expansion of recycling stem cells in cultures of plastic-adherent cells from human bone marrow. *Proc Natl Acad Sci USA 97*:3213-8.

Duncan, S.A., Navas, M.A., Dufor, D., Rossant, J., and Stoffel, M. (1998). Regulation of a transcription factor network required for differentiation and metabolism. *Science 281*: 692-5.

Ema, H., Takano, H., Sudo, K., and Nakauchi, H. (2000). In vitro self-renewal division of hematopoietic stem cells. *J Exp Med 192*:1281-8.

Ferrari, G., Cusella-DeAngelis, G., Coletta, M., Paolucci, E., Stornaiuolo, A., Cossu, G., and Mavilio, F. (1998). Muscle regeneration by bone marrow-derived myogenic progenitors. *Science 279*:1528-30.

Friedenstein AJ, Gorskaja JF, Kulagina NN. Fibroblast precursors in normal and irradiated mouse hematopoietic organs. *Exp Hematol* 1976; 4: 267.

Fox, I. J., Chowdhury, J. R., Kaufman, S. S., Goertzen, T. C., Chowdhury, N. R., Warkentin, P. I. et al: Treatment of the Crigler-Najjar syndrome type I with hepatocyte transplantation. *N Engl J Med, 338*: 1422, 1998.

Gage, F. H.: Mammalian neural stem cells. *Science, 287*: 1433, 2000.

Gage, F. H., Coates, P. W., Palmer, T. D., Kuhn, H. G., Fisher, L. J., Suhonen, J. O. et al: Survival and differentiation of adult neuronal progenitor cells transplanted to the adult brain. *Proc Natl Acad Sci USA, 92*: 11879, 1995.

Gaiano, N. and Fishell, J.: Transplantation as a tool to study progenitors within the vertebrate nervous system. *J Neurobiol, 36*: 152, 1998.

Gimble JM, Kats AJ, Bunnel BA. Adipose-Derived Stem Cells for Regenerative Medicine. *Circulation Research, Circ Res.* 2007;100:1249-1260.

Goldring, K., Parttridge, T. and Watt, D.: Muscle stem cells. *J Pathol, 197*: 457, 2002.

Greenfield JP, Ayuso-Sacido A, Schwartz, Theodore H, Pannullo S, Souweidane M, Stieg PE, Boockvar JA. Use of human neural tissue for the generation of progenitors. *Neurosurgery. 62*(1):21-30, 2008.

Lagasse, E., Connors, H., Al-Dhalimy, M., Reitsma, M., Dohse, M., Osborne, L., Wang, X., Finegold, M., Weissman, I.L., and Grompe, M. (2000). Purified hematopoietic stem cells can differentiate into hepatocytes in vivo. *Nat Med 6*:1229-34.

Li, M. Pevny, L., Lovell-Badge, R., and Snith, A. (1998). Generation of purified neural precursors from embryonic stem cells by lineage selection. *Curr Biol 8*:971-4.

Lo KC, Chuang WW, Lamb DJ. Stem cell research: the facts, the myths and the promises. *The Journal of Urology*, 2003, 170, 2453–2458.

Lumelsky, N., Blondel, O., Laeng, P., Velasco, I., Ravin, R., and McKay, R. (2001). Differentiation of embryonic stem cells to insulin-secreting structures similar to pancreatic islets. *Science 292*:1389-94.

Ma Y, Xu Y, Xiao Z, et al. Reconstruction of chemically burned rat corneal surface by bone marrow-derived human mesenchymal stem cells. *Stem Cells* 2005; 24: 315.

Makino S, Fukuda K, Miyoshi S, et al. Cardiomyocytes can be generated from marrow stromal cells in vitro. *J Clin Invest* 1999; 103: 697.

Mezey, E., Chandross, K.J., Harta, G., Maki, R.A., and McKercher, S.R. (2000). Turning blood into brain: cells bearing neuronal antigens generated in vivo from bone marrow. *Science 290*:1779-82.

Morrison, S. J., Wandycz, A. M., Hemmati, H. D., Wright, D. E. and Weissman, I. L.: Identification of a lineage of multipotent hematopoietic progenitors. *Development, 124*: 1929, 1997.

Odorico, J.S., Kaufman, D.S., and Thomson, J.A. (2001). Multilineage differentiation from human embryonic stem cell lines. *Stem Cells 19*:193-204.

Oswald J, Boxberger S, Jorgensen B, et al. Mesenchymal stem cells can be differentiated into endothelial cells in vitro. *Stem Cells* 2004; 22: 377.

Peterson, B. E., Zajac, A. F. and Michalopoulos, G. K.: Hepatic oval cell activation in response to injury following chemically induced periportal or pericentral damage in rats. *Hepatology, 27*: 1030, 1998.

Pittenger MF, Mackay AM, Beck SC, et al. Multilineage potential of adult human mesenchymal stem cells. *Science* 1999; 284: 143.

Planat-Bernard V, Menard C, Andrè M, et al. Spontaneous cardiomyocyte differentiation from adipose tissue stroma cells. *Circ Res* 2004; 94: 223.

Poulsom, R., Alison, M. R., Forbes, S. J. and Wright, N. A.: Adult stem cell plasticity. *J Pathol, 197*: 441, 2002.

Prockop DJ. Marrow stromal cells as stem cells for nonhematopoietic tissues. *Science* 1997; 276: 71.

Prunet-Marcassus B, Cousin B, Caton D, Andre M, Penicaud L, Casteilla L. From heterogeneity to plasticity in adipose tissues: sitespecific differences. *Exp Cell Res.* 2006;312:727–736.

Rando, T. A. and Blau, H. M.: Methods for my oblast transplantation. *Methods Cell Biol, 52*: 261, 1997.

Rodbell M. Metabolism of isolated fat cells. 2. The similar effects of phospholipase c (clostridium perfringens alpha toxin) and of insulin on glucose and amino acid metabolism. *J Biol Chem.* 1966;241:130 –139.

Rodbell M, Jones AB. Metabolism of isolated fat cells. 3. The similar inhibitory action of phospholipase c (clostridium perfringens alpha toxin) and of insulin on lipolysis stimulated by lipolytic hormones and theophylline. *J Biol Chem.* 1966;241:140 –142.

Schuldiner, M., Yanuka, O., Istkovitz-Eldor, J. Melton, D., and Benvenistry, N. (2000). Effects of eight growth factors on the differentiation of cells derived from human embryonic stem cells. *Proc Natl Acad Sci USA 907*:11307-12.

Sensebè L and Bourin P. Mesenchymal Stem Cells for Therapeutic Purposes *Transplantation.* 2009, 87, 49-53.

Spees JL, Olson SD, Ylostlo J, et al. Differentiation, cell fusion, and nuclear fusion during ex vivo repair of epithelium by human adult stem cells from bone marrow stroma. *Proc Natl Acad Sci USA* 2003; 100: 2397.

Strom, S. C., Fisher, R. A., Thompson, M. T., Sanyal, A. J., Cole, P. E., Ham, J. M. et al: Hepatocyte transplantation as a bridge to orthotopic liver transplantation in terminal liver failure. *Tranplantation, 63*: 559, 1997.

Suhonen, J. O., Peterson, D. A., Ray, J. and Gage, F. H.: Differentiation of adult hippocampus-derived progenitors into olfactory neurons in vivo. *Nature, 383*: 624, 1996.

Taniguchi, H., Toyoshima, T., Fukau, K., and Nakauchi, H. (1996). Presence of hematopoietic stem cells in the adult liver. *Nat Med 2*:198-203.

Theise, N.D., Nimmakayalu, M., Gardner, R., Illei, P.B., Morgan, G., Teperman, L., Henegariu, O., and Krause, D.S. (2001). Liver from bone marrow in humans. *Hepatology 32*:11-16.

Thomson, J.A., Itskovitz-Eldor, J., Shapiro, S.S., Waknitz, M.A., Swiergiel, J.J., Marshall, V.S., and Jones, J.M. (1998). Embryonic stem cell lines derived from human blastocysts. *Science 282*:1145-7.

Torres, M. (1998). The use of embryonic stem cells for the genetic manipulation of the mouse. *Curr Top Dev Biol 36*:99-114.

Wakitani, S., Saito, T. and Caplan, A.I.: Myogenic cells derived from rat bone marrow mesenchymal stem cells exposed to 5-azacytidine. *Muscle Nerve, 18*: 1417, 1995.

Wobus, A.M., and Boheler, K.R. (1999). Embryonic stem cells as a developmental model in vitro. *Cells Tissues Organs 165*:3-4.

Woodbury D, Schwarz EJ, Prockop DJ, et al. Adult rat and human bone marrow stromal cells differentiate into neurons. *J Neurosci Res* 2000; 61: 364.

Uchida, N., Tsukamoto, A., He, D., Friera, A. M., Scollay, R. and Weissman, I. L.: High doses of purified stem cells cause early hematopoietic recovery in syngeneic and allogeneic hosts. *J Clin Invest, 101*: 961, 1998.

Uchida, N., Buck, D. W., He, D., Reitsma, M. J., Masek, M., Phan, T. V. et al: Direct isolation of human central nervous system stem cells. *Proc Natl Acad Sci USA, 97*: 14720, 2000.

Van Harmelen V, Rohrig K, Hauner H. Comparison of proliferation and differentiation capacity of human adipocyte precursor cells from the omental and subcutaneous adipose tissue depot of obese subjects. *Metabolism.* 2004;53:632– 637.

Vessay, C. J. and de la Hall, P. M.: Hepatic stem cells: a review. *Pathology, 33*: 130, 2001.

Zuk, P.A., Zhu, M., Mizuno, H., Huang, J., Futrell, J.W., Katz, A.J., Benhaim, P., Lorenz, P., and Hedrick, M.H. (2001). Multilineage cells from human adipose tissue: implications for cell-based therapies. *Tiss Eng 7*(2):211-28.

In: Daughter Cells
Editors: A. Hitomi et al., pp. 293-304
ISBN: 978-1-60876-790-8

Chapter X

Unique Behavior of Daughter Cells in Chlorarachniophytes

Shuhei Ota*

Station Biologique de Roscoff, Centre National de la Recherche Scientifique et Université Pierre et Marie Curie, Roscoff, France.

Abstract

Chlorarachniophytes constitute a group of photosynthetic cercozoans (one of the ecologically significant protists) living in marine environments. They possess plastids of secondary endosymbiotic origin, which have a remnant nucleus of the green algal endosymbiont (nucleomorph) located in the periplastidial compartment, between the inner and the outer pairs of the plastid membranes. With respect to the plastid evolution, the chlorarachniophytes constitute therefore one of the most interesting unicellular algal groups. Since this algal group was found in 1930, little is known about their cell cycle as well as their taxonomy and geographical distribution. During our taxonomic studies of the chlorarachniophytes, an interesting difference in behavior between the daughter cells has been observed: once the cell has divided, one of the amoeboid daughter cells constricts till it is as narrow as one of its pseudopodia; then, the cytoplasm of this new cell flows in a thin trickle along the pseudopodium, collecting to form another cell body some distance from the stay-at-home nucleus. This recently-discovered phenomenon raises some questions about the mechanism and the evolution of the unique behavior. Here I describe in detail the daughter cells' behavior of the chlorarachniophytes, and discuss about the evolution and biological implications.

* Correspondence concerning this article should be addressed to: (Present address) Shuhei Ota, Graduate School of Science and Engineering, Yamaguchi University, Yoshida 1677-1, Yamaguchi, 753-8512, Japan. e-maill: ohtashuhei@gmail.com, or ota@sb-roscoff.fr

1. INTRODUCTION

1.1. Evolutionary and Taxonomic Background of Chlorarachniophytes

Chlorarachniophytes are marine phototrophic protists which while cosmopolitan, are rarely encountered in natural habitats. The chlorarachniophytes possess relatively high diversity of life cycle, because they are adapted in various habitats such as costal or open ocean waters (Ishida et al. 2007). The life cycles are basically composed of amoeboid, coccoid and flagellate stages. What stage is dominant in the life cycles depend on the species; for example, in some species the amoeboid stage is dominant (Figure 1a), and in other species the coccoid is dominant (Figure 1b). The amoeboid cells are star- or irregular-shaped and usually possess several filopodia (very fine pseudopodia) (Figure 1a). The coccoid cells are spherical with a non-cellulosic cell wall. The flagellate cells are spherical or ellipsoid possessing a single flagellum.

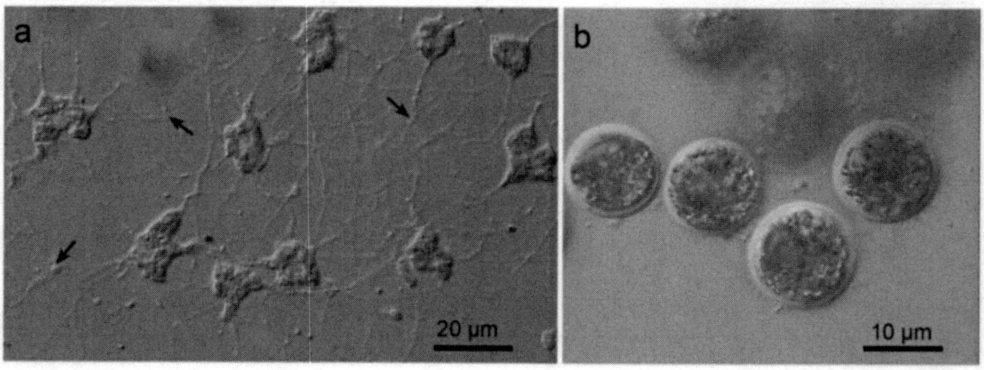

Figure 1. Examples of chlorarachniophytes. (a) Amoeboid species, *Chlorarachnion reptans*. Anastomosing filopodia (meroplasmodium) are seen between cells (arrows). (b) Coccoid species, *Norrisiella sphaerica*.

The chlorarachniophyte cell possesses a nucleus and several four-membrane-bound chloroplasts containing chlorophylls *a* and *b* (Hibberd and Norris 1984). The chloroplast possesses a tiny double membrane-bound organelle, called the "nucleomorph", in a periplastidial compartment (i. e., the space between inner and outer membrane pairs that envelop the chloroplast, red space in Figure 2). Ludwig and Gibbs (1989) showed that the nucleomorph contains DNA, and McFadden et al. (1994) demonstrated that it is a vestigial nucleus of an endosymbiont and the chlorarachniophyte chloroplast was acquired by secondary endosymbiosis. Recently Gilson et al. (2006) published the complete nucleomorph genome sequence of the chlorarachniophyte *Bigelowiella natans*, showing that it has only 373 kbp in size. The nucleomorph is an interesting organelle for understanding the process of the endosymbiont-turned-organelle in the context of secondary endosymbiosis (see Archibald 2007).

The first species of the chlorarachniophyte *Chlorarachnion reptans* (Figure 1a) was described by Geitler (1930). About half a century later, Hibberd and Norris re-examined *C. reptans* by light and electron microscopy and pigment analysis, and established the phylum

Chlorarachniophyta. In the current taxonomic study of the chlorarachniophytes, 12 species in 7 genera have been described to date. In the zoological classification system, the class **"Chlorarachnea"** (= Chlorarachniophyta) belongs to the subphylum Filosa in the phylum Cercozoa which are major protists with cosmopolitan distribution and therefore ecologically significant in soil, freshwater and marine environments (Bass et al. 2004, 2005, Cavalier-Smith & Chao 2003; see Figure 3).

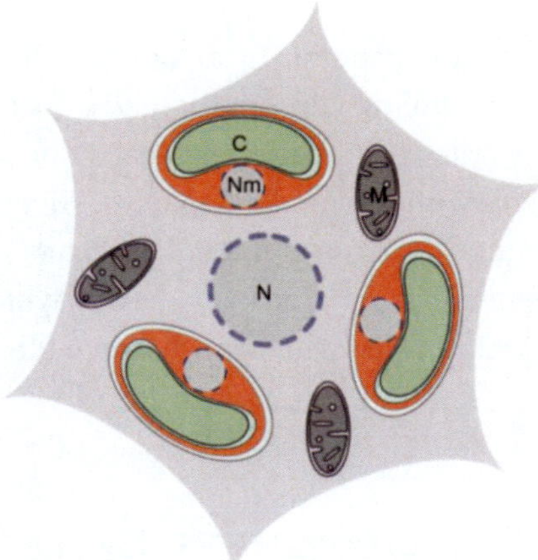

Figure 2. Schematic drawing of generic cell structure of a chlorarachniophyte. C, chloroplast; M, mitochondrion; N, nucleus; Nm, nucleomorph. Periplastidial compartment are show in red.

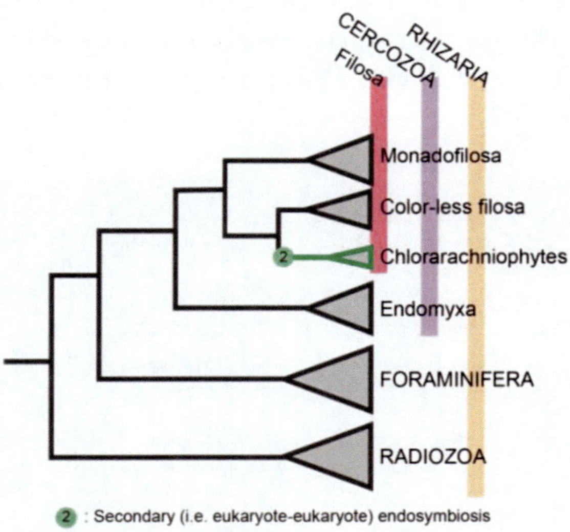

Figure 3. Schematic tree of rhizarians (Rhizaria Cavalier-Smith 2002). The tree topology is based on the published paper (Bass et al. 2005) without some minor groups.

Research on this algal group has mainly developed into two directions, taxonomy and evolutionally cell biology. Little is yet known about the biodiversity of the chlorarachniophytes, and taxonomic studies are needed in this area. Because they are so evolutionally interesting, the chlorarachniophytes are also used in the field of cellular biology, molecular biology or evolutionary biology (e.g., Gilson et al. 2006; Hirakawa et al. 2009). Thus, this algal group has been recognized as an excellent model organism for understanding how the plastid acquisition occurred via secondary endosymbiosis.

1.2. Daughter Cells' Behavior

Recently, interesting behavioral patterns in daughter cells have been revealed during taxonomic studies of the chlorarachniophytes, and initially on *Lotharella polymorpha.* They have been found to show the similar post-cytokinesis behaviors (Dietz et al. 2003, Ota et al. 2005). In another taxonomic study of chlorarachniophytes, *Bigelowiella longifila*, genetically distinct species from *Lotharella*, also presents the unique daughter cells' behavior (Ota et al. 2007). In this chapter, I will show examples of the behavior with time-lapse video sequences in *L. vacuolata* and *B. longifila*, and then discuss the evolutionary aspects and biological implications of this behavior.

2. Examples of Doughter Cells' Behavior

2.1. Lotharella Vacuolata

L. vacuolata is one of the chlorarachniophytes possessing coccoid (walled) cells as main vegetative stage (Figure 4a), but amoeboid cells (Figure 4b) are partially observed in its life cycle (Ota et al. 2005). This species, strain CCMP240, is maintained at the Provasoli-Guillard National Center for Culture of Marine Phytoplankton at Bigelow Laboratory for Ocean Sciences, USA (Ota et al. 2005).

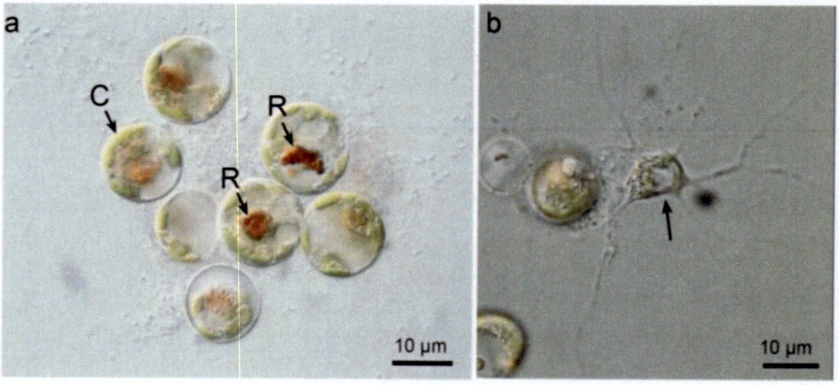

Figure 4. Light micrographs of *Lotharella vacuolata*. (a) Coccoid cells (vegetative cells). Red particles (R) are often observed in cytoplasm which may be excretion products. C, chloroplast . (b) Amoeboid cell (arrow).

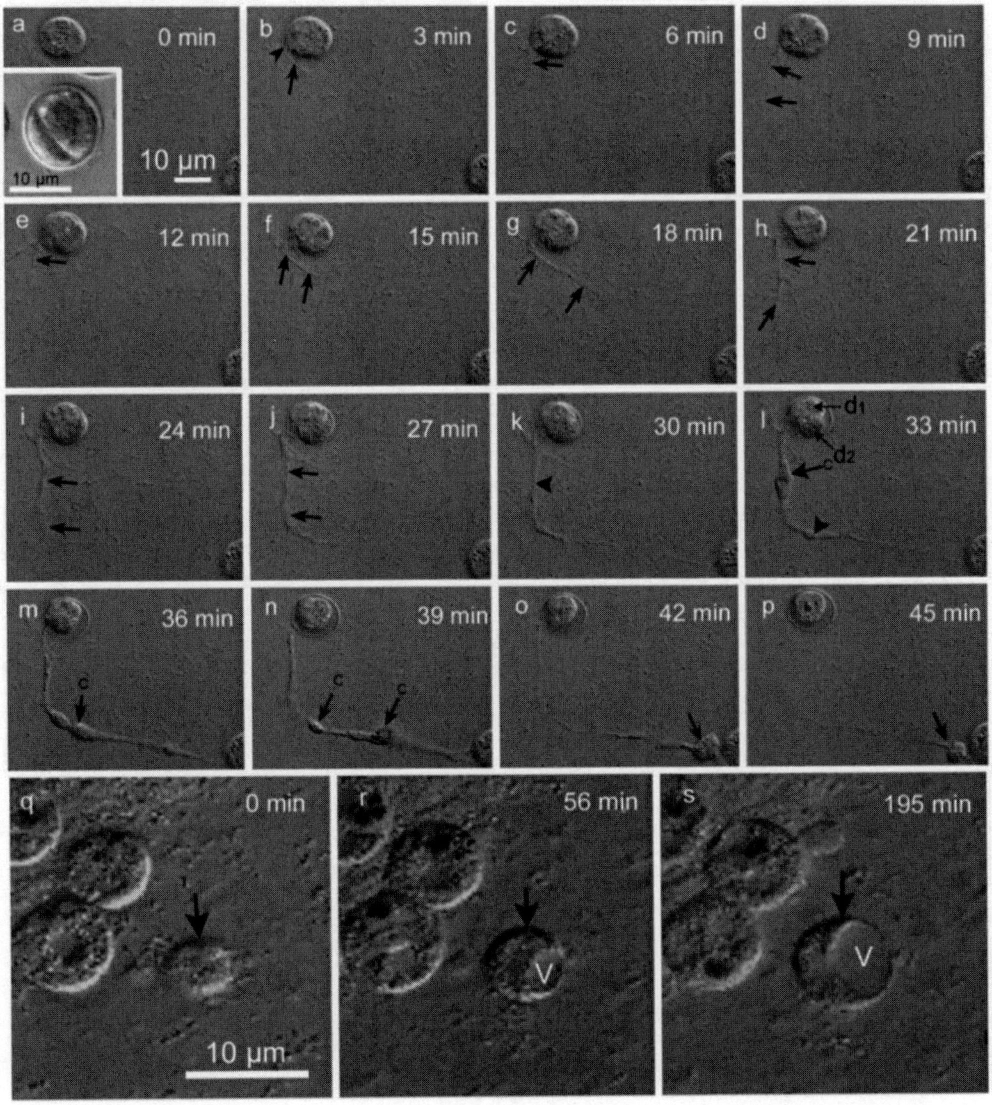

Figure 5. Two time-lapse video sequences (a-p and q-s) showing the post-cytokinesis behavior of *Lotharella vacuolata*. (a) Cell divided into two daughter cells. Scale bar is shown in a. Same magnification in b–p. Inset: high magnification view of a cell just after cytokinesis. (b-j) Filopodium (arrows) of one daughter cell emerging through a pore in the parental cell wall (arrowhead in b). (k-n) First, nucleus (arrowhead) is transported through the filopodium, subsequently, other organelles including chloroplasts (C) are transported to the distal end of the filopodium. d_1 indicates a daughter cell staying within the parental cell wall, d_2 indicates a daughter cell becoming an amoeboid cell. (o-p) The amoeboid daughter cell (arrow) is now completely outside the parental cell wall. The other daughter cell remained within the parental cell wall. (q) The amoeboid daughter cell has withdrawn its filopodia and rounded off (arrow). Scale bar is shown in q. Same magnification in r, s. (r) The amoeboid daughter cell (arrow) has started to develop a vacuole (V) and to form a cell wall. (s) The young coccoid cell (arrow) developing a vacuole (V).

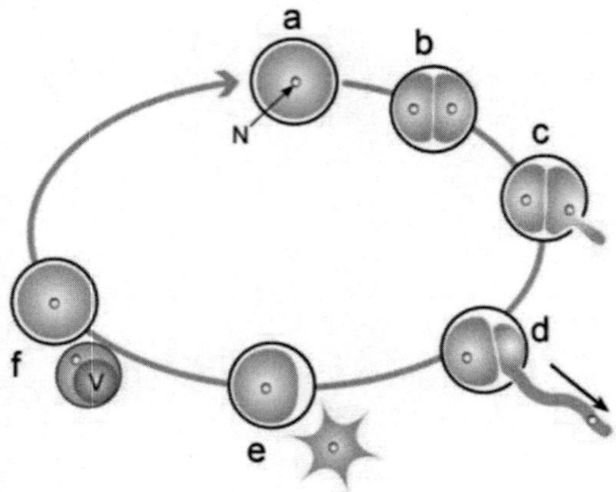

Figure 6. A diagrammatic representation of the daughter cells' behavior in *Lotharella vacuolata*. (a) Coccoid cell (vegetative cell). N, nucleus. (b) Cytokinesis is occurred within the cell wall. (c) One of the two daughter cells extending a filopodium. (d) Organelles (e.g., nucleus, chloroplasts) transported through the filopodium to the distal end of filopodium. Arrow indicates a direction of cytoplasmic streaming. (e) The amoeboid daughter cell has formed completely outside the parental cell wall. The other daughter cell remained within the parental wall. (f) The young coccoid cell is developing a vacuole (V) and forming a new cell wall.

2.2. Bigelowiella Longifila

Bigelowiella longifila is another chlorarachniophyte possessing the unique daughter cells' behavior. It is genetically distinct from *Lotharella vacuolata* based on 18S rDNA phylogeny (Ota et al. 2009, Silver et al. 2007). This species was obtained as strain CCMP242 from the Provasoli-Guillard National Center for Culture of Marine Phytoplankton at Bigelow Laboratory for Ocean Sciences, USA (Ota et al. 2007). *B. longifila* has amoeboid cells as its vegetative stage (Figure 7a, b); however flagellate, coccoid and walled amoeboid stages are also observed under different culture conditions (Ota et al. 2007).

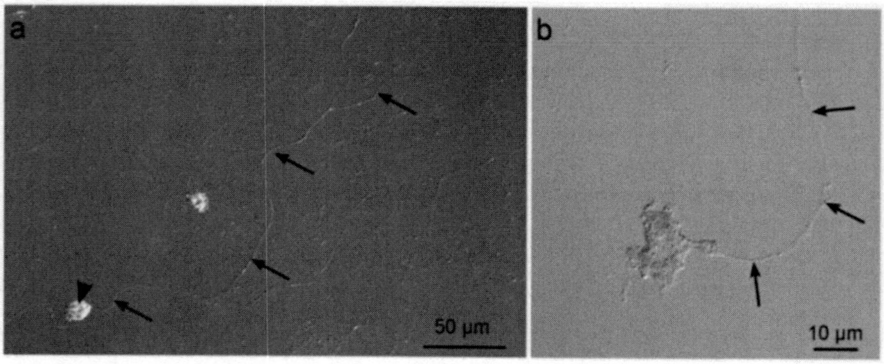

Figure 7. Light micrographs of *Bigelowiella longifila*. (a) Low magnification of amoeboid cells. Arrowhead indicates main cell body, and arrows indicate very fine long filopodium. (b) High magnification of the main cell body with the unilateral filopodium (arrows).

A series of micrographs from a time-lapse video sequence is shown in Figure 8a-d and summarized in Figure 9. This behavior in the daughter cells was seen in those amoeboid cells which possess one (sometimes two) long filopodium (Figure 8a). Cytokinesis occurred at the main body of the amoeboid cell; one of the daughter cell inherited the long filopodium from the parental cell while the other daughter cell did not have a filopodium just after the cytokinesis (Figure 8b). A few hours after cytokinesis, unidirectional cytoplasmic streaming began in the daughter cell possessing the long filopodium, that is, organelles such as the chloroplasts, nucleus and mitochondria slowly moved distally (Figure 8c). The transport of cell content usually took 2-3 hours to complete (Figure 8d). (The movie of this scene is available at Pickett-Heaps and Pickett-Heaps (2006).) The other daughter cell showed almost no movement and eventually developed a new filopodium (Ota et al. 2007). This sequence of events was commonly seen in the amoeboid cells, but the duration of the process and details of filopodium development varied.

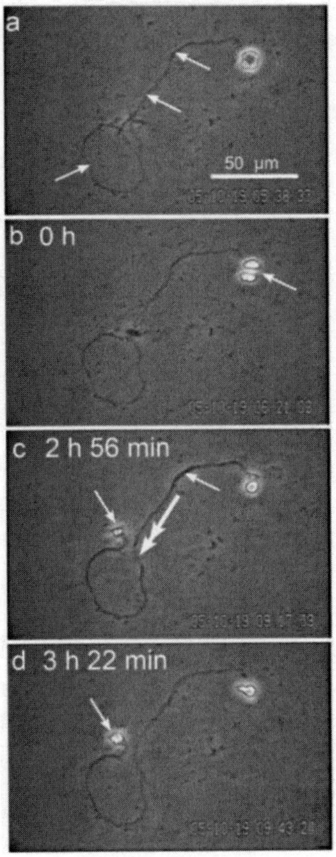

Figure 8. Time-lapse video sequence showing the post-cytokinesis behavior of amoeboid daughter cells in *Bigelowiella longifila*. (a) An amoeboid cell with a long filopodium. Scale bar is shown in a. Same magnification in b–d. (b) Cell divide into two daughter cells. (c) One of the daughter cell possessing a filopodium transport the organelles to the distal end of the filopodium (arrowheads). Double-arrow indicates the direction of cytoplasmic streaming. (d) The nearly completed migration of the amoeboid cell (arrow).

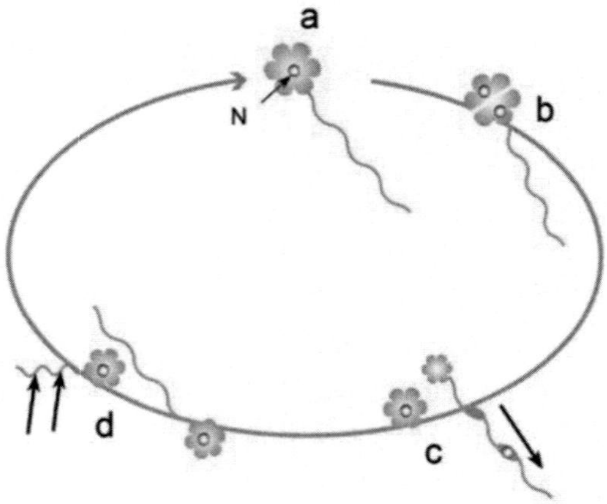

Figure 9. A diagrammatic representation of the daughter cells' behavior in *Bigelowiella longifila.* (a) An amoeboid cell. N; nucleus. (b) Cytokinesis is occurred in main body of the cell. (c) The daughter cell that inherited the filopodium transports its cell contents (e.g., nucleus, chloroplasts) to the distal end of the filopodium. Arrow indicates a direction of cytoplasmic streaming. (d) Another daughter cell extends a new filopodium (arrows).

3. DISCUSSION

3.1. Unique daughter Cells' Behavior in other Organisms

First I will discuss similar daughter cells' behavior in other organisms. In the chlorarachniophyte *Gymnochlora dimorpha,* a new species in the Chlorarachniophyta, also possess this unique daughter cells' behavior (S. Ota, Phycologia, submitted). *G. dimorpha* has an amoeboid life cycle devoid of coccoid or flagellate stages. In contrast, *Bigelowiella longifila*, one of the daughter cells of *G. dimorpha* has the capability of amoeboid locomotion after cytokinesis, whereas the other daughter cell remains sessile (Unpubl. observ.).

Besides the chlorarachniophytes as mentioned above, a testate amoeba, *Paulinella chromatophora* has been reported to show the similar daughter cells' behavior after cytokinais (Hoogenraad 1927). *Paulinella* cell is covered with a test of silica scales, and the mother cell forms a new daughter test before the cytokinesis and one daughter cell moves into the new daughter test after cytokinesis, while the other inherits the mother test (Hoogenraad 1927). Although the chlorarachniophytes and *P. chromatophora* are members of the Cercozoa, they are not closely related each other (Bass et al. 2005, Cavalier-Smith and Chao 2003).

Some prasinophyte algae (one of the green algal groups: see Lewis and McCourt (2004) for details), show similar behavior after cytokinesis (Miyashita et al. 1993; Hasegawa et al. 1996). For example, Hasegawa et al. (1996) reported the post-cytokinesis daughter cells' behavior in *Prasinoderma coloniale* which is a small coccoid prasinophyte from the Pacific Ocean. In the vegetative reproduction of *P. coloniale*, one of the two daughter cells is excluded through a pore of the parental cell wall. A new cell wall is formed around both daughter cells before one daughter cell is released. As a result, the released daughter cell has a single layered cell wall, whereas the other daughter cells has multi-layered cell wall (Hasegawa et al. 1996).

Recently, Horn et al. (2007) reported that an amoeboid heterokont (photosynthetic stramenopiles) *Synchroma grande* had similar daughter cells' behavior. The heterokonts are one of the largest algal groups including uni- and multi-cellular organisms such as brown algae and diatoms. The amoeboid *S. grande* which is found in marine environments (Horn et al. 2007), possesses a filopodium. After binary cell division of the sessile amoeboid cell, one daughter cell stays within a lorica, whereas the second daughter cell becomes a migrating amoeboid cell (Horn et al. 2007).

3.2. Similarity and Difference of the Behavior

Not only chlorarachniophytes but also other cercozoan species *Paulinella chromatophora*, some green algae and a heterokont *Synchroma grande* show asymmetrical daughter cells' behavior. Among those species, a common manner of the behavior is that two daughter cells always show different movement (asymmetrical movement) after cytokinesis. However, there are some minor differences among those species. For example, *P. chromatophora* build up a new daughter test prior to releasing of one of the daughter cells, whereas in *Lotharella polymorpha* and *L. vacuolata*, a one of the daughter cells form a new cell wall after releasing from the mother cell. In *Bigelowiella longifila*, daughter cells do not show amoeboid locomotion, whereas in *Gymnochlora dimorpha* and two *Lotharella* species, one daughter cell shows amoeboid locomotion, while the other remains sessile after cytokinesis. In *Prasinoderma coloniale* a new cell wall is formed around daughter cells before one daughter cell is released, whereas in *Lotharella polymorpha* and *L. vacuolata,* a new cell wall is formed around daughter cells after one daughter cell is released.

3.3. Biological Implications of the Behavior

In *Lotharella vacuolata* and *L. polymorpha*, the unique daughter cells' behavior occurs during a life cycle possessing the shorter amoeboid stage and the longer walled stage. The life cycle might be an adaptation to habitats with highly variable environmental marine conditions, because naked (wall-less) amoeboid cells tend to be sensitive to osmotic stress in littoral environment. The formation of cell walls thus might help resist such environmental stresses. Amoeboid locomotion in one daughter cell might play a role in dispersal of its colonies over solid substrata such as macro-algae or sea sands on the other hand.

In contrast, *Bigelowiella longifila* is a planktonic species living in open oceanic waters, and complete loss of amoeboid locomotion is not important (Ota et al. 2007). In this species the cytoplasmic streaming may also lead to form low-density planktonic gelatinous colonies (Figure 10), which remain suspended in the water column more easily. This is because one of the daughter cells consistently migrates the almost fixed distance via a filopodium, allowing the daughter cells to keep a certain distance apart in the colony. Thus, the daughter cells' behavior in *B. longifila* would be adapted to open ocean waters.

It is still unclear that the exact and general roles of the daughter cells' behavior in other organisms. More observations are needed to understand the true biological meaning of the behavior in eukaryotes as a whole.

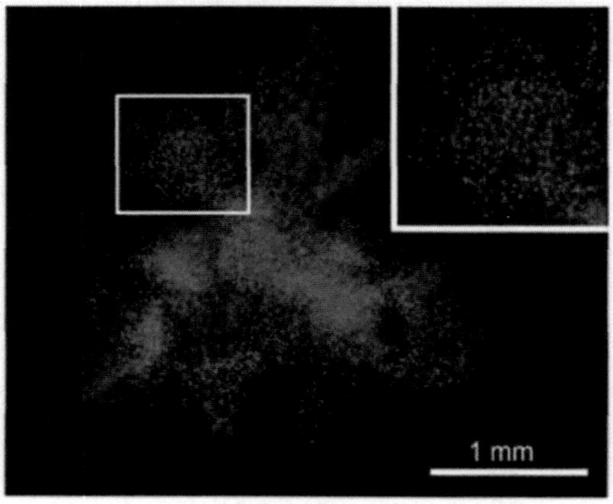

Figure 10. Low-density planktonic colonies of *Bigelowiella longifila*. The boxed area is enlarged in the inset.

3.4. Current Questions

Here, I will mention two interesting issues that arise from this behavior of daughter cells.

3.4.1. When does Daughter Cell fate become Determined?

The first is an issue of cell differentiation. In *Lotharella vacuolata*, two identical daughter cells are formed just after cytokinesis; one becomes amoeboid and exits the parental cell wall through the pore in the wall, whereas the other daughter cell stays within the parental cell wall. In this case, when is the fate of the two cells determined? Time-lapse observations revealed that slightly morphological differences between two daughter cells in *L. vacuolata* (Ota et al. 2005). In *L. vacuolata*, red particles (see Figure 4a) are often observed in cytoplasm, which may be excretion products suggested by Calderon-Saenz and Schnetter (1989). In time-lapse video observations of Ota et al. (2005), the reddish particles were always left within the daughter cells that remain in the parental cell wall, and the amoeboid daughter cells never carried the particles. This observation suggests that the determination of which daughter cell becomes amoeboid is not random.

3.4.2. The Behaviors in the Context of Eukaryote Evolution

As mentioned above, similar behaviors are observed in various eukaryotic groups, i.e., the behaviors are seen over the Kingdom-level groups. From the perspective of evolution of cell behavior, a key question is whether they are homologous or not. In my opinion, the daughter cells' behaviors among the chlorarachniophytes appear to be homologous because the manner of the daughter cell movement (cytoplasmic streaming) is basically same among the chlorarachniophytes. For example, a nucleus is always transported at first through a filopodium. Considering the similarity of the cytoplasmic streaming manner, it is reasonable to conclude that the common ancestor of the chlorarachniophytes already possessed the capability of the behavior. If so, the next question is whether the behaviors are homologues in other eukaryotes. Or the behaviors were evolving independently? Because the data on the daughter cells' behavior are so limited, I cannot answer this question now. More observations are needed to look the real picture of the behavior in the context of eukaryote evolution.

ACKNOWLEDGMENTS

Dr. J. D. Pickett-Heaps thanked for critical review of this manuscript. S.O. was supported by fellowship from the French government (2008-2009).

REFERENCES

Archibald J.M. (2007). Nucleomorph genomes: structure, function, origin and evolution. *BioEssays, 29,* 392–402.

Bass, D. & Cavalier-Smith, T. (2004). Phylum-specific environmental DNA analysis reveals remarkably high global biodiversity of Cercozoa (Protozoa). *Int. J. Syst. Evol. Microbiol., 54,* 2393–2404.

Bass, D., Moreira, D., López-García, P., Polet, S., Chao, E.E., von der Heyden, S., Pawlowski, J. & Cavalier-Smith, T. (2005). Polyubiquitin Insertions and the Phylogeny of Cercozoa and Rhizaria, *Protist, 156,* 149–161.

Calderon-Saenz, E. & Schnetter, R. (1989). Morphology, biology, and systematics of *Cryptochlora perforans* (Chlorarachniophyta), a phagotrophic marine alga. *Pl. Syst. Evol., 163,* 165–76.

Cavalier-Smith, T. 2002. The phagotrophic origin of eukaryotes and phylogenetic classification of Protozoa. *Int. J. Syst. Evol. Microbiol., 52,* 297–354.

Cavalier-Smith, T. and Chao, E.E. (2003) Phylogeny and classification of phylum Cercozoa (Protozoa). *Protist, 154,* 341–358.

Dietz, C., Ehlers, K.,Wilhelm, C., Gil-Rodrı´guez,M. C. & Schnetter, R. (2003). *Lotharella polymorpha* sp. nov. (Chlorarachniophyta) from the coast of Portugal. *Phycologia, 42,* 582–93.

Geitler, L. (1930). Ein grünes Filarplasmodium und andere neue Protisten. *Arc. Protistenkd., 69,* 615–636.

Gilson, P.R., Su, V., Slamovits, C.H., Reith, M.E., Keeling, P.J., & McFadden, G.I. (2006). Complete nucleotide sequence of the chlorarachniophyte nucleomorph: nature's smallest nucleus. *Proc Natl Acad Sci USA, 103,* 9566–9571.

Hasegawa, T., Miyashita, H., Kawachi, M., Ikemoto, H., Kurano, N., Miyachi, S. & Chihara, M. (1996). *Prasinoderma coloniale* gen. et sp. nov., a new pelagic coccoid prasinophyte from the western Pacific Ocean, *Phycologia, 35,* 170-176.

Hibberd, D.J. & Norris, R. E. (1984). Cytology and ultrastructure of *Chlorarachnion reptans* (Chlorarachniophyta divisio nova, Chlorarachniophyceae classis nova). *J. Phycol., 20,* 310–330.

Hirakawa Y., Nagamune K. & Ishida K. (2009). Protein targeting into secondary plastids of chlorarachniophytes. *Proc. Natl. Acad. Sci. USA, 106,* 12820– 12825.

Hoogenraad, H.R. (1927). Zur Kenntnis der Fortpflanzung von *Paulinella chromatophora* Lauterb. *Zool. Anz., 72,* 140–50.

Horn, S., Ehlers, K., Fritzsch, G., Gil-Rodriguez, M.C., Wilhelm, C. & Schnetter, R. (2007). *Synchroma grande* spec. nov. (Synchromophyceae class.nov.,Heterokontophyta):an amoeboid marine alga with unique plastid complexes. *Protist, 158,* 277–293.

Ishida, K., Yabuki, A. & Ota, S. (2007). The chlorarachniophytes: evolution and classification. In: Brodie J. and Lewis J. (eds) *Unravelling the Algae: The Past, Present, and Future of Algal Systematics.* CRC Press, Boca Raton, Florida, pp.171-182.

Lewis, L.A. & McCourt, R.M. (2004). Green algae and the origin of land plants. *Am. J. Bot., 91,* 1535–1556.

Ludwig, M. & Gibbs, S.P. (1989). Evidence that the nucleomorphs of *Chlorarachnion reptans* (Chlorarachniophyceae) are vestigial nuclei: morphology, division and DNA-DAPI fluorescence. *J. Phycol., 25,* 385–394.

McFadden, G.I., Gilson, P.R., Hofmann, C.J., Adcock, G.J. & Maier, U.G. (1994). Evidence that an amoeba acquired a chloroplast by retaining part of an engulfed eukaryotic alga. *Proc. Natl. Acad. Sci. USA, 91,* 3690–3694.

Miyashita, H., Ikemoto, H., Kurano, N., Miyachi, S. & Chihara, M. (1993). *Prasinococcus capsulatus* gen. et sp. nov., a new marine coccoid prasinophyte. *J Gen Appl. Microbiol., 39,* 571-582.

Ota, S., Ueda, K. & Ishida, K. (2005). *Lotharella vacuolata* sp. nov., a new species of chlorarachniophyte algae, and time-lapse video observations on its unique post-cell division behavior. *Phycol. Res., 53,* 275–286.

Ota, S., Ueda, K. & Ishida, K. (2007). Taxonomic study of *Bigelowiella longifila* sp. nov. (Chlorarachniophyta) and a time-lapse video observation on the unique migration of amoeboid cells. *J. Phycol., 43,* 333–343.

Ota, S., Daniel, V., Le Gall, F., Yabuki, A. & Ishida, K. (2009). *Partenskyella glossopodia* gen. et sp. nov., the first report of a chlorarachniophyte that lacks a pyrenoid, *Protist, 160,* 137–150.

Pickett-Heaps, J.D. & Picket-Heaps, J.F. (eds.) (2006). *The Kingdom Protista: The Dazzling World of Living Cells.* Cytographics. Australia. DVD. Available from: cytographics.com

Silver, T.D., Koike, S., Yabuki, A., Kofuji, R., Archibald, J.M. & Ishida, K. (2007). Phylogeny and nucleomorph karyotype diversity of chlorarachniophyte algae. *J. Eukaryot. Microbiol., 54,* 403–410.

INDEX

A

B

C

D

E

F

G

H

I

K

L

M

N

O

P

Q

R

S

T

U

V

W

X

Y

Z